INFLAMMATION IN HEART FAILURE

INFLAMMATION IN HEART FAILURE

Edited by

W. Matthijs Blankesteijn
Department of Pharmacology,
Cardiovascular Research Institute Maastricht,
Maastricht University,
The Netherlands

AND

Raffaele Altara
Department of Physiology and Biophysics,
University of Mississippi Medical Center,
Jackson, MS, USA

Department of Pharmacology,
Cardiovascular Research Institute Maastricht,
Maastricht University,
The Netherlands

AMSTERDAM • BOSTON • HEIDELBERG • LONDON
NEW YORK • OXFORD • PARIS • SAN DIEGO
SAN FRANCISCO • SINGAPORE • SYDNEY • TOKYO
Academic Press is an imprint of Elsevier

Academic Press is an imprint of Elsevier
32 Jamestown Road, London NW1 7BY, UK
525 B Street, Suite 1800, San Diego, CA 92101-4495, USA
225 Wyman Street, Waltham, MA 02451, USA
The Boulevard, Langford Lane, Kidlington, Oxford OX5 1GB, UK

© 2015 Elsevier Inc. All rights reserved.

No part of this publication may be reproduced or transmitted in any form or by any means, electronic or mechanical, including photocopying, recording, or any information storage and retrieval system, without permission in writing from the publisher. Details on how to seek permission, further information about the Publisher's permissions policies and our arrangements with organizations such as the Copyright Clearance Center and the Copyright Licensing Agency, can be found at our website: www.elsevier.com/permissions

This book and the individual contributions contained in it are protected under copyright by the Publisher (other than as may be noted herein).

Notices
Knowledge and best practice in this field are constantly changing. As new research and experience broaden our understanding, changes in research methods, professional practices, or medical treatment may become necessary.

Practitioners and researchers must always rely on their own experience and knowledge in evaluating and using any information, methods, compounds, or experiments described herein. In using such information or methods they should be mindful of their own safety and the safety of others, including parties for whom they have a professional responsibility.

To the fullest extent of the law, neither the Publisher nor the authors, contributors, or editors, assume any liability for any injury and/or damage to persons or property as a matter of products liability, negligence or otherwise, or from any use or operation of any methods, products, instructions, or ideas contained in the material herein.

British Library Cataloguing in Publication Data
A catalogue record for this book is available from the British Library

Library of Congress Cataloging-in-Publication Data
A catalog record for this book is available from the Library of Congress

ISBN: 978-0-12-800039-7

For information on all **Academic Press** publications
visit our website at **store.elsevier.com**

Printed and bound in the United States
14 15 16 17 10 9 8 7 6 5 4 3 2 1

Contents

3 TARGETING OF THE INFLAMMATORY RESPONSE

14. The Role of Cytokines in Clinical Heart Failure

DOUGLAS L. MANN

Contributors

Raffaele Altara Department of Pharmacology, Cardiovascular Research Institute Maastricht (CARIM), Maastricht University, Maastricht, The Netherlands, and Department of Physiology and Biophysics, University of Mississippi Medical Center, Jackson, MS, USA

Jonathan Beaudoin Massachusetts General Hospital, Boston, Massachusetts, USA

W. Matthijs Blankesteijn Department of Pharmacology, Cardiovascular Research Institute Maastricht (CARIM), Maastricht, The Netherlands

Hans-Peter Brunner-La Rocca Department of Cardiology, Maastricht University Medical Centre, Maastricht, The Netherlands

Emmanuel Buys Massachusetts General Hospital, Boston, Massachusetts, USA

Federico Carbone Division of Cardiology, Department of Medical Specialties, Foundation for Medical Researches, University of Geneva, Geneva, Switzerland; Department of Internal Medicine, University of Genoa School of Medicine, and IRCCS Azienda Ospedaliera Universitaria San Martino–IST Istituto Nazionale per la Ricerca sul Cancro, Genoa, Italy

Arrigo F.G. Cicero Medical and Surgical Sciences Department, University of Bologna, Bologna, Italy

Evangelos P. Daskalopoulos Department of Pharmacology, Cardiovascular Research Institute Maastricht (CARIM), Maastricht, The Netherlands

Lisandra E. de Castro Brás San Antonio Cardiovascular Proteomics Center, and Mississippi Center for Heart Research, Department of Physiology and Biophysics, University of Mississippi Medical Center, Jackson, MS, USA

Kristine Y. Deleon-Pennell San Antonio Cardiovascular Proteomics Center, and Mississippi Center for Heart Research, Department of Physiology and Biophysics, University of Mississippi Medical Center, Jackson, MS, USA

Uli L.M. Eisel Department of Molecular Neurobiology, University of Groningen, and Department of Psychiatry, University Medical Centre Groningen, Groningen, The Netherlands

Elda Favari Pharmaceutical Sciences Department, University of Parma, Parma, Italy

Nikolaos G. Frangogiannis Department of Medicine (Cardiology), The Wilf Family Cardiovascular Research Institute, Albert Einstein College of Medicine, Bronx, New York, USA

Stefan Frantz Department of Internal Medicine I, University Hospital Würzburg, Comprehensive Heart Failure Center, University of Würzburg, Würzburg, Germany

Olga Frunza Departm ent of Medicine (Cardiology), The Wilf Family Cardiovascular Research Institute, Albert Einstein College of Medicine, Bronx, New York, USA

Michael E. Hall San Antonio Cardiovascular Proteomics Center; Mississippi Center for Heart Research, Department of Physiology and Biophysics, University of Mississippi Medical Center, and Cardiology Division, University of Mississippi Medical Center, Jackson, MS, USA

Kevin C.M. Hermans Department of Pharmacology, Cardiovascular Research Institute Maastricht (CARIM), Maastricht, The Netherlands

Stephane Heymans Center for Heart Failure Research, Cardiovascular Research Institute Maastricht, Maastricht University, Maastricht, The Netherlands

Rugmani Padmanabhan Iyer San Antonio Cardiovascular Proteomics Center, and Mississippi Center for Heart Research, Department of Physiology and Biophysics, University of Mississippi Medical Center, Jackson, MS, USA

Dennis H.M. Kusters Department of Biochemistry, Cardiovascular Research Institute Maastricht (CARIM), Maastricht University, Maastricht, The Netherlands

Richard A. Lange San Antonio Cardiovascular Proteomics Center, Jackson, MS, and Paul L. Foster School of Medicine, Texas Tech University Health Sciences Center El Paso, El Paso, TX, USA

Merry L. Lindsey San Antonio Cardiovascular Proteomics Center; Mississippi Center for Heart Research, Department of Physiology and Biophysics, University of Mississippi Medical Center, and Research Services, G.V. (Sonny) Montgomery Veterans Affairs Medical Center, Jackson, MS, USA

Yonggang Ma San Antonio Cardiovascular Proteomics Center, and Mississippi Center for Heart Research, Department of Physiology and Biophysics, University of Mississippi Medical Center, Jackson, MS, USA

Douglas L. Mann Cardiovascular Division, Department of Medicine, Center for Cardiovascular Research, Washington University School of Medicine, St. Louis, MO, USA

Fabrizio Montecucco Department of Internal Medicine, University of Genoa School of Medicine; IRCCS Azienda Ospedaliera Universitaria San Martino–IST Istituto Nazionale per la Ricerca sul Cancro, Genoa, Italy, and Division of Laboratory Medicine, Department of Genetics and Laboratory Medicine, Geneva University Hospitals, Geneva, Switzerland

Anna Planavila Departament de Bioquímica i Biologia Molecular, Institut de Biomedicina de la Universitat de Barcelona (IBUB), Universitat de Barcelona and CIBER Fisiopatología de la Obesidad y Nutrición (CIBEROBN), Barcelona, Spain

Chris P.M. Reutelingsperger Department of Biochemistry, Cardiovascular Research Institute Maastricht (CARIM), Maastricht University, Maastricht, The Netherlands

Nicoletta Ronda Pharmaceutical Sciences Department, University of Parma, Parma, Italy

Marielle Scherrer-Crosbie Massachusetts General Hospital, Boston, Massachusetts, USA

Regien G. Schoemaker Department of Cardiology, University Medical Centre Groningen, and Department of Molecular Neurobiology, University of Groningen, Groningen, The Netherlands

Blanche Schroen Center for Heart Failure Research, Cardiovascular Research Institute Maastricht, Maastricht University, Maastricht, The Netherlands

Leon J. Schurgers Department of Biochemistry, Cardiovascular Research Institute Maastricht (CARIM), Maastricht, The Netherlands

Jan Tegtmeier Department of Biochemistry, Cardiovascular Research Institute Maastricht (CARIM), Maastricht University, Maastricht, The Netherlands

Robrecht Thoonen Massachusetts General Hospital, Boston, Massachusetts, USA

Hiroe Toba San Antonio Cardiovascular Proteomics Center, Jackson; Mississippi Center for Heart Research, Department of Physiology and Biophysics, University of Mississippi Medical Center, Jackson, MS, USA, and Department of Clinical Pharmacology, Division of Pathological Sciences, Kyoto Pharmaceutical University, Kyoto, Japan

Marc van Bilsen Department of Physiology, Cardiovascular Research Institute Maastricht (CARIM), Maastricht University, Maastricht, The Netherlands

Lieke van Delft Department of Pharmacology, Cardiovascular Research Institute Maastricht (CARIM), Maastricht University, Maastricht, The Netherlands

Vanessa van Empel Department of Cardiology, Maastricht University Medical Centre, Maastricht, The Netherlands

Sara Vandenwijngaert Massachusetts General Hospital, Boston, Massachusetts, USA

Johannes Weirather Department of Internal Medicine I, University Hospital Würzburg, Comprehensive Heart Failure Center, University of Würzburg, Würzburg, Germany

Andriy Yabluchanskiy San Antonio Cardiovascular Proteomics Center, Jackson, and Mississippi Center for Heart Research, Department of Physiology and Biophysics, University of Mississippi Medical Center, Jackson, MS, USA

Francesca Zimetti Pharmaceutical Sciences Department, University of Parma, Parma, Italy

Preface

Heart failure is a progressive condition that affects an increasing number of patient worldwide and severely impairs their physical capabilities and quality of life. Despite large scientific efforts, the molecular mechanisms that lead to heart failure are still far from elucidated. Therefore, diagnosis of this condition is difficult unless the patient has reached a progressed state, accompanied with clinical symptoms. A better understanding of the molecular mechanisms contributing to the earlier phases of heart failure development would therefore help to improve the diagnosis and therapy. The drugs that are currently used can slow down heart failure progression but cannot cure the patient; moreover, the effectiveness of these interventions may very much depend on the subtype of heart failure, as many patients suffering from heart failure with preserved ejection fraction show little benefit from therapies with proven efficacy in heart failure with reduced ejection fraction.

An example of a molecular mechanism that is involved in the development and progression of heart failure is inflammation. It was originally observed in the wound-healing response that takes place in the area of injury after myocardial infarction. There, the inflammatory response is crucial for the removal of the necrotic debris from the area of injury and helps to attract the cells involved in the formation of a scar. In the meantime, inflammation has been described in cardiac remodeling due to other causes, for example, hypertension, and is already activated early on in its development. This highlights the importance of inflammation as a common molecular pathway of heart failure, providing potentially interesting options for diagnosis and therapy. However, the clinical results of interventions in inflammatory pathways have been disappointing so far, underscoring the complexity of the inflammatory response and the need for a better understanding of its molecular mechanisms. Therapeutic targeting of inflammation will therefore likely require careful patient selection and precise timing of the intervention to become successful.

The purpose of this book is to provide the latest information on the role of inflammation in heart failure to researchers and advanced students in the cardiovascular diseases. To this end, we have invited experts in the field to provide a comprehensive and timely overview of their research areas. The book is structured into three sections, providing the reader with easy access to the information. In Section 1, which focuses on the *pathophysiology of the inflammatory response in heart failure*, an overview is provided of the extensive literature on the role of inflammation in heart failure, with a distinction between ischemia-induced heart failure and heart failure due to other causes. Specific emphasis is put on the role of the innate immune system and the interaction between the extracellular matrix and the inflammatory mediators. The cross talk between the inflammatory response in the heart and the brain is highlighted and the section is finalized with an overview of different animal models of heart failure and their advantages and restrictions for the study of this condition.

In Section 2, the focus is on *inflammatory biomarkers*. The section starts with an overview of multiple inflammatory mediators as biomarkers for adverse remodeling and heart failure. Next, the pros and cons of different analytical techniques for measuring panels of inflammatory biomarkers in a single sample are discussed. In the last chapter of this section, an overview of imaging modalities to visualize the inflammatory response is provided.

Targeting of the inflammatory response is the subject of the third section of this book. Here, we focus on the experimental and clinical evidence for the beneficial effects of interventions on mineralocorticoid receptor and peroxisome proliferator-activated receptors. The modulating effects of statins and the involvement of miRNAs in the control of the inflammatory response and their therapeutic potential are discussed. Finally, the results of clinical trials with anti-inflammatory agents are presented and interpreted in light of our current understanding of the inflammatory response in heart failure.

W. Matthijs Blankesteijn
Raffaele Altara

S E C T I O N 1

PATHOPHYSIOLOGY OF THE INFLAMMATORY RESPONSE IN HEART FAILURE

CHAPTER

1

Inflammation in Heart Failure with Preserved Ejection Fraction

Vanessa van Empel, Hans-Peter Brunner-La Rocca

Department of Cardiology, Maastricht University Medical Centre, Maastricht, The Netherlands

1.1 INTRODUCTION

Until recently, heart failure was considered as *one* syndrome with different underlying causes. Treatment, apart from underlying cause, was uniform, despite acknowledging that no treatment trials had been done in patients with heart failure with preserved left-ventricular ejection fraction LVEF (HFpEF) [1]. Heart failure was basically considered as a final common pathway with uniform pathophysiology, irrespective of underlying disease. However, all the studies supporting this concept have been performed in patients with heart failure and reduced LVEF (HFrEF), because it was thought to be the more advanced stage of heart failure in general. Apart from diseases such as hypertrophic or restrictive cardiomyopathies, HFpEF was considered as a less advanced stage of heart failure and to be a relatively rare disease. This, however, was to a large extent related to the lack of knowledge and relatively rare use of echocardiography to measure LVEF in elderly patients who comprise the majority of HFpEF patients. Over the years, it became more and more evident that nearly half of the patients with the clinical syndrome of heart failure have normal or preserved LVEF, thus HFpEF [2]. Still, it took years before it became evident that HFpEF and HFrEF may have quite distinct underlying pathophysiology. For more details on the pathophysiology of HFrEF, see Chapters 2 and 3 of this book.

1.2 CONSEQUENCES OF LIMITED UNDERSTANDING OF PATHOPHYSIOLOGY IN HFpEF

Treatment that was shown to be highly efficacious in HFrEF patients [3] failed to show beneficial effects in HFpEF patients [4,5]. Even more so, intensifying similar therapy in HFrEF and HFpEF led to completely different results. Thus, whereas intensifying heart failure medication based on NT-proBNP guidance resulted in significant reduction of heart failure hospitalization-free survival, the opposite was the case in patients with HFpEF [6]. The question, therefore, that arises is which underlying pathophysiolocial pathways are most important in HFpEF that obviously are less or not important in HFrEF; information in this regard is, however, limited.

The limited understanding in the pathophysiology of HFpEF not only makes treatment difficult, but also makes diagnosis difficult (Figure 1.1). The current diagnostic algorithm of HFpEF focuses on symptoms and signs of HF in addition to evidence of diastolic dysfunction on imaging, usually echocardiography [3]. Yet, symptoms in HFpEF are nonspecific and echocardiographic measures correlate poorly with actual LV (left ventricle) filling pressure and lack accuracy [7,8]. Moreover, an important assumption was the fact that the important pathophysiological pathways are equal in all HFpEF patients as it is the case in HFrEF—at least to a large extent. So far, this has not yet been proven in basic research or in clinical studies. Better characterization may be helpful in this regard, which may include determination of underlying cause and investigation of myocardial and systemic consequences of

Inflammation in Heart Failure
http://dx.doi.org/10.1016/B978-0-12-800039-7.00001-3

© 2015 Elsevier Inc. All rights reserved.

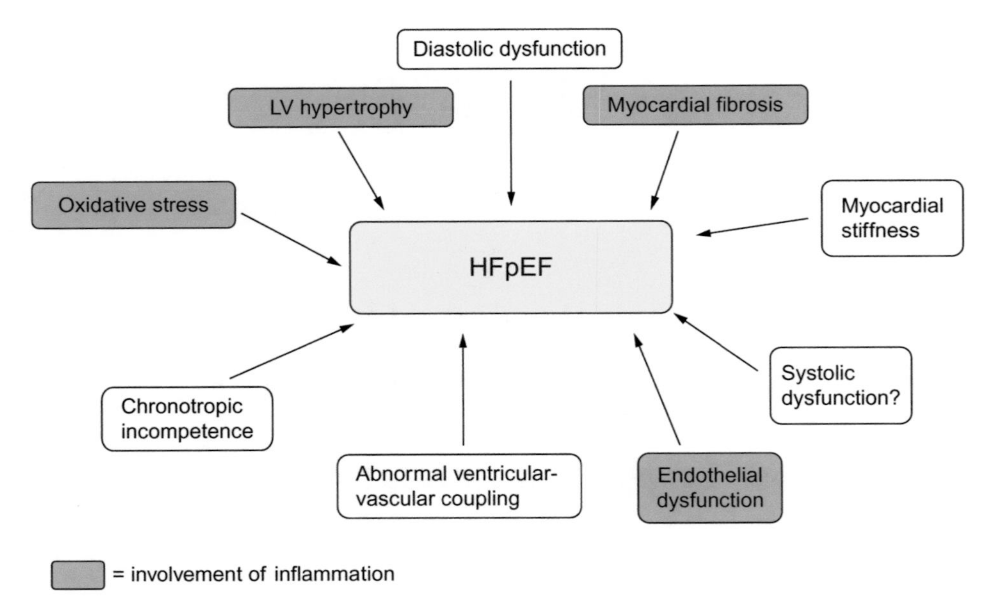

FIGURE 1.1 Pathophysiology of HFpEF. Blue background indicates involvement of inflammation. HFpEF, heart failure with preserved ejection fraction; LV, left ventricular.

these causes. Biomarkers may be helpful in this regard. Levels of brain natriuretic peptide (BNP) or its N-terminal propeptide (NT-proBNP) were suggested as additional diagnostic aids in the HFpEF workup [3]. Their value is proven for diagnosis in acute decompensated heart failure [9,10] as well as risk stratification [11] and guiding therapy [12,13] in chronic HFrEF, but the use of BNP and NT-proBNP is less well established in HFpEF [14]. Several other biomarkers may be helpful in this regard and may provide information on the pathogenesis of HFpEF as compared to HFrEF and may, therefore, help in a better understanding as well as in improving diagnosis of HFpEF and risk stratification.

1.3 UNDERLYING CAUSES OF HFpEF

Different etiologies lead to heart failure and may differ between HFpEF and HFrEF. Patient characteristics and risk factors of HFpEF differ significantly from those of HFrEF (Figure 1.2). HFpEF patients are likely to be older and more often are female compared to HFrEF patients [15]. Furthermore, cardiovascular and noncardiovascular comorbidities are highly present in HFpEF patients and may significantly contribute to the patients' limitations [16]. In a population study in Olmsted County, hypertension was prevalent in 63% of HFpEF compared to 48% in HFrEF patients and atrial fibrillation in 41% of HFpEF compared to 28% in HFrEF [2]. Interestingly, the presence of atrial fibrillation may identify a HFpEF cohort with more advanced disease and a significantly reduced exercise capacity compared to HFpEF patients without atrial fibrillation [17]. The prevalence of cardiovascular comorbidities varied in the different studies, depending on the type of study and the different criteria used to diagnose HFpEF. Overall, data from population-based studies, registries, and randomized control trials reported in HFpEF a prevalence of coronary artery disease of 20-76%, diabetes mellitus of 13-70%, atrial fibrillation of 15-41%, and hypertension of 25-88% [18]. However, studies comparing HFpEF with HFrEF all reported an increased prevalence of hypertension and atrial fibrillation in HFpEF and a decreased prevalence of coronary artery disease compared with HFrEF.

There are additional cardiac and cardiovascular mechanisms that could contribute to the clinical picture of HFpEF. Thus, not only tachyarrhythmias (i.e., usually atrial fibrillation), but also bradyarrhythmia may cause symptoms of heart failure, such as exercise intolerance and dyspnoea [19]. In severe bradycardia, this is obvious clinically, although little prospective studies are done in this regard. However, chronotropic incompetence—lack of increase in heart rate during exercise—may play an often unrecognized role [20,21]. Moreover, pulmonary hypertension (PH) may also play a role in such patients. It is known that pulmonary artery pressure is often slightly elevated in patients with HFpEF [22]. As most data come from echocardiographic studies, it is difficult to distinguish if this increase is purely passive, that is, caused by increased pressure in the left atrium with consecutive increase in pulmonary venous pressure, or (additionally) caused by an increase in pulmonary vascular resistance. Invasive measurement in small trials suggested that increase in left-ventricular filling pressure, particularly during exercise, causes symptoms,

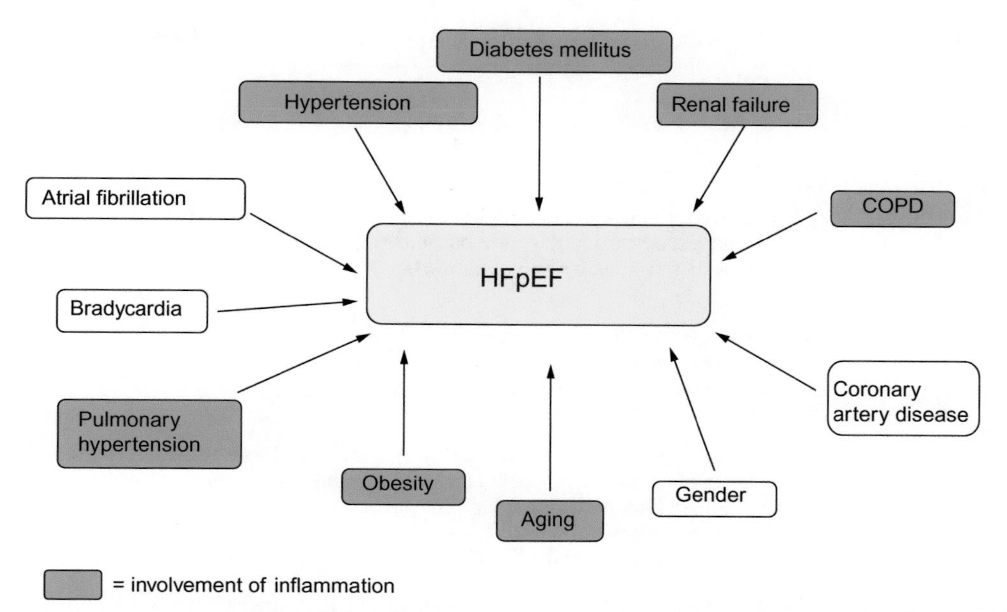

FIGURE 1.2 Comorbidities involved in HFpEF. Blue background indicates involvement of inflammation. Although one can discuss whether inflammation is involved in aging since aging is associated with increased comorbidities itself. HFpEF, heart failure with preserved ejection fraction.

although pulmonary vascular resistance was also slightly above the normal range [23]. However, invasive measurements at rest in a relatively large cohort of HFpEF patients revealed a substantial number of these patients having both increased pressure in the left atrium and significantly elevated pulmonary vascular resistance [24]. Finally, coronary artery disease is the more common cause of HFrEF, but myocardial ischemia also causes diastolic dysfunction. Thus, coronary artery disease can be a reason for HFpEF.

Although also common in patients with HFrEF, noncardiovascular comorbidities are more often associated with HFpEF, such as renal impairment, liver disease, peptic ulcer disease, and hypothyroidism [25]. Additionally, HFpEF patients typically have a higher body mass index (BMI) and are more likely to be obese [2]. Obesity is paradoxically associated with higher survival rates in heart failure. When comparing HFrEF with HFpEF, the obesity paradox was present in both with the highest survival rates in patients with BMI between 30.0 and 34.9 $kg\,m^{-2}$ [26]. Because HFpEF is highly associated with comorbidities, many of which significantly contribute to symptoms, this triggered a discussion whether HFpEF was merely a combination of comorbidities or a distinct disease. When comparing a community-based cohort of HFpEF patients and control patients without HF, fundamental cardiovascular structural and functional abnormalities were seen, however, even after accounting for body size and comorbidities, demonstrating that HFpEF is more than just a compilation of comorbidities [27].

The increased prevalence of comorbidities in HFpEF was also demonstrated when calculating the Charlson Comorbidity Index, a method of predicting mortality by classifying or weighting comorbid conditions [28]. Thus, the Charlson index was 3 or more in 70% of HFpEF patients [18]. Ather *et al.* studied the impact of noncardiac comorbidities on prognosis and mortality [29]. Although they studied a cohort of veterans, which was predominantly males, the HFpEF population was older, and had a high prevalence of diabetes, hypertension, obesity, and chronic obstructive lung disease. The overall hospitalization rates were similar for those with HFpEF compared with HFrEF, however there was a higher noncardiac hospitalization rate in HFpEF compared with HFrEF [29]. In general, HFpEF patients have a higher noncardiovascular cause of death, and HFrEF patients more often die of cardiovascular cause [30,31]. Although a majority of studies report a similar mortality rate in HFpEF and HFrEF, a meta-analysis by Somaratne and associates, including 7.688 HFpEF patients and 16.831 HFrEF patients from 17 studies, reported a 50% lower hazard mortality in HFpEF compared to HFrEF [32]. However, it is worth noting that concerning HFpEF, community-based studies reported a higher noncardiovascular death and, in contrast, clinical trials reported a higher percentage of cardiovascular deaths. This could be due to the inclusion criteria of controlled clinical trials, where relative "healthier" patients are included compared to the total population. Moreover, diagnosis may be less accurate and more difficult in HFpEF as discussed above, which may be particularly true in community-based studies. Table 1.1 summarizes important underlying diseases in HFpEF as compared to HFrEF.

TABLE 1.1 Selection of Pathways Involved in the Pathophysiology of Heart Failure, Comparing HFpEF and HFrEF

	HFpEF	HFrEF
Inflammation/oxidative stress	Increased, probably at early stage contributing to disease [33]. Inhibition possibly helpful in small human and in many animal studies [34,35]	Increased, but probably more as a consequence of disease. Inhibition has no impact on disease in human [36,37]
Renin-angiotensin system (RAS)	Increased, but probably not uniformly involved in pathophysiology. Animal work with evidence, but large studies in human not successful [4,5,38]	Increased and crucial pathway in remodeling and progression of disease [39]. Much evidence in human that inhibition uniformly improves outcome in many large trials [40]
Aldosterone	Increased and thought to be crucial pathway (related to fibrosis) based on animal work [41]. Recent large trial in human overall negative, but possibly helpful in subgroup [42]	Increased and important pathway in remodeling and progression of disease. Inhibition results in improved outcome in humans based on large trials [43,44]
Sympathetic nervous system	Increased probably in most instances [45]. Probably not uniform (e.g. atrial fibrillation vs. bradycardia) [46]. Limited human studies with some positive findings by β-blockers [47], but less effective in hypertension than other drugs [48]	Increased and important pathway in progression and complications (arrhythmias, SCD) of disease. Inhibition results in improved outcome based on many large trials in humans [49]
Endothelin-1	Increased, but not entirely clear if directly related to HFpEF or to pulmonary hypertension. Small studies suggest potential benefit in some patients [50]	Increased, but pathophysiology not entirely clear. Probably contributing to progression of disease, but inhibition does not result in improved outcome in human [51,52]
Natriuretic peptides	Increased, but less than in HFrEF [14]. Thought to be beneficial, possibly by inhibiting inflammation [34]	Increased, probably beneficial as it is counterpart to systems mentioned above. Used in acute heart failure, results not convincing [53,54]
NO system/cGMP	Inhibited, thought to play a crucial role in the pathophysiology of HFpEF, i.e., endothelial dysfunction/inflammation [33]. Preliminary data in humans promising [55], but not uniform [56].	Inhibited, not entirely clear if causal or consequence of HFrEF. Preliminary data suggest beneficial effect if cGMP is increased [57]
Kinins	Limited data, probably increased. Possibly inhibiting inflammation [58]	Increased, thought to be counterpart of RAS; but some evidence of increase in inflammation [59]. Bradykinin thought to be responsible for part of effects of ACE-inhibition [60]

SCD, sudden cardiac death.

Therefore, the important key question is whether HFpEF really is a uniform disease. Until recently, this has rarely been questioned even if HFpEF is increasingly recognized as significantly different from HFrEF. Recently, a paradigm shift was suggested, but this was again an attempt to explain HFpEF with one central underlying pathophysiological mechanism (see below), where inflammation plays the crucial role [33]. Therefore, the questions that arise are whether increased inflammation is independent of the etiology of heart failure, whether there are disease-specific mechanisms, and in what way is inflammation of particular importance in HFpEF.

1.4 ADAPTIVE MECHANISMS IN HFpEF

Recently, there is emerging evidence that the adaptive mechanisms may differ between patients with HFpEF as compared to patients with HFrEF. Table 1.1 summarizes various pathways that have been found to be of importance in patients with heart failure, irrespective of reduced or preserved LVEF. Although not all these pathways are explored in detail in both HFpEF and HFrEF, and particularly comparing the two, it is obvious that many pathophysiological pathways are not equally affected in HFpEF as compared to HFrEF. Thus, response to therapy suggests that activation of the renin-angiotensin-aldosterone system as well as the sympathetic nervous system is much more important in HFrEF than in HFpEF. On the other hand, there is increasing evidence that inflammation as well as pathways closely related to inflammation such as oxidative stress, endothelial dysfunction, and microvascular dysfunction may play a different role in HFpEF as compared to HFrEF.

Still, it is important to keep in mind that clinically it seems likely that the different causes of HFpEF not only trigger common pathways in all patients with HFpEF, but they also significantly influence the disease process as a whole. Moreover, results in HFpEF are quite diverse for many pathways, which seem less to be the case in HFrEF (Table 1.1). Here, we focus on inflammation and pathways related to inflammation and discuss the potential relevance of them in HFpEF and some differences with HFrEF.

1.5 INFLAMMATION IN HFpEF

For many years, HFpEF was considered mainly to be a consequence of chronically increased afterload. The fact that left-ventricular hypertrophy (LVH) is significantly related to HFpEF, and reduction in LVH and clinical events act in parallel, to a large extent and independently of the type of antihypertensive drug used—with some exceptions—was support for a very prominent role of afterload increase in the development of HFpEF. Still, questions remained why not all patients with hypertension develop LVH and poor outcome, irrespective of the treatment of hypertension. This concept also includes the development of myocardial fibrosis and increased myocardial stiffness, resulting in treatment suggestions that are not different from those in HFrEF. However, large trials addressing this failed to show positive results [4,5,61]. Very recently, even spironolactone, which is thought to more specifically reduce myocardial fibrosis than blockade of the renin-angiotensin system, failed to show convincing benefit in HFpEF although some methodological problems may have negatively influenced the results [42]. Other factors have been proposed as outlined above, but none were sufficiently convincing to explain HFpEF.

More recently, a new concept has been proposed that poses inflammation as a result of multiple comorbidities central in the pathophysiology of HFpEF [33]. Indeed, the concept is attractive because it provides an explanation for the fact that comorbidities are not only very common in patients with HFpEF, but also seem to significantly influence the presence of HFpEF and outcome [62]. In fact, as discussed above, comorbidities are very common in patients with HFpEF because the average HFpEF population is elderly. Moreover, it has been shown that patients with significant obesity and diabetes mellitus, even at younger age, have a significantly increased risk to develop HFpEF, also in the absence of macrovascular disease. There is increasing evidence that this form of diabetic cardiomyopathy is related to microvascular disease, which is known to be a significant problem in diabetes mellitus in other organs and systems. Microvascular disease in turn may be triggered by inflammation. Inflammation takes place in basically all chronic diseases.

The basis for the new concept comes both from human and from animal studies. Moreover, independently of this new concept, inflammation has been considered for quite sometime as an important factor in the pathophysiology of HFpEF.

1.5.1 Inflammation in HFpEF Animal Studies

In the 1990s, inflammation was found to be increased in animal models of LVH in hypertensive models. Obviously, this was not yet named HFpEF. Still, in these animal studies, a cause-effect relationship between inflammation and fibrosis in LVH was suggested [63]. In models of spontaneous (SHR) and renovascular hypertension rats, inflammation (macrophages) and fibrosis was found co-localized in the perivascular region in these animals with pressure overload [63,64]. Alteration of adhesion molecules were suggested to play an important role in this. Thus, altered expression of Intercellular Adhesion Molecule 1 (ICAM-1) were found in chronic SHR, which was related to pressure overload [65]. In the renovascular hypertension model, the potential positive effect of angiotensin-blockade and mineralocorticoid receptor antagonism has also been suggested [66]. This seems important because blockade of the renin-angiotensin system as well as mineralocorticoid receptor antagonism have been found to be of limited value in HFpEF patients, at least if these drugs are used in an unselected cohort of HFpEF patients [4,5,42,61], as discussed elsewhere in this chapter. Later, changes in different inflammatory and fibrotic markers were found in a model of rapid increase of arterial blood pressure by suprarenal aortic banding. Interestingly, expression of mediators of macrophages and fibrosis varied significantly over time. Thus, whereas activation was seen early, suppression was present later (i.e., 28 days after banding) [67]. This was also related to diastolic dysfunction. Myocyte chemoattractant protein-1 (MCP-1) was activated early (peak day 3), whereas transforming growth factor (TGF-)β remained elevated also late (28 days). Whereas LVH was seen already after day 7 with increased fibrosis and myocyte hypertrophy, diastolic dysfunction with normal LVEF was present on day 28. All of these effects could be prevented by inhibition of MCP-1, suggesting that inflammation may play an important role in the early (pre-clinical) stage of LVH with diastolic dysfunction [67]. This is in line with a more recent study that investigated the early cellular mechanisms linking interstitial fibrosis with the onset of the tissue inflammatory response in a cardiac hypertrophy and failure model of angiotensin-II infusion with nonadaptive fibrosis [68]. This nonadaptive fibrosis seen in hypertrophy could be prevented by genetic depletion of MCP-1, whereas the development of hypertension, cardiac hypertrophy, and increased systolic function was seen in both wild-type and MCP-1 KO hearts, suggesting a specific role of the inflammatory response on the fibrotic response considered to be a central underlying mechanism of diastolic dysfunction in LVH. Chemokine receptor CCR2 seems to play an important role in the development of cardiac fibrosis in this model, resulting from accumulation of bone marrow-derived fibroblast precursors [69]. Moreover, overexpression of the murine renin transgene in a transgenic rat model, which depicts insulin resistance, results in salt-sensitive cardiac inflammation and oxidative stress, accompanied with myocardial fibrosis and diastolic dysfunction. This seems to play a particularly important role in female rats [70]. On the contrary, calorie restriction in DahlS.Z-Lepr(fa)/Lepr(fa) (DS/obese) rats, derived from a cross of Dahl salt-sensitive and Zucker rats, showed downregulation of ACE and angiotensin-II type 1 receptor as well as reduced inflammation. These obese rats have phenotype resembling HFpEF and calorie reduction attenuates obesity, hypertension, LVH, and diastolic dysfunction [71].

There is more evidence of the crucial role of inflammation in the development of HFpEF from different models. Thus, in DOCA-salt hypertensive rats, hypertension increased leukocyte extravasation into cardiac tissue, resulting in increased collagen deposition and ventricular stiffness [72]. In this, the anaphylatoxin C5a generated by activation of the innate immunity complement system, which is a potent inflammatory peptide mediator through the G-protein-coupled receptor C5aR (CD88) present in immune-inflammatory cells, including monocytes, macrophages, neutrophils, T cells, and mast cells, is critically involved. Thus, inhibition with the selective C5aR antagonist PMX53 attenuated inflammatory cell infiltration and reduced collagen deposition and ventricular stiffness [72]. Interestingly, new links between complement signaling and metabolism were found and demonstrated that aberrant immune responses may exacerbate obesity and metabolic dysfunction [73]. This is in line with the above-discussed link between obesity, inflammation, LVH and diastolic dysfunction in the DS/obese rate model [71], and Dahl-SS rat models of LVH with diastolic dysfunction that can be improved by calorie restriction [74]. On the contrary, exercise training, which is known to improve metabolic syndrome, has similar beneficial effects [75]. Altered metabolic homeostasis in adipose tissue promotes insulin resistance, type 2 diabetes, hypertension, and cardiovascular disease. Inflammatory and metabolic processes are mediated by certain proteolytic enzymes that share a common cellular target, protease-activated receptor 2 (PAR2), which was shown to be an important contributor to metabolic and inflammatory dysfunction and inhibition of it attenuated not only metabolic, but also cardiovascular dysfunction [76].

Interleukins (IL) were also found to be involved in the process of inflammation and diastolic dysfunction. Thus, IL-16 was found to be elevated in a rat model of HFpEF and positively correlated with LV end-diastolic pressure, lung weight, and LV myocardial stiffness constant. The cardiac expression of IL-16 was upregulated in this model. In transgenic mice, enhanced cardiac expression of IL-16 induced cardiac fibrosis and LV myocardial stiffening accompanied by increased macrophage infiltration. Treatment with anti-IL-16 neutralizing antibody ameliorated cardiac fibrosis [77]. Similarly, IL-18 overexpression in rats fed by fructose to induce metabolic syndrome using adenovirus encoding rat IL-18 had comparable effects [78]. A blockade inhibited the development of fibrosis and diastolic dysfunction in spontaneous hypertensive rats [79]. However, the simple assumption that inflammatory cytokines only mediate adverse effects in models of HFpEF has recently been challenged [80].

1.5.1.1 Interactions with Other Systems

There are significant interactions with other systems. On the one hand, various factors may stimulate or suppress inflammation. On the other hand, inflammation has effects on the heart by different pathways, of which fibrosis and changes in the extracellular matrix (ECM) are very prominent ones as described above. Another important result of inflammation is the inhibition of NO bioavailability, which may in turn decrease protein kinase G (PKG) activity, thereby inducing concentric remodeling of the left ventricle, as recently described in detail [33]. The mechanisms of peroxynitrite production, reduced NO availability, and lower soluble guanylate cyclase (sGC) activity are described below.

There is, however, evidence that reduced NO, cGMP, and sGC may not only be a consequence of inflammation and oxidative stress, respectively, but that they may also lack inhibition of inflammation. Thus, the natriuretic peptide receptor A (NPRA), which if stimulated increases cGMP, has an important role in the regulation of fibrotic and inflammatory pathways in LVH. NPRA deletion in KO mice causes salt-resistant hypertension, LVH, and fibrosis [81] and increased expression of fibrotic genes such as collagen, metalloproteinases, transforming growth factor-β (TGF-β), and tumor necrosis factor-α (TNF-α) [82]. Deletion of the BNP gene (Nppb−/−) may cause focal cardiac ventricular fibrotic lesions and increase ventricular expression of profibrotic genes, including ACE, TGF-β3, and pro-α1-collagen [83]. BNP can upregulate the production of pro- and anti-inflammatory molecules such as reactive oxygen and nitrogen species, leukotriene B4, and prostaglandin E2; increase IL10 levels; and affect cell motility of monocytic THP1 cells [84]. Moreover, co-culture of peripheral blood mononuclear cells (PBMC) from cardiac transplant recipients with BNP caused a reduction in pro-inflammatory cytokines (TNF-α, interleukin-6 (IL-6), IL-1a), while expression of anti-inflammatory and regulatory cytokines (IL-4, IL-5, IL-13) was preserved [85]. BNP was also able to directly oppose human monocyte migration to MCP1, but its ability to block MCP1-induced chemotaxis was attenuated in monocytes from HTN and HFPEF patients suggesting that this potentially beneficial anti-inflammatory function of BNP is likely compromised in chronic pressure overload and HFpEF [86]. Taken together, these studies suggest that natriuretic peptides and consequently sGC and cGMP play an important regulatory role in inflammation in HFpEF, which might also provide the opportunity for therapeutic intervention in HFpEF.

These studies from animal models fit very nicely in the above-mentioned new concept of inflammation-induced HFpEF [33]. However, the therapies tested so far in humans should have much more prominent, beneficial effects as there is a large amount of experimental evidence that inhibition of renin-angiotensin-aldosterone system has anti-inflammatory effects (e.g., Refs. [38,87,88]). Moreover, it is unclear which of all the pathways described is most important in mediating cardiac inflammation, fibrosis, and diastolic dysfunction as well as which interactions between all of these pathways, in relation to myocardial fibrosis and diastolic dysfunction, are pivotal. Many other pathways not (directly) related to inflammation have been described as playing an important role in the development of HFpEF. Finally, the impact of the chosen model is not clear, and it is also unclear which of these models are most relevant to the human situation. Obviously, this is particularly true for a controversial disease such as HFpEF.

1.5.2 Inflammation in HFpEF Human Studies

There are significant studies showing increased inflammation in patients with HFpEF. Obviously, most of these studies investigated systemic inflammation as measured by biomarkers of inflammation in peripheral blood [89] or inflammation in peripheral tissue, but not directly myocardial inflammation. There are, however, a few studies that also obtained human tissue from HFpEF patients [90]. These studies revealed structural and functional alterations in the myocardium that may be relevant to LVH, diastolic dysfunction, and HFpEF. Thus, systemic inflammation was suggested to gradually affect the cardiac vascular endothelium resulting in increased expression of endothelial adhesion molecules including VCAM1 in the heart [90]. VCAM and other endothelial adhesion molecules may lead to the activation and subendothelial migration of circulating leukocytes. HFpEF patients had high numbers of CD3, CD11, and CD45-positive leukocytes in the myocardium, increased inflammatory cell TGF-β expression, and increased levels of collagen I and III [90]. TGF-β is a very strong inducer of collagen production and stimulates the differentiation of fibroblasts into myofibroblasts. Moreover, activated myofibroblasts may themselves induce inflammation by producing cytokines and chemokines, which stimulate inflammatory cell recruitment and activation [91]. Therefore, it impacts the cardiac homeostasis of the ECM and intensifies fibrosis, predisposing to diastolic dysfunction and subsequent HFpEF, basically confirming findings from animal studies relevant to LVH and diastolic dysfunction.

There is a significant amount of data showing increased biomarkers of inflammation in patients with both HFrEF and HFpEF. Comparing biomarker levels in HFpEF versus HFrEF may help to uncover pathways that are of specific importance in HFpEF patients. There are, however, limited studies in this regard, particularly with respect to focus on inflammation. Whereas several studies have investigated biomarkers in HFrEF, the data in HFpEF is much more limited [92]. Additionally, most studies that have been performed in HFpEF are either cross-sectional or do not have a consecutive HF population with both preserved and reduced LVEF included. Furthermore, many studies only investigated a single biomarker or a group of markers with similar pathophysiological background. An important shortcoming of measuring biomarkers in the plasma, however, is the fact that it remains unclear, whether a given marker is causally involved in cardiac remodeling, whether it is upregulated in a compensatory manner, or whether it is simply an epiphenomenon of a catabolic state caused by HF. Furthermore, we do not know whether the increase in circulating levels of many of these biomarkers in the setting of HF reflects increased local cardiac synthesis *per se*, or whether it just reflects a systemic inflammatory state [93]. All these facts make interpretation of (circulating) biomarkers in human studies difficult. Nevertheless, such studies may be helpful for hypothesis generation, for identification of pathological processes in individual patients, for prognostic assessment, and possibly in the future to guide therapy in individual patients.

In hypertensive patients with metabolic syndrome, increased markers of inflammation (urinary albumin, C-reactive protein (CRP), TNF-α, and TGF-β) and fibrosis (procollagen type 1 carboxy-terminal propeptide (PICP)) were found to be independently associated with asymptomatic diastolic dysfunction as compared to patients with hypertension but without metabolic syndrome [94]. In other studies, increased inflammation as evident from increased CRP levels [95], platelet activation measured by soluble P-selectin, and endothelial dysfunction as assessed by plasma van Willebrand factor (vWf) [96] were closely related to diastolic dysfunction in patients with stable coronary artery disease. Increased inflammation measured by circulating IL-6, TNF-α, IL-8, and MCP1 and fibrosis measured by fibrotic signals (PIIINP and CITP) and matrix turnover signals (matrix metalloproteinases; TIMP-1) were also detected in a cross-sectional study of 275 stable hypertensive patients with and without HFpEF, defining varying fibro-inflammatory profiles throughout different stages of hypertensive heart disease (HHD) [97]. Two medium-sized, cross-sectional studies—one in patients with acute dyspnoea and preserved LVEF [98] and one in HFpEF patients [99]—identified the independent systemic inflammatory markers soluble ST2 (s-ST2; member of the IL1 receptor family) and pentraxin 3 (PTX3), respectively, to correlate with the presence of left-ventricular diastolic dysfunction and HFpEF and to be independent predictors of mortality in these patients.

Comparing biomarkers profiles of HFpEF and HFrEF patients is limited as mentioned above. Still, there are interesting first studies, suggesting important differences in the biomarker profiles of these two diseases, particularly with respect of inflammatory biomarkers. Thus, in a recent study comparing physiologically distinct circulating biomarkers in HFpEF patients, HFrEF patients, and community controls, there seems to be a distinguishing role for myocardial injury (high-sensitivity troponin T) with increased wall stress (N-terminal pro-BNP) in the pathophysiology of HFrEF, but a different pattern in HFpEF patients [89]. In the latter patients, systemic inflammation, as assessed by high levels of growth differentiation factor 15 (GDF15), seemed to play a crucial role specifically in the progression of HFpEF. In a recent analysis of TIME-CHF (Trial of Intensified versus standard Medical therapy in Elderly patients with Congestive Heart Failure [6,100,101]), different biomarker patterns could be found in patients with HFpEF versus HFrEF (unpublished data). We found biomarkers of cardiac damage and overload (hs-TnT, NT-proBNP) to be significantly more elevated in HFrEF patients, but markers of inflammation and fibrosis (hsCRP, s-ST2) to be significantly higher in HFpEF patients. It needs to be noted, however, that not all previous studies found the same regarding hsCRP and s-ST2 [89,99,102,103], some of which found other markers of inflammation to be elevated.

Important evidence comes from a large prospective cohort on incident heart failure. Thus, systemic inflammatory state, as evident from high circulating levels of IL-6 and TNF-α, has recently been shown to be predictive of incident HFpEF, but not of incident HFrEF [104]. This risk was independent of other known risk factors of incident HF, including ankle-arm index and incident coronary artery disease. More recently, this association was also found for soluble TNF type 1 receptor in the same cohort, further supporting the important role of inflammation regarding incident HFpEF [105].

As described in Chapter 14 of this book, specific anti-inflammatory treatment in HFrEF was disappointing. This was true not only for anti-TNF-α therapy [36,37], but also for statin treatment even in patients with coronary artery disease [106,107]. Various reasons have been discussed why this was the case. In HFpEF, such a study has not yet been performed. Therefore, the answer cannot be given if differences between HFrEF and HFpEF in this regard are of importance. Still, there is some preliminary data that anti-inflammatory therapy might well be beneficial in HFpEF patients, in contrast to HFrEF as discussed below.

1.6 OXIDATIVE STRESS, ENDOTHELIAL DYSFUNCTION AND MICROVASCULAR DISEASE

Markers of inflammation and oxidative stress have been associated with incident heart failure suggesting that inflammation has a direct effect on the myocardium, possibly by increased oxidative stress. Chronic inflammation causes excessive production of free radicals and depletion of antioxidants [108]. Generation of reactive oxygen species (ROS) is significantly increased in heart failure, as has been shown unequivocally in experimental and clinical studies [109]. Important sources of ROS in heart failure include the mitochondrial electron chain transport, xanthine oxidase, uncoupled endothelial nitric oxide synthases (eNOS), and nicotinamide adenine dinucleotide phosphate (NADPH) oxidases [110]. Additionally, inflammatory cytokines stimulate NADPH oxidases. Increased activation of NADPH oxidases, a catalyst of the one-electron transfer from NADPH to O^- and ROS, results in increased endothelial production of ROS. In experimental diabetic cardiomyopathy, NADPH oxidases were the primary source of ROS and inhibition of NADPH oxidases diminished oxidative stress and proved beneficial on systolic function [111].

The majority of ROS in heart failure originates from the mitochondria, and impaired electron chain transport has been pointed out as the main source of intracellular ROS [112]. How ROS induces myocardial remodeling and contributes to the development of heart failure remains, however, a topic of discussion [113]. It has been suggested that ROS has a direct effect on the myocardium, by causing physical damage of cellular and mitochondrial structures, such as sarcomeric proteins, which results in impaired cardiac function [113]. In addition, several hypertrophic signaling pathways are regulated by ROS, involving protein kinase C, Jun-N-terminal kinase, and Ras signaling [114]. Furthermore, ROS facilitates ECM remodeling, either directly or indirectly through activation of the nuclear factor kB pathwa [115]. Lastly, ROS causes DNA strand breaks, which lead to activation of nuclear enzyme poly(ADP-ribose) polymerase-1 (PARP-1), which in turn regulates the expression of a variety of inflammatory mediators. In a recent study, plasma levels of inflammatory and oxidative stress biomarkers (TBARS and 8-epi-prostaglandin F2α) were increased in HFpEF patients compared to healthy control subjects [116]. There was a significant relationship between the 8-epi-prostaglandin F2α levels and peak VO_2, suggesting a relationship between oxidative stress and functional capacity in HFpEF.

Endothelial dysfunction is an important link between the pro-inflammatory state in HFpEF and increased levels of oxidative stress [117]. Cytokines enhance expression of adhesion molecules and induce inflammatory cytokines in endothelial cells resulting in increased inflammation within the vessel wall. More importantly, cytokines modulate the balance between endogenous vasodilators (e.g., nitric oxide (NO)) and vasoconstriction (e.g., endothelin-1) [118]. Increase of oxidative stress and endothelial dysfunction has been described in both HFpEF and HFrEF, but is also implicated in other diseases, including diabetes, hypertension (both of which are highly prevalent in HFpEF), and in atherosclerosis (which is highly prevalent in HFrEF) [119]. However, in a study by Chiang *et al.*, both HFpEF and HFrEF patients have decreased circulating endothelial progenitor cells, an indicator for impaired endothelial turnover, compared to age-, gender-, and comorbidity-matched controls. This suggests that increased oxidative stress and endothelial dysfunction is not merely a result of a myriad of comorbidities, but is also linked specifically to the pathophysiology of heart failure [120].

Exercise intolerance is the key symptom of HFpEF, yet its underlying pathophysiological mechanism remains a topic of discussion [121]. Several studies have demonstrated a blunted exercise-induced vasodilatation. This may be explained in part by systemic endothelial dysfunction [46,122]. In a study comparing HFpEF patients with age- and sex-matched hypertensives without symptoms as well as age- and sex-matched healthy controls, global vascular function (Ea and systemic vascular resistance index (SVRI)) was not significantly different between groups, but endothelial function was impaired in both HFpEF and hypertensive subjects compared to the controls, even after adjusting for history of coronary disease [123]. Hyperemic increase in peripheral arterial tonometry (PAT) amplitude after cuff occlusion was augmented in HFpEF and hypertensives, as was depression of endothelium-dependent vasodilation. Interestingly, HFpEF patients had an impaired endothelial function accompanied by a reduction in exercise capacity, but hypertensive controls only displayed an impaired endothelial function, suggesting endothelial dysfunction does not directly lead to reduced exercise capacity. In contrast, Akiyama *et al.* reported a poorer NYHA class in those with impaired endothelial function [124]. HFpEF patients demonstrated a higher prevalence of endothelial dysfunction even after matching for age, gender, diabetes, and hypertension. Additionally, endothelial dysfunction predicted cardiovascular events, independently of age, diabetes, hospitalization, NYHA class, E/e′, ejection fraction, and BNP. Another study demonstrated flow-mediated arterial dilation as expression of endothelial dysfunction of the femoral artery, measured using MRI, was impaired in HFrEF, but not in HFpEF [125]. A potential explanation for this at first sight discrepant findings is that flow-mediated vasodilation may be different in the larger arteries compared to the microvasculature.

Inflammation does not only affect the systemic microvasculature but also the coronary microvasculature. Myocardial biopsy samples of HFpEF patients showed increased inflammatory cells, marked by CD3, CD11a, and CD45, but also abundant expression of endothelial adhesion molecules such as vascular cell adhesion molecule (VCAM) accompanied by accumulation of collagen and presence of ROS [90]. Furthermore, a double staining in these samples showed secretion of the profibrotic growth factor; TGF-β by inflammatory cells. These findings suggest inflammation plays an important role in the regulation of the ECM and thus may contribute to diastolic dysfunction in HFpEF. In patients with cardiac parvovirus B19 infection, coronary endothelial function was impaired [126]. After intracoronary application of acetylcholine, the coronary microvasculature displayed an impaired vasodilation response. In these patients, coronary endothelial dysfunction was correlated with diastolic dysfunction.

NO is considered a key player in endothelial dysfunction. NO promotes LV relaxation through cyclic guanosine monophosphate (cGMP)—PKG dependent and independent mechanisms. The effects of NO depend on where and by which NO synthase isoform NO is produced [127]. Inflammation induces endothelial production of ROS, which leads to the formation of peroxynitrite and reduces NO bioavailability. As a result, it decreases sGC activity in cardiomyocytes adjacent to the dysfunctional endothelium. Low sGC activity in turn reduces cGMP and consequently PKG activity. Myocardial samples from HpEF patients have reduced levels of PKG activity and lower cGMP concentrations, related to increased cardiomyocyte stiffness, measured using passive tension (Fpassive) [128]. The downregulation of cGMP-PKG was likely related to low myocardial NO bioavailability, demonstrated by high nitrosative/oxidative stress assessed by immunohistochemical determination of nitrotyrosine.

PKG dependent phosphorylation of the sarcomeric protein titin seems to play an important role in this process [129]. Expression and phosphorylation of titin isoforms were analyzed in LV biopsies of heart failure patients, aortic stenosis patients, and controls. Titin expression shifted in heart failure compared to the aortic stenosis and control patients, with lower expression of its compliant N2BA isoform and higher expression of its stiff N2B isoform in heart failure [130]. High diastolic stiffness was correlated to relative hypophosphorylation of the stiff N2B titin isoform, and in HFpEF the N2BA:N2B expression ratio was decreased compared to HFrEF patients [131].

An additional downstream effect of PKG involves reuptake of Ca^{2+} into the sarcoplasmic reticulum (SR), inhibition of calcium influx, and suppression of hypertrophic and fibrotic signaling pathways [131,132]. The lack of PKG in the cardiomyocyte enhances hypertrophy, as seen in experimental and clinical studies [133]. In patients with diabetic cardiomyopathy and concentric LV remodeling treatment with sildenafil, which increases myocardial PKG activity through inhibited breakdown of cGMP by phosphodiesterase 5, reduced hypertrophy, e.g., LV mass/volume ratio, and improved cardiac kinetics [134]. In a study comparing HFpEF with HFrEF, HFpEF patients exhibited lower myocardial PKG activity, which correlated with elevated cardiomyocyte diameter compared to HFrEF [128].

Endothelial dysfunction is not just confined to the systemic or the coronary arteries, but also includes the pulmonary arteries. PH is a common feature in HFpEF. In a community-based study, PH was present in 83% of HFpEF patients [22]. This population demonstrated an increased pulmonary capillary wedge pressure; however, the severity of PH suggests an additional pre-capillary component contributes as well. It is possible that endothelial dysfunction plays a role in that aspect. In HFpEF patients with PH, sildenafil treatment resulted in improvement of diastolic stiffness and reduction of pulmonary pressure [135].

1.6.1 Potential Implications for Treatment of HFpEF

As mentioned above, there is no specific treatment available for HFpEF patients apart from treating potentially underlying diseases such as hypertension and comorbidities that are important in HFpEF. As inflammation, oxidative stress, and endothelial dysfunction seem to play an important, though probably not exclusive, role in HFpEF, inhibition of inflammation and oxidative stress and improvement of endothelial are potentially promising targets for these patients with significant morbidity and mortality.

Although failed in HFrEF [106,107], statins are interesting agents in this regard if inflammation played a much more causal role in HFpEF as compared to HFrEF. Statins are believed to have pleiotropic effects independent of cholesterol lowering. Thus, direct effects on the vascular wall in human arteries were found by the use of atorvastatin in improved endothelial redox balance and reduced superoxide anion production and restored NO bioavailability [136]. These effects are not limited to the vascular wall, but also reach cardiomyocytes and fibroblasts thereby preventing myocardial fibrosis and diastolic dysfunction in experimental hypertension or hypercholesterolemia as summarized by Ramasubbu *et al.* [137]. Therefore, it may be seen as no big surprise that treatment of patients without overt cardiovascular disease selected based on slightly elevated CRP levels had significantly less cardiovascular events than those untreated [138]. Although these patients had no HFpEF and diastolic dysfunction was not measured, the significant reduction in cardiovascular events is promising for the patient population described here because it is likely that at least some of the patients included were also at significant risk for developing HFpEF. This is in line with the positive effect on clinical outcome in a relatively small not randomized study on the effects of statins in HFpEF patients [139]. Still, subgroup analysis of the GISSI-HF trial did not find positive effects in patients with LVEF of >40% [107]. It must be noted, however, that this comprised only about 10% of all patients, and patients with slightly reduced LVEF and HFpEF are probably two different types of patient populations.

As nicely summarized in a recent review article [34], there is quite a substantial number of experimental studies that provide promising results of targeting immunomodulatory and inflammatory pathways in hypertensive HFpEF-relevant animal models. They include targeting cytokines and chemokines (MCP1 [67,140], MCP3 [141], IL10 [142], IL-1 receptors [143]), matrix-modulating enzymes (MMP) [144-146], pentraxins (PTX3) [147,148], and inflammatory signal transduction mediators (phosphatidylinositol 3-kinase gamma (PI3 Kc)) [149]. Still, it needs to be mentioned that these factors were investigated to a large extent in isolation. Thus, the interaction between the various pathways has not been investigated in detail, and it is not clear which of them is most promising as potential therapeutic targets or if a less specific inhibition of inflammation is most important.

Direct evidence in humans is very limited so far. A preliminary study testing inhibition of IL-1 in patients with HFpEF provides interesting results [35]. Thus, IL-1 blockade with anakinra for 14 days significantly reduced the systemic inflammatory response and improved the aerobic exercise capacity of patients with HFpEF and elevated plasma CRP levels in a cross-over, randomized, double-blind design. Still, this was a very small study, including only 12 patients [35] with a significant risk of a chance finding. Another pilot study investigates the effects of vitamin D (paricalcitol) in patients with HFpEF (NCT01630408). An animal study in Dahl salt-sensitive rats was promising, and was also found retrospectively in patients on hemodialysis [150]. Moreover, low vitamin D is associated with poor outcome related to RAAS activation and, importantly, inflammation [151]. Whether vitamin D can be seen as directly targeting inflammation or whether there are indirect effects is not yet clear.

PDE-5 inhibition, which may increase cGMP by inhibiting its breakdown is another potential therapeutic option that directly targets one of the important suggested consequences of inflammation, as discussed above. Indeed, preliminary findings in small studies have been promising [55,135], but the first larger trial (phase IIb), unfortunately, did not show any benefit in HFpEF patients regarding improvement of exercise capacity and clinical status [56]. In fact, there was not even a trend toward improvement by the PDE-5 inhibitor sildenafil. The reason for this finding is, obviously, not clear. Still, it cannot be claimed that a large proportion of patients included did not really have HFpEF as in other studies since exercise capacity was significantly reduced and there was echocardiographic evidence for increased filling pressure. Obviously, more studies are required to get a sufficient answer to the question of whether PDE-5 inhibition is a tool to improve outcome in HFpEF patients. Given the diversity of patients considered to have HFpEF, it may well be that only a subgroup may profit. This obviously includes those with some increase in pulmonary vascular resistance, but proper studies are required to answer these questions.

1.7 CONCLUSIONS

Undoubtedly, inflammation is increased in heart failure. This is true for both HFrEF and HFpEF. In this chapter, we specifically focused on HFpEF because inflammation in HFrEF is largely covered in other chapters of this book. There is substantial evidence that inflammation and the processes related to it such as oxidative stress, endothelial dysfunction and microvascular disease are not only activated and are important bystanders of HFpEF, but that they also play a pathophysiologically important and causative role. However, as to whether a "simple" paradigm shift in HFpEF from afterload excess to coronary microvascular inflammation due to pro-inflammatory state related to multiple comorbidities provides identification of the most important central link in HFpEF [33], remains to be determined [62]. Importantly, this paradigm shift does not explain why previous attempts in treating patients with HFpEF failed. Thus, blockade of the renin-angiotensin system, either by ACE-inhibitors or by ARBs, has anti-inflammatory effects and may also reduce structural alterations in the myocardium proposed to be of importance—namely, fibrosis [152]. Thus, an effect that is larger than the effects caused purely by antihypertensive action of these drugs, would be expected. Still, the opposite is the case. The same is also true for inhibition of PDE-5, which is believed to address a central part of the cascade affected by the initial trigger of inflammation [153]. However, results from the largest trial are disappointing despite earlier promising results in smaller mechanistic trials. Even for statins, the same picture is incomplete and resembles that of the PDE-5 inhibitors, where some studies are promising, but others are not. Therefore, it might be that the central mechanism of HFpEF is still not yet determined or HFpEF is not a uniform disease, but a clinical syndrome of different diseases where different pathways play differently important roles. The latter hypothesis is favored, but this needs to be further investigated. Among these potential mechanisms, inflammation may well be an important one.

To address these uncertainties, it seems important not to easily adopt or reject a certain concept. All of the pathways studied so far, despite none providing convincing data on interventions in patients, may still play a role in certain (sub)groups of patients in the HFpEF population. Moreover, potentially crucial mechanisms need to be considered without neglecting the possibility that results may not apply to all HFpEF patients. From what is known so far, inflammation is a very interesting potential target in this regard. There is a significant number of emerging anti-inflammatory targets with potential benefit for HFpEF therapy [34]. Numerous trials are ongoing or planned. However, it might be wise to do good phenotyping of the patients included in the trials, or to even target such agents in subgroups of HFpEF patients only. One possibility would be to use biomarkers of inflammation for selection of patients for inclusion in trials.

References

[1] Remme WJ, Swedberg K. Guidelines for the diagnosis and treatment of chronic heart failure. Eur Heart J 2001;22(17):1527–60.
[2] Owan TE, Hodge DO, Herges RM, Jacobsen SJ, Roger VL, Redfield MM. Trends in prevalence and outcome of heart failure with preserved ejection fraction. N Engl J Med 2006;355(3):251–9.
[3] McMurray JJ, Adamopoulos S, Anker SD, Auricchio A, Bohm M, Dickstein K, et al. ESC Guidelines for the diagnosis and treatment of acute and chronic heart failure 2012: the Task Force for the Diagnosis and Treatment of Acute and Chronic Heart Failure 2012 of the European Society of Cardiology. Developed in collaboration with the Heart Failure Association (HFA) of the ESC. Eur Heart J 2012;33(14):1787–847.

[4] Massie BM, Carson PE, McMurray JJ, Komajda M, McKelvie R, Zile MR, et al. Irbesartan in patients with heart failure and preserved ejection fraction. N Engl J Med 2008;359(23):2456–67.
[5] Cleland JG, Tendera M, Adamus J, Freemantle N, Polonski L, Taylor J. The perindopril in elderly people with chronic heart failure (PEP-CHF) study. Eur Heart J 2006;27(19):2338–45.
[6] Maeder MT, Rickenbacher P, Rickli H, Abbuhl H, Gutmann M, Erne P, et al. N-terminal pro brain natriuretic peptide-guided management in patients with heart failure and preserved ejection fraction: findings from the Trial of Intensified versus standard medical therapy in elderly patients with congestive heart failure (TIME-CHF). Eur J Heart Fail 2013;15(10):1148–56.
[7] Brunner-La Rocca HP, Attenhofer CH, Jenni R. Can the extent of change of the left ventricular Doppler inflow pattern during the Valsalva maneuver predict an elevated left ventricular end-diastolic pressure? Echocardiography 1998;15(3):211–8.
[8] Swanson KL, Utz JP, Krowka MJ. Doppler echocardiography-right heart catheterization relationships in patients with idiopathic pulmonary fibrosis and suspected pulmonary hypertension. Med Sci Monit 2008;14(4):CR177–82.
[9] Maisel AS, Krishnaswamy P, Nowak RM, McCord J, Hollander JE, Duc P, et al. Rapid measurement of B-type natriuretic peptide in the emergency diagnosis of heart failure. N Engl J Med 2002;347(3):161–7.
[10] Mueller C, Scholer A, Laule-Kilian K, Martina B, Schindler C, Buser P, et al. Use of B-type natriuretic peptide in the evaluation and management of acute dyspnea. N Engl J Med 2004;350(7):647–54.
[11] Anand IS, Fisher LD, Chiang YT, Latini R, Masson S, Maggioni AP, et al. Changes in brain natriuretic peptide and norepinephrine over time and mortality and morbidity in the Valsartan Heart Failure Trial (Val-HeFT). Circulation 2003;107(9):1278–83.
[12] Pfisterer M, Buser P, Rickli H, Gutmann M, Erne P, Rickenbacher P, et al. BNP-guided vs symptom-guided heart failure therapy: the Trial of Intensified vs Standard Medical Therapy in Elderly Patients With Congestive Heart Failure (TIME-CHF) randomized trial. JAMA 2009;301(4):383–92.
[13] Troughton RW, Frampton CM, Brunner-La Rocca HP, Pfisterer M, Eurlings LW, Erntell H, et al. Effect of B-type natriuretic peptide guided treatment of chronic heart failure on total mortality and hospitalisation: an individual patient meta-analysis. Eur Heart J 2014;35(23):1559–67.
[14] Bishu K, Deswal A, Chen HH, Lewinter MM, Lewis GD, Semigran MJ, et al. Biomarkers in acutely decompensated heart failure with preserved or reduced ejection fraction. Am Heart J 2012;164(5):763–70.
[15] Hummel SL, Kitzman DW. Update on heart failure with preserved ejection fraction. Curr Cardiovasc Risk Rep 2013;7(6):495–502.
[16] Edelmann F, Stahrenberg R, Gelbrich G, Durstewitz K, Angermann CE, Dungen HD, et al. Contribution of comorbidities to functional impairment is higher in heart failure with preserved than with reduced ejection fraction. Clin Res Cardiol 2011;100(9):755–64.
[17] Zakeri R, Borlaug BA, McNulty SE, Mohammed SF, Lewis GD, Semigran MJ, et al. Impact of atrial fibrillation on exercise capacity in heart failure with preserved ejection fraction: a RELAX trial ancillary study. Circ Heart Fail 2014;7(1):123–30.
[18] Lam CS, Donal E, Kraigher-Krainer E, Vasan RS. Epidemiology and clinical course of heart failure with preserved ejection fraction. Eur J Heart Fail 2011;13(1):18–28.
[19] Caliskan K, Balk AH, Jordaens L, Szili-Torok T. Bradycardiomyopathy: the case for a causative relationship between severe sinus bradycardia and heart failure. J Cardiovasc Electrophysiol 2010;21(7):822–4.
[20] Mohammed SF, Borlaug BA, McNulty S, Lewis GD, Lin G, Zakeri R, et al. Resting ventricular-vascular function and exercise capacity in heart failure with preserved ejection fraction: a RELAX trial ancillary study. Circ Heart Fail 2014;7:580–9.
[21] Bhuiyan T, Maurer MS. Heart failure with preserved ejection fraction: persistent diagnosis, therapeutic enigma. Curr Cardiovasc Risk Rep 2011;5(5):440–9.
[22] Lam CS, Roger VL, Rodeheffer RJ, Borlaug BA, Enders FT, Redfield MM. Pulmonary hypertension in heart failure with preserved ejection fraction: a community-based study. J Am Coll Cardiol 2009;53(13):1119–26.
[23] Maeder MT, Thompson BR, Brunner-La Rocca HP, Kaye DM. Hemodynamic basis of exercise limitation in patients with heart failure and normal ejection fraction. J Am Coll Cardiol 2010;56(11):855–63.
[24] Thenappan T, Shah SJ, Gomberg-Maitland M, Collander B, Vallakati A, Shroff P, et al. Clinical characteristics of pulmonary hypertension in patients with heart failure and preserved ejection fraction. Circ Heart Fail 2011;4(3):257–65.
[25] Bursi F, Weston SA, Redfield MM, Jacobsen SJ, Pakhomov S, Nkomo VT, et al. Systolic and diastolic heart failure in the community. JAMA 2006;296(18):2209–16.
[26] Padwal R, McAlister FA, McMurray JJ, Cowie MR, Rich M, Pocock S, et al. The obesity paradox in heart failure patients with preserved versus reduced ejection fraction: a meta-analysis of individual patient data. Int J Obes 2014;38:1110–4.
[27] Mohammed SF, Borlaug BA, Roger VL, Mirzoyev SA, Rodeheffer RJ, Chirinos JA, et al. Comorbidity and ventricular and vascular structure and function in heart failure with preserved ejection fraction: a community-based study. Circ Heart Fail 2012;5(6):710–9.
[28] Charlson ME, Pompei P, Ales KL, MacKenzie CR. A new method of classifying prognostic comorbidity in longitudinal studies: development and validation. J Chronic Dis 1987;40(5):373–83.
[29] Ather S, Chan W, Bozkurt B, Aguilar D, Ramasubbu K, Zachariah AA, et al. Impact of noncardiac comorbidities on morbidity and mortality in a predominantly male population with heart failure and preserved versus reduced ejection fraction. J Am Coll Cardiol 2012;59(11):998–1005.
[30] Hamaguchi S, Kinugawa S, Sobirin MA, Goto D, Tsuchihashi-Makaya M, Yamada S, et al. Mode of death in patients with heart failure and reduced vs. preserved ejection fraction: report from the registry of hospitalized heart failure patients. Circ J 2012;76(7):1662–9.
[31] Rickenbacher P, Pfisterer M, Burkard T, Kiowski W, Follath F, Burckhardt D, et al. Why and how do elderly patients with heart failure die? Insights from the TIME-CHF study. Eur J Heart Fail 2012;14(11):1218–29.
[32] Somaratne JB, Berry C, McMurray JJ, Poppe KK, Doughty RN, Whalley GA. The prognostic significance of heart failure with preserved left ventricular ejection fraction: a literature-based meta-analysis. Eur J Heart Fail 2009;11(9):855–62.
[33] Paulus WJ, Tschope C. A novel paradigm for heart failure with preserved ejection fraction: comorbidities drive myocardial dysfunction and remodeling through coronary microvascular endothelial inflammation. J Am Coll Cardiol 2013;62(4):263–71.
[34] Glezeva N, Baugh JA. Role of inflammation in the pathogenesis of heart failure with preserved ejection fraction and its potential as a therapeutic target. Heart Fail Rev 2014;19:681–94.

[35] Van Tassell BW, Arena R, Biondi-Zoccai G, McNair CJ, Oddi C, Abouzaki NA, et al. Effects of interleukin-1 blockade with anakinra on aerobic exercise capacity in patients with heart failure and preserved ejection fraction (from the D-HART pilot study). Am J Cardiol 2014;113(2):321–7.
[36] Mann DL, McMurray JJ, Packer M, Swedberg K, Borer JS, Colucci WS, et al. Targeted anticytokine therapy in patients with chronic heart failure: results of the Randomized Etanercept Worldwide Evaluation (RENEWAL). Circulation 2004;109(13):1594–602.
[37] Chung ES, Packer M, Lo KH, Fasanmade AA, Willerson JT. Randomized, double-blind, placebo-controlled, pilot trial of infliximab, a chimeric monoclonal antibody to tumor necrosis factor-alpha, in patients with moderate-to-severe heart failure: results of the anti-TNF Therapy Against Congestive Heart Failure (ATTACH) trial. Circulation 2003;107(25):3133–40.
[38] Sciarretta S, Paneni F, Palano F, Chin D, Tocci G, Rubattu S, et al. Role of the renin-angiotensin-aldosterone system and inflammatory processes in the development and progression of diastolic dysfunction. Clin Sci (Lond) 2009;116(6):467–77.
[39] Wright JW, Mizutani S, Harding JW. Pathways involved in the transition from hypertension to hypertrophy to heart failure. Treatment strategies. Heart Fail Rev 2008;13(3):367–75.
[40] Garg R, Yusuf S. Overview of randomized trials of angiotensin-converting enzyme inhibitors on mortality and morbidity in patients with heart failure. Collaborative Group on ACE Inhibitor Trials. JAMA 1995;273(18):1450–6.
[41] Mohammed SF, Ohtani T, Korinek J, Lam CS, Larsen K, Simari RD, et al. Mineralocorticoid accelerates transition to heart failure with preserved ejection fraction via "nongenomic effects". Circulation 2010;122(4):370–8.
[42] Pitt B, Pfeffer MA, Assmann SF, Boineau R, Anand IS, Claggett B, et al. Spironolactone for heart failure with preserved ejection fraction. N Engl J Med 2014;370(15):1383–92.
[43] Pitt B, Zannad F, Remme WJ, Cody R, Castaigne A, Perez A, et al. The effect of spironolactone on morbidity and mortality in patients with severe heart failure. Randomized Aldactone Evaluation Study Investigators. N Engl J Med 1999;341(10):709–17.
[44] Zannad F, McMurray JJ, Krum H, van Veldhuisen DJ, Swedberg K, Shi H, et al. Eplerenone in patients with systolic heart failure and mild symptoms. N Engl J Med 2011;364(1):11–21.
[45] Schlaich MP, Kaye DM, Lambert E, Sommerville M, Socratous F, Esler MD. Relation between cardiac sympathetic activity and hypertensive left ventricular hypertrophy. Circulation 2003;108(5):560–5.
[46] Borlaug BA, Melenovsky V, Russell SD, Kessler K, Pacak K, Becker LC, et al. Impaired chronotropic and vasodilator reserves limit exercise capacity in patients with heart failure and a preserved ejection fraction. Circulation 2006;114:2138–47.
[47] van Veldhuisen DJ, McMurray JJ. Pharmacological treatment of heart failure with preserved ejection fraction: a glimpse of light at the end of the tunnel? Eur J Heart Fail 2013;15(1):5–8.
[48] Kjeldsen SE, Dahlof B, Devereux RB, Julius S, Aurup P, Edelman J, et al. Effects of losartan on cardiovascular morbidity and mortality in patients with isolated systolic hypertension and left ventricular hypertrophy: a Losartan Intervention for Endpoint Reduction (LIFE) substudy. JAMA 2002;288(12):1491–8.
[49] Packer M, Coats AJ, Fowler MB, Katus HA, Krum H, Mohacsi P, et al. Effect of carvedilol on survival in severe chronic heart failure. N Engl J Med 2001;344(22):1651–8.
[50] Zile MR, Bourge RC, Redfield MM, Zhou D, Baicu CF, Little WC. Randomized, double-blind, placebo-controlled study of sitaxsentan to improve impaired exercise tolerance in patients with heart failure and a preserved ejection fraction. JACC Heart Fail 2014;2(2):123–30.
[51] Kiowski W, Suetsch G, Oechslin E, Schalcher C, Brunner-La Rocca HP, Bertel O. Rationale and perspective of endothelin-1 antagonism in acute heart failure. J Cardiovasc Pharmacol 2001;38(Suppl. 2):S53–7.
[52] Packer M, McMurray J, Massie BM, Caspi A, Charlon V, Cohen-Solal A, et al. Clinical effects of endothelin receptor antagonism with bosentan in patients with severe chronic heart failure: results of a pilot study. J Card Fail 2005;11(1):12–20.
[53] Sackner-Bernstein JD, Kowalski M, Fox M, Aaronson K. Short-term risk of death after treatment with nesiritide for decompensated heart failure: a pooled analysis of randomized controlled trials. JAMA 2005;293(15):1900–5.
[54] O'Connor CM, Starling RC, Hernandez AF, Armstrong PW, Dickstein K, Hasselblad V, et al. Effect of nesiritide in patients with acute decompensated heart failure. N Engl J Med 2011;365(1):32–43.
[55] Guazzi M, Vicenzi M, Arena R, Guazzi MD. Pulmonary hypertension in heart failure with preserved ejection fraction: a target of phosphodiesterase-5 inhibition in a 1-year study. Circulation 2011;124(2):164–74.
[56] Redfield MM, Chen HH, Borlaug BA, Semigran MJ, Lee KL, Lewis G, et al. Effect of phosphodiesterase-5 inhibition on exercise capacity and clinical status in heart failure with preserved ejection fraction: a randomized clinical trial. JAMA 2013;309(12):1268–77.
[57] Guazzi M, Samaja M, Arena R, Vicenzi M, Guazzi MD. Long-term use of sildenafil in the therapeutic management of heart failure. J Am Coll Cardiol 2007;50(22):2136–44.
[58] Tschope C, Walther T, Escher F, Spillmann F, Du J, Altmann C, et al. Transgenic activation of the kallikrein-kinin system inhibits intramyocardial inflammation, endothelial dysfunction and oxidative stress in experimental diabetic cardiomyopathy. FASEB J 2005;19(14):2057–9.
[59] Wei CC, Chen Y, Powell LC, Zheng J, Shi K, Bradley WE, et al. Cardiac kallikrein-kinin system is upregulated in chronic volume overload and mediates an inflammatory induced collagen loss. PLoS One 2012;7(6):e40110.
[60] Liu YH, Yang XP, Sharov VG, Nass O, Sabbah HN, Peterson E, et al. Effects of angiotensin-converting enzyme inhibitors and angiotensin II type 1 receptor antagonists in rats with heart failure. Role of kinins and angiotensin II type 2 receptors. J Clin Invest 1997;99(8):1926–35.
[61] Yusuf S, Pfeffer MA, Swedberg K, Granger CB, Held P, McMurray JJ, et al. Effects of candesartan in patients with chronic heart failure and preserved left-ventricular ejection fraction: the CHARM-preserved trial. Lancet 2003;362(9386):777–81.
[62] Desai AS. Heart failure with preserved ejection fraction: time for a new approach? J Am Coll Cardiol 2013;62(4):272–4.
[63] Nicoletti A, Heudes D, Mandet C, Hinglais N, Bariety J, Michel JB. Inflammatory cells and myocardial fibrosis: spatial and temporal distribution in renovascular hypertensive rats. Cardiovasc Res 1996;32(6):1096–107.
[64] Hinglais N, Heudes D, Nicoletti A, Mandet C, Laurent M, Bariety J, et al. Colocalization of myocardial fibrosis and inflammatory cells in rats. Lab Invest 1994;70(2):286–94.
[65] Komatsu S, Panes J, Russell JM, Anderson DC, Muzykantov VR, Miyasaka M, et al. Effects of chronic arterial hypertension on constitutive and induced intercellular adhesion molecule-1 expression in vivo. Hypertension 1997;29(2):683–9.

[66] Nicoletti A, Heudes D, Hinglais N, Appay MD, Philippe M, Sassy-Prigent C, et al. Left ventricular fibrosis in renovascular hypertensive rats. Effect of losartan and spironolactone. Hypertension 1995;26(1):101–11.
[67] Kuwahara F, Kai H, Tokuda K, Takeya M, Takeshita A, Egashira K, et al. Hypertensive myocardial fibrosis and diastolic dysfunction: another model of inflammation? Hypertension 2004;43(4):739–45.
[68] Haudek SB, Cheng J, Du J, Wang Y, Hermosillo-Rodriguez J, Trial J, et al. Monocytic fibroblast precursors mediate fibrosis in angiotensin-II-induced cardiac hypertrophy. J Mol Cell Cardiol 2010;49(3):499–507.
[69] Xu J, Lin SC, Chen J, Miao Y, Taffet GE, Entman ML, et al. CCR2 mediates the uptake of bone marrow-derived fibroblast precursors in angiotensin II-induced cardiac fibrosis. Am J Physiol Heart Circ Physiol 2011;301(2):H538–47.
[70] Whaley-Connell AT, Habibi J, Aroor A, Ma L, Hayden MR, Ferrario CM, et al. Salt loading exacerbates diastolic dysfunction and cardiac remodeling in young female Ren2 rats. Metabolism 2013;62(12):1761–71.
[71] Takatsu M, Nakashima C, Takahashi K, Murase T, Hattori T, Ito H, et al. Calorie restriction attenuates cardiac remodeling and diastolic dysfunction in a rat model of metabolic syndrome. Hypertension 2013;62(5):957–65.
[72] Iyer A, Woodruff TM, Wu MC, Stylianou C, Reid RC, Fairlie DP, et al. Inhibition of inflammation and fibrosis by a complement C5a receptor antagonist in DOCA-salt hypertensive rats. J Cardiovasc Pharmacol 2011;58(5):479–86.
[73] Lim J, Iyer A, Suen JY, Seow V, Reid RC, Brown L, et al. C5aR and C3aR antagonists each inhibit diet-induced obesity, metabolic dysfunction, and adipocyte and macrophage signaling. FASEB J 2013;27(2):822–31.
[74] Seymour EM, Parikh RV, Singer AA, Bolling SF. Moderate calorie restriction improves cardiac remodeling and diastolic dysfunction in the Dahl-SS rat. J Mol Cell Cardiol 2006;41(4):661–8.
[75] Nunes RB, Alves JP, Kessler LP, Dal LP. Aerobic exercise improves the inflammatory profile correlated with cardiac remodeling and function in chronic heart failure rats. Clinics (Sao Paulo) 2013;68(6):876–82.
[76] Lim J, Iyer A, Liu L, Suen JY, Lohman RJ, Seow V, et al. Diet-induced obesity, adipose inflammation, and metabolic dysfunction correlating with PAR2 expression are attenuated by PAR2 antagonism. FASEB J 2013;27(12):4757–67.
[77] Tamaki S, Mano T, Sakata Y, Ohtani T, Takeda Y, Kamimura D, et al. Interleukin-16 promotes cardiac fibrosis and myocardial stiffening in heart failure with preserved ejection fraction. PLoS One 2013;8(7):e68893.
[78] Xing SS, Bi XP, Tan HW, Zhang Y, Xing QC, Zhang W. Overexpression of interleukin-18 aggravates cardiac fibrosis and diastolic dysfunction in fructose-fed rats. Mol Med 2010;16(11–12):465–70.
[79] Liu W, Wang X, Feng W, Li S, Tian W, Xu T, et al. Lentivirus mediated IL-17R blockade improves diastolic cardiac function in spontaneously hypertensive rats. Exp Mol Pathol 2011;91(1):362–7.
[80] Garcia AG, Wilson RM, Heo J, Murthy NR, Baid S, Ouchi N, et al. Interferon-gamma ablation exacerbates myocardial hypertrophy in diastolic heart failure. Am J Physiol Heart Circ Physiol 2012;303(5):H587–96.
[81] Kuhn M, Holtwick R, Baba HA, Perriard JC, Schmitz W, Ehler E. Progressive cardiac hypertrophy and dysfunction in atrial natriuretic peptide receptor (GC-A) deficient mice. Heart 2002;87(4):368–74.
[82] Vellaichamy E, Khurana ML, Fink J, Pandey KN. Involvement of the NF-kappa B/matrix metalloproteinase pathway in cardiac fibrosis of mice lacking guanylyl cyclase/natriuretic peptide receptor A. J Biol Chem 2005;280(19):19230–42.
[83] Tamura N, Ogawa Y, Chusho H, Nakamura K, Nakao K, Suda M, et al. Cardiac fibrosis in mice lacking brain natriuretic peptide. Proc Natl Acad Sci U S A 2000;97(8):4239–44.
[84] Chiurchiu V, Izzi V, D'Aquilio F, Carotenuto F, Di NP, Baldini PM. Brain Natriuretic Peptide (BNP) regulates the production of inflammatory mediators in human THP-1 macrophages. Regul Pept 2008;148(1–3):26–32.
[85] Shaw SM, Critchley WR, Puchalka CM, Williams SG, Yonan N, Fildes JE. Brain natriuretic peptide induces CD8+ T cell death via a caspase 3 associated pathway–implications following heart transplantation. Transpl Immunol 2012;26(2–3):119–22.
[86] Glezeva N, Collier P, Voon V, Ledwidge M, McDonald K, Watson C, et al. Attenuation of monocyte chemotaxis—a novel anti-inflammatory mechanism of action for the cardio-protective hormone B-type natriuretic peptide. J Cardiovasc Transl Res 2013;6(4):545–57.
[87] Westermann D, Rutschow S, Jager S, Linderer A, Anker S, Riad A, et al. Contributions of inflammation and cardiac matrix metalloproteinase activity to cardiac failure in diabetic cardiomyopathy: the role of angiotensin type 1 receptor antagonism. Diabetes 2007;56(3):641–6.
[88] Usui M, Egashira K, Tomita H, Koyanagi M, Katoh M, Shimokawa H, et al. Important role of local angiotensin II activity mediated via type 1 receptor in the pathogenesis of cardiovascular inflammatory changes induced by chronic blockade of nitric oxide synthesis in rats. Circulation 2000;101(3):305–10.
[89] Santhanakrishnan R, Chong JP, Ng TP, Ling LH, Sim D, Leong KT, et al. Growth differentiation factor 15, ST2, high-sensitivity troponin T, and N-terminal pro brain natriuretic peptide in heart failure with preserved vs. reduced ejection fraction. Eur J Heart Fail 2012;14(12):1338–47.
[90] Westermann D, Lindner D, Kasner M, Zietsch C, Savvatis K, Escher F, et al. Cardiac inflammation contributes to changes in the extracellular matrix in patients with heart failure and normal ejection fraction. Circ Heart Fail 2011;4(1):44–52.
[91] Souders CA, Bowers SL, Baudino TA. Cardiac fibroblast: the renaissance cell. Circ Res 2009;105(12):1164–76.
[92] Cheng JM, Akkerhuis KM, Battes LC, van Vark LC, Hillege HL, Paulus WJ, et al. Biomarkers of heart failure with normal ejection fraction: a systematic review. Eur J Heart Fail 2013;15(12):1350–62.
[93] Kaess BM, Vasan RS. Heart failure: pentraxin 3-a marker of diastolic dysfunction and HF? Nat Rev Cardiol 2011;8(5):246–8.
[94] Sciarretta S, Ferrucci A, Ciavarella GM, De PP, Venturelli V, Tocci G, et al. Markers of inflammation and fibrosis are related to cardiovascular damage in hypertensive patients with metabolic syndrome. Am J Hypertens 2007;20(7):784–91.
[95] Williams ES, Shah SJ, Ali S, Na BY, Schiller NB, Whooley MA. C-reactive protein, diastolic dysfunction, and risk of heart failure in patients with coronary disease: Heart and Soul Study. Eur J Heart Fail 2008;10(1):63–9.
[96] Lee KW, Blann AD, Lip GY. Impaired tissue Doppler diastolic function in patients with coronary artery disease: relationship to endothelial damage/dysfunction and platelet activation. Am Heart J 2005;150(4):756–66.
[97] Collier P, Watson CJ, Voon V, Phelan D, Jan A, Mak G, et al. Can emerging biomarkers of myocardial remodelling identify asymptomatic hypertensive patients at risk for diastolic dysfunction and diastolic heart failure? Eur J Heart Fail 2011;13(10):1087–95.

[98] Shah KB, Kop WJ, Christenson RH, Diercks DB, Henderson S, Hanson K, et al. Prognostic utility of ST2 in patients with acute dyspnea and preserved left ventricular ejection fraction. Clin Chem 2011;57(6):874–82.
[99] Matsubara J, Sugiyama S, Nozaki T, Sugamura K, Konishi M, Ohba K, et al. Pentraxin 3 is a new inflammatory marker correlated with left ventricular diastolic dysfunction and heart failure with normal ejection fraction. J Am Coll Cardiol 2011;57(7):861–9.
[100] van Empel VP, Kaufmann BA, Bernheim AM, Goetschalckx K, Min SY, Muzzarelli S, et al. Interaction between pulmonary hypertension and diastolic dysfunction in an elderly heart failure population. J Card Fail 2014;20(2):98–104.
[101] Brunner-La Rocca HP, Buser PT, Schindler R, Bernheim A, Rickenbacher P, Pfisterer M. Management of elderly patients with congestive heart failure—design of the Trial of Intensified versus standard Medical therapy in Elderly patients with Congestive Heart Failure (TIME-CHF). Am Heart J 2006;151(5):949–55.
[102] Michowitz Y, Arbel Y, Wexler D, Sheps D, Rogowski O, Shapira I, et al. Predictive value of high sensitivity CRP in patients with diastolic heart failure. Int J Cardiol 2008;125(3):347–51.
[103] Kerzner R, Gage BF, Freedland KE, Rich MW. Predictors of mortality in younger and older patients with heart failure and preserved or reduced left ventricular ejection fraction. Am Heart J 2003;146(2):286–90.
[104] Kalogeropoulos A, Georgiopoulou V, Psaty BM, Rodondi N, Smith AL, Harrison DG, et al. Inflammatory markers and incident heart failure risk in older adults: the Health ABC (Health, Aging, and Body Composition) study. J Am Coll Cardiol 2010;55(19):2129–37.
[105] Marti CN, Khan H, Mann DL, Georgiopoulou VV, Bibbins-Domingo K, Harris T, et al. Soluble tumor necrosis factor receptors and heart failure risk in older adults: Health, Aging, and Body Composition (Health ABC) Study. Circ Heart Fail 2014;7(1):5–11.
[106] Kjekshus J, Apetrei E, Barrios V, Bohm M, Cleland JG, Cornel JH, et al. Rosuvastatin in older patients with systolic heart failure. N Engl J Med 2007;357(22):2248–61.
[107] Tavazzi L, Maggioni AP, Marchioli R, Barlera S, Franzosi MG, Latini R, et al. Effect of rosuvastatin in patients with chronic heart failure (the GISSI-HF trial): a randomised, double-blind, placebo-controlled trial. Lancet 2008;372(9645):1231–9.
[108] Hold GL, El-Omar EM. Genetic aspects of inflammation and cancer. Biochem J 2008;410(2):225–35.
[109] Tsutsui H, Kinugawa S, Matsushima S. Oxidative stress and heart failure. Am J Physiol Lung Cell Mol Physiol 2011;301(6):H2181–90.
[110] Octavia Y, Brunner-La Rocca HP, Moens AL. NADPH oxidase-dependent oxidative stress in the failing heart: from pathogenic roles to therapeutic approach. Free Radic Biol Med 2012;52(2):291–7.
[111] Roe ND, Thomas DP, Ren J. Inhibition of NADPH oxidase alleviates experimental diabetes-induced myocardial contractile dysfunction. Diabetes Obes Metab 2011;13(5):465–73.
[112] Ide T, Tsutsui H, Kinugawa S, Utsumi H, Kang D, Hattori N, et al. Mitochondrial electron transport complex I is a potential source of oxygen free radicals in the failing myocardium. Circ Res 1999;85(4):357–63.
[113] Bayeva M, Ardehali H. Mitochondrial dysfunction and oxidative damage to sarcomeric proteins. Curr Hypertens Rep 2010;12(6):426–32.
[114] Kwon SH, Pimentel DR, Remondino A, Sawyer DB, Colucci WS. H(2)O(2) regulates cardiac myocyte phenotype via concentration-dependent activation of distinct kinase pathways. J Mol Cell Cardiol 2003;35(6):615–21.
[115] Siwik DA, Colucci WS. Regulation of matrix metalloproteinases by cytokines and reactive oxygen/nitrogen species in the myocardium. Heart Fail Rev 2004;9(1):43–51.
[116] Vitiello D, Harel F, Touyz RM, Sirois MG, Lavoie J, Myers J, et al. Changes in cardiopulmonary reserve and peripheral arterial function concomitantly with subclinical inflammation and oxidative stress in patients with heart failure with preserved ejection fraction. Int J Vasc Med 2014;2014:917271. doi:10.1155/2014/917271.
[117] Tousoulis D, Charakida M, Stefanadis C. Inflammation and endothelial dysfunction as therapeutic targets in patients with heart failure. Int J Cardiol 2005;100(3):347–53.
[118] Colombo PC, Banchs JE, Celaj S, Talreja A, Lachmann J, Malla S, et al. Endothelial cell activation in patients with decompensated heart failure. Circulation 2005;111(1):58–62.
[119] Bhimaraj A, Tang WH. Role of oxidative stress in disease progression in Stage B, a pre-cursor of heart failure. Heart Fail Clin 2012;8(1):101–11.
[120] Chiang CH, Huang PH, Leu HB, Hsu CY, Wang KF, Chen JW, et al. Decreased circulating endothelial progenitor cell levels in patients with heart failure with preserved ejection fraction. Cardiology 2013;126(3):191–201.
[121] van Empel VP, Kaye DM. Integration of exercise evaluation into the algorithm for evaluation of patients with suspected heart failure with preserved ejection fraction. Int J Cardiol 2013;168(2):716–22.
[122] Ennezat PV, Malendowicz SL, Testa M, Colombo PC, Cohen-Solal A, Evans T, et al. Physical training in patients with chronic heart failure enhances the expression of genes encoding antioxidative enzymes. J Am Coll Cardiol 2001;38(1):194–8.
[123] Borlaug BA, Olson TP, Lam CS, Flood KS, Lerman A, Johnson BD, et al. Global cardiovascular reserve dysfunction in heart failure with preserved ejection fraction. J Am Coll Cardiol 2010;56(11):845–54.
[124] Akiyama E, Sugiyama S, Matsuzawa Y, Konishi M, Suzuki H, Nozaki T, et al. Incremental prognostic significance of peripheral endothelial dysfunction in patients with heart failure with normal left ventricular ejection fraction. J Am Coll Cardiol 2012;60(18):1778–86.
[125] Hundley WG, Bayram E, Hamilton CA, Hamilton EA, Morgan TM, Darty SN, et al. Leg flow-mediated arterial dilation in elderly patients with heart failure and normal left ventricular ejection fraction. Am J Physiol Lung Cell Mol Physiol 2007;292(3):H1427–34.
[126] Tschope C, Bock CT, Kasner M, Noutsias M, Westermann D, Schwimmbeck PL, et al. High prevalence of cardiac parvovirus B19 infection in patients with isolated left ventricular diastolic dysfunction. Circulation 2005;111(7):879–86.
[127] Seddon M, Shah AM, Casadei B. Cardiomyocytes as effectors of nitric oxide signalling. Cardiovasc Res 2007;75(2):315–26.
[128] van Heerebeek L, Hamdani N, Falcao-Pires I, Leite-Moreira AF, Begieneman MP, Bronzwaer JG, et al. Low myocardial protein kinase G activity in heart failure with preserved ejection fraction. Circulation 2012;126(7):830–9.
[129] Borbely A, van Heerebeek L, Paulus WJ. Transcriptional and posttranslational modifications of titin: implications for diastole. Circ Res 2009;104(1):12–4.
[130] Borbely A, Falcao-Pires I, van Heerebeek L, Hamdani N, Edes I, Gavina C, et al. Hypophosphorylation of the Stiff N2B titin isoform raises cardiomyocyte resting tension in failing human myocardium. Circ Res 2009;104(6):780–6.
[131] van Heerebeek L, Franssen CP, Hamdani N, Verheugt FW, Somsen GA, Paulus WJ. Molecular and cellular basis for diastolic dysfunction. Curr Heart Fail Rep 2012;9(4):293–302.

[132] Kruger M, Kotter S, Grutzner A, Lang P, Andresen C, Redfield MM, et al. Protein kinase G modulates human myocardial passive stiffness by phosphorylation of the titin springs. Circ Res 2009;104(1):87–94.
[133] Takimoto E, Champion HC, Li M, Belardi D, Ren S, Rodriguez ER, et al. Chronic inhibition of cyclic GMP phosphodiesterase 5A prevents and reverses cardiac hypertrophy. Nat Med 2005;11(2):214–22.
[134] Giannetta E, Isidori AM, Galea N, Carbone I, Mandosi E, Vizza CD, et al. Chronic Inhibition of cGMP phosphodiesterase 5A improves diabetic cardiomyopathy: a randomized, controlled clinical trial using magnetic resonance imaging with myocardial tagging. Circulation 2012;125(19):2323–33.
[135] Guazzi M, Vicenzi M, Arena R, Guazzi MD. PDE5 inhibition with sildenafil improves left ventricular diastolic function, cardiac geometry, and clinical status in patients with stable systolic heart failure: results of a 1-year, prospective, randomized, placebo-controlled study. Circ Heart Fail 2011;4(1):8–17.
[136] Antoniades C, Bakogiannis C, Leeson P, Guzik TJ, Zhang MH, Tousoulis D, et al. Rapid, direct effects of statin treatment on arterial redox state and nitric oxide bioavailability in human atherosclerosis via tetrahydrobiopterin-mediated endothelial nitric oxide synthase coupling. Circulation 2011;124(3):335–45.
[137] Ramasubbu K, Estep J, White DL, Deswal A, Mann DL. Experimental and clinical basis for the use of statins in patients with ischemic and nonischemic cardiomyopathy. J Am Coll Cardiol 2008;51(4):415–26.
[138] Ridker PM, Danielson E, Fonseca FA, Genest J, Gotto Jr AM, Kastelein JJ, et al. Rosuvastatin to prevent vascular events in men and women with elevated C-reactive protein. N Engl J Med 2008;359(21):2195–207.
[139] Fukuta H, Sane DC, Brucks S, Little WC. Statin therapy may be associated with lower mortality in patients with diastolic heart failure: a preliminary report. Circulation 2005;112(3):357–63.
[140] Hayashidani S, Tsutsui H, Shiomi T, Ikeuchi M, Matsusaka H, Suematsu N, et al. Anti-monocyte chemoattractant protein-1 gene therapy attenuates left ventricular remodeling and failure after experimental myocardial infarction. Circulation 2003;108(17):2134–40.
[141] Westermann D, Savvatis K, Lindner D, Zietsch C, Becher PM, Hammer E, et al. Reduced degradation of the chemokine MCP-3 by matrix metalloproteinase-2 exacerbates myocardial inflammation in experimental viral cardiomyopathy. Circulation 2011;124(19):2082–93.
[142] Nishio R, Matsumori A, Shioi T, Ishida H, Sasayama S. Treatment of experimental viral myocarditis with interleukin-10. Circulation 1999;100(10):1102–8.
[143] Suzuki K, Murtuza B, Smolenski RT, Sammut IA, Suzuki N, Kaneda Y, et al. Overexpression of interleukin-1 receptor antagonist provides cardioprotection against ischemia-reperfusion injury associated with reduction in apoptosis. Circulation 2001;104(12 Suppl 1):I308–13.
[144] Cox MJ, Hawkins UA, Hoit BD, Tyagi SC. Attenuation of oxidative stress and remodeling by cardiac inhibitor of metalloproteinase protein transfer. Circulation 2004;109(17):2123–8.
[145] Li YY, Kadokami T, Wang P, McTiernan CF, Feldman AM. MMP inhibition modulates TNF-alpha transgenic mouse phenotype early in the development of heart failure. Am J Physiol Heart Circ Physiol 2002;282(3):H983–9.
[146] Lindeman JH, Abdul-Hussien H, van Bockel JH, Wolterbeek R, Kleemann R. Clinical trial of doxycycline for matrix metalloproteinase-9 inhibition in patients with an abdominal aneurysm: doxycycline selectively depletes aortic wall neutrophils and cytotoxic T cells. Circulation 2009;119(16):2209–16.
[147] Salio M, Chimenti S, De AN, Molla F, Maina V, Nebuloni M, et al. Cardioprotective function of the long pentraxin PTX3 in acute myocardial infarction. Circulation 2008;117(8):1055–64.
[148] Latini R, Masson S, Anand I, Salio M, Hester A, Judd D, et al. The comparative prognostic value of plasma neurohormones at baseline in patients with heart failure enrolled in Val-HeFT. Eur Heart J 2004;25(4):292–9.
[149] Patrucco E, Notte A, Barberis L, Selvetella G, Maffei A, Brancaccio M, et al. PI3Kgamma modulates the cardiac response to chronic pressure overload by distinct kinase-dependent and -independent effects. Cell 2004;118(3):375–87.
[150] Bodyak N, Ayus JC, Achinger S, Shivalingappa V, Ke Q, Chen YS, et al. Activated vitamin D attenuates left ventricular abnormalities induced by dietary sodium in Dahl salt-sensitive animals. Proc Natl Acad Sci U S A 2007;104(43):16810–5.
[151] Liu LC, Voors AA, van Veldhuisen DJ, van der Veer E, Belonje AM, Szymanski MK, et al. Vitamin D status and outcomes in heart failure patients. Eur J Heart Fail 2011;13(6):619–25.
[152] Funabiki K, Onishi K, Dohi K, Koji T, Imanaka-Yoshida K, Ito M, et al. Combined angiotensin receptor blocker and ACE inhibitor on myocardial fibrosis and left ventricular stiffness in dogs with heart failure. Am J Physiol Heart Circ Physiol 2004;287(6):H2487–92.
[153] Westermann D, Becher PM, Lindner D, Savvatis K, Xia Y, Frohlich M, et al. Selective PDE5A inhibition with sildenafil rescues left ventricular dysfunction, inflammatory immune response and cardiac remodeling in angiotensin II-induced heart failure in vivo. Basic Res Cardiol 2012;107(6):308, 0308.

CHAPTER

2

Role of the Innate Immune System in Ischemic Heart Failure

Johannes Weirather, Stefan Frantz

Department of Internal Medicine I, University Hospital Würzburg, Comprehensive Heart Failure Center, University of Würzburg, Würzburg, Germany

2.1 INTRODUCTION

Cardiac remodeling refers to cellular, molecular, and interstitial variations in the myocardium that clinically manifest as changes in the structure, shape, and physiology of the heart. Geometric and histopathological changes include myocyte apoptosis, myocardial fibrosis, and left-ventricular (LV) enlargement associated with a decline of LV performance. As the heart remodels, the LV chamber becomes progressively spherical and increases the LV luminal volume [1].

Cardiac remodeling has been recognized as an important aspect in the development of heart failure, and an interference with this process has emerged as an appealing therapeutic approach to prevent disease progression. Therefore, considerable efforts have been taken to decipher pathological mechanisms driving cardiac remodeling. Pathological remodeling can develop in response to myocardial infarction (MI), pressure overload, myocarditis, or volume overload. Although exhibiting different etiologies, remodeling in the context of all these diseases correlates with an activation of cross-talking inflammatory and profibrotic pathways that are mediated and modulated by components of innate immunity [1].

In heart failure patients, MI constitutes the most frequent trigger causative for the development of cardiac remodeling [2]. Therefore, mechanistic studies focusing on remodeling have been predominantly conducted in experimental models of MI, or in patients sustaining a heart attack, that is, after sterile ischemic tissue injury. In this regard, the mechanistic link between an activation of innate immunity and the progression of remodeling is paradigmatic for remodeling in the context of different etiologies. Given the prevailing clinical relevance and the extensive mechanistic understanding, the interrelation between remodeling and the activation of innate immunity will be exemplified in the context of MI.

MI and reperfusion injury are characterized by a damage of cardiomyocytes and nonmyocytes including the extracellular collagen matrix (ECM) [3]. Myocardial necrosis triggers an inflammatory reaction that clears the infarct zone from ECM and cell debris, representing an important physiological mechanism that paves the way for the healing phase [4]. During early infarct healing, invading innate immune cells and angiogenic capillary sprouts comprise a *de novo* formed granulation tissue. In parallel, reparative pathways become activated by the inflammatory reaction and promote the formation of a collagenous scar that ultimately replaces the irreversibly injured myocardium [5].

Remodeling after MI is a complex and dynamic process that parallels healing over several days to weeks. LV remodeling is characterized by changes in LV structure and topography comprising differential changes in the infarct and noninfarct zone. After an early infarct expansion and disproportionate thinning of the LV chamber wall, progressive LV dilatation is followed by a compensatory hypertrophy in the noninfarct zone that develops over weeks to maintain both blood pressure and output [3,6,7].

Inflammation in Heart Failure
http://dx.doi.org/10.1016/B978-0-12-800039-7.00002-5

© 2015 Elsevier Inc. All rights reserved.

In parallel to LV remodeling, the extracellular matrix (ECM) undergoes substantial changes that, in turn, pivotally influence cardiac global shape [3,6]. An initial ECM disruption in the infarct zone contributes to an early expansion of the infarct region, followed by successive deposition of consolidating collagen and *de novo* ECM construction. Collagen-producing fibroblasts gradually acquire contractile function and differentiate into myofibroblasts that additionally mediate scar contraction. After large MI, early collagen degradation even in the noninfarct zone contributes to the progressive LV dilatation that develops over weeks. Ultimately, during late stage remodeling after large MI, interstitial fibrosis in the noninfarct zone emerges preserving cardiac tensile strength while impairing systolic function [8].

The quality of wound healing and the excess of cardiac remodeling are inevitably linked and both pivotally modulated by immunity involving inflammatory and anti-inflammatory reactions [9]. Despite the importance of inflammation for proper infarct healing, an accentuated or prolonged inflammatory reaction is regarded as detrimental. An exuberant inflammatory response may provoke necrosis of surviving cardiomyocytes or elicit an overreaching ECM degradation, resulting in adverse expansion of the infarct zone. Infarct size, in turn, constitutes a major determinant for the degree of cardiac remodeling and correlates directly with the extent of LV dilatation and wall thinning [10]. In contrast, spatiotemporally restricted inflammation and efficacious scarring in the healing myocardium are considered to attenuate LV dilatation by preventing an early infarct expansion [9]. During scar tissue formation, innate immunity modulates fibroblast function and stimulates collagen synthesis by soluble mediators. Subsequently, immunity is involved in regulating inflammation and collagen deposition preventing further parenchymal injury or excessive fibrosis in the noninjured remote zone [11]. Conclusively, a fine balance between inflammation and inflammation resolution is necessary to ensure proper infarct healing without promoting detrimental remodeling that may finally lead to fatal cardiac dysfunction. Both inflammatory and anti-inflammatory reactions during healing and remodeling are predominantly mediated by components of innate immunity. This chapter highlights the contribution of soluble and cellular effectors to cardiac remodeling in terms of cardiac damage—namely, after MI or reperfusion injury.

2.2 INITIATION OF THE IMMUNE RESPONSE

2.2.1 Receptors

After prolonged coronary artery occlusion, the preponderant mechanism of cardiomyocyte death is coagulative necrosis resulting in the release of intracellular content. The released constituents and the ECM fragments engage receptors on the local parenchymal cells, the vasculature, fibroblasts, or infiltrating leukocytes. Receptor ligation, in turn, transduces the recognition of these danger-associated molecular patterns (DAMPs) into activation signals provoking an inflammatory response [12]. DAMPs thereby comprise all molecules with capacity to initiate or perpetuate the inflammatory reaction in a noninfectious and sterile condition. MI-associated DAMPs include heat shock proteins, high-mobility group box (HMGB)-1, low molecular hyaluronic acid, fibronectin fragments, or components of the coagulation system such as fibrinogen [12–14].

Receptors involved in DAMP sensing comprise pattern recognition receptors (PRRs) such as toll-like receptors (TLRs), NOD-like receptors, C-type lectin receptors, or the receptor for advanced glycation end-products (RAGE) [12]. After PRR engagement, inflammatory pathways become initiated, converging in the activation of transcription factors such as nuclear factor 'kappa-light-chain-enhancer' of activated B-cells (NFkB), or the activator protein 1 (AP-1) that drive the expression of inflammatory effectors. In response to inflammatory cytokines or DAMPs, the endothelium upregulates adhesion molecules and initiates the secretion of chemokines facilitating leukocyte recruitment into the myocardium [15,16].

So far, TLRs are the best-characterized receptors in terms of healing and remodeling. Generally, TLRs serve as pathogen recognition receptors recognizing conserved molecular pathogen-intrinsic motifs, so-called pathogen-associated molecular patterns (PAMPs) [17]. *Inter alia*, PAMPs comprise the lipopolysaccharides of Gram-negative bacteria, the teichoic acids of Gram-positive microbes, the glycolipids of mycobacteria, the yeast zymosan, or double-stranded RNAs of certain virus families. TLRs constitute, therefore, an integral component of immunity capacitating the host to recognize potentially harmful microorganisms [17,18].

The similarity of TLR-induced immune reactions secondary to tissue injury and infection is likely phylogenetically conserved, insofar as healing is commonly accompanied by an infection. DAMPs with ability to trigger TLRs include proteins such as fibronectin or HMGB1, nucleic acids as mitochondrial DNA, or purine metabolites such as ATP [19]. Generally, TLR engagement triggers signaling pathways that end up activating the transcription factors NFκB, AP-1, and IRF3 that drive the expression of inflammatory mediators such as tumor necrosis factor alpha (TNFα), interleukin (IL)-6, or type I interferons [20].

At present, 13 mammalian TLR paralogs have been identified, and TLR1 to TLR9 are highly conserved between human and mouse [21,22]. According to the subcellular location, TLRs are divided into two subgroups—intracellular and cell surface TLRs [22,23]. The spectrum of TLRs is expressed in different combinations in all types of immune cells, and also in nonimmune cells such as fibroblasts, endothelial cells, or keratinocytes [20]. TLR2 to TLR5, TLR7, and TLR9 expression has been evidenced in murine cardiomyocytes, and mRNAs covering the entire TLR repertoire are detectable in human hearts [24].

In terms of cardiac injury, classic loss-of-function studies demonstrate that TLR signaling aggravates healing and remodeling. Targeted disruption of TLR2 or TLR4 in mice results in alleviated neutrophil recruitment into the myocardium and both a reduction of inflammatory cytokines and diminished proapoptotic signaling after ischemia/reperfusion. Coherently, infarct size and remodeling are attenuated in these mice, in line with an improved survival [25–30]. From the mechanistic point of view, the advantageous outcome in TLR2-deficient mice is abrogated after transplantation of bone marrow from TLR2-sufficient animals, suggesting that TLR2-mediated deleterious effects are attributable to a modulation of leukocyte functions in response to receptor ligation [26].

TLR signaling exacerbates the outcome after MI spurred efforts to disrupt these pathways by the use of TLR-specific drugs. Instancing, treatment with a TLR4 antagonist or a TLR2-specific inhibitory monoclonal antibody prior to reperfusion attenuates cardiac inflammation and abates infarct size, along with improved cardiac function and geometry [26,31].

In conclusion, DAMPs efficiently trigger the inflammatory response by engaging innate receptors such as TLRs. Antagonizing receptor ligation or a targeted disruption of receptor signaling pathways constitutes, therefore, an appealing modality to attenuate remodeling after MI.

2.2.2 Complement

The complement system consists of a number of soluble serum proteins that circulate freely as inactive precursors. The cardinal function of complement is to support phagocyte-mediated clearance of pathogens from the host [32]. In general, three pathways of complement activation can be distinguished and all converge in the activation of a C3 convertase, which catalyzes the generation of downstream complement effector molecules. The effectors comprise proteins that mediate pathogen cell lysis, but also anaphylatoxins supporting both leukocyte recruitment and function [32].

The classical pathway of complement activation becomes initiated by antigen-antibody aggregates, that is, immunoglobulins that have previously labeled a target cell in terms of an infection [32]. Additionally, C1 can become activated by binding the C-reactive protein (CRP), which recognizes constituents released from dying cells in terms of cell or tissue injury [33]. Moreover, polyanions such as nucleic acids have been demonstrated to bear the potential for direct C1 binding and, thus, complement activation in response to cell necrosis [34]. Ligand engagement by C1, in turn, induces a conformational change in the C1 molecule imparting proteolytic activity. Subsequent C1-mediated cleavage of substrates downstream in the activation cascade ultimately leads to generation of an active C3 convertase [32,33].

Analogous to the classical pathway, the lectin pathway becomes activated by attachment of the mannose-binding lectin to sugar moieties, for example, on the cell surface of pathogens in terms of bacterial infections. The alternative pathway of complement activation, in contrast, results from a spontaneous C3 hydrolysis resulting in the formation of an enzymatically active C3 convertase [33].

Complement activation in the wrong context has the potential to inflict extreme damage on the host. In 1971, Hill and Ward were the first to show that complement becomes activated in the wake of MI; virtually all components of the classical pathway increase within the infarcted myocardium [35]. Later, studies have suggested that myocardial necrosis results in the release of mitochondria-derived membrane fragments capable of C1-binding [36]. However, apart from a direct C1 ligation to cellular constituents, circulating antibodies mediate an activation of the classical pathway following ischemic tissue damage. In models for skeleton muscle and intestinal reperfusion injury, natural monoclonal antibodies recognizing nonmuscle myosin heavy chain type II are involved in the activation of complement [37–39]. In accordance with this observation, natural immunoglobulin M-mediated complement activation has been recognized in reperfused hearts [40]. More recent evidence underscores the contribution of complement to cardiac injury in clinical and experimental studies showing that CRP is involved in C1 activation post-MI [41,42]. Furthermore, a role of the lectin pathway is implicated in postinfarction remodeling. High serum levels of lectin pathway components in combination with high CRP is associated with increased infarct size as well as deteriorated remodeling in patients with MI [43]. However, evidence for a direct complement-mediated injury of cardiomyocytes is missing and may therefore play a minor pathomechanistical role.

Aside from direct damage by complement effectors, anaphylatoxins crucially mediate the phlogistic functions of the system [33]. During the early phase of reperfusion, cardiac lymph is chemotactically active due to the presence of C5a [44,45]. Consistently, antibody-mediated C5 neutralization *in vivo* results in infarct size reduction and attenuated neutrophil recruitment after ischemia/reperfusion [46].

From the clinical point of view, a targeted interference with complement activation or effector molecules may constitute an effective treatment modality to prevent remodeling after MI. Although a direct evidence for complement-mediated aggravation of remodeling is missing, an early mitigation of complement function likely attenuates LV enlargement. In rodents, treatment with a C5-specific inhibitory antibody reduces infarct size after ischemia/reperfusion [46]. Consistently, similar results have been provided in experiments with anaphylatoxin receptor blockade [47]. Because infarct size correlates directly with the severity of remodeling, a complement-mediated deterioration of LV dilatation is implicated. However, in a clinical trial, interference with C5 function by neutralizing antibodies did not improve infarct size or clinical outcome in patients undergoing fibrinolysis [48]. Nevertheless, C5 neutralization resulted in significantly improved survival rates in MI patients undergoing percutaneous coronary intervention [49].

In summary, the complement system becomes activated in response to cardiac injury and facilitates the recruitment of leukocytes. Mitigation of complement activation leads to a reduction of leukocyte recruitment and infarct size, indicating a role of complement in aggravating remodeling.

2.2.3 Oxidative Stress

Reactive oxygen species (ROS) are chemically reactive, oxygen-containing molecules exhibiting unpaired valence electrons or an open electron shell. ROS can wreak havoc directly by oxidizing macromolecules such as nucleic acids, unsaturated fatty acids, or proteins [50]. The infliction of cumulative damage to cell structures is defined as oxidative stress that can trigger apoptosis, or necrosis in the case of overwhelming oxidative damage [51]. ROS are endogenously produced in the cell metabolism. During oxidative phosphorylation in mitochondria, electrons are transported across the inner mitochondrial membrane and finally transferred to oxygen molecules producing water by reacting with protons. Sporadically, the oxygen molecule is incompletely reduced, leading to the formation of superoxide radical that, subsequently, dismutates to hydrogen peroxide [52,53]. The hydrogen peroxide molecule contains no unpaired electrons and is therefore no ROS *stricto sensu*. However, hydrogen peroxide can diffuse freely over lipid membranes and constitutes a potent oxidizing agent that may generate free radicals in the presence of metal cations [51].

2.2.3.1 *ROS Generation Post-MI*

To prevent radical formation during homeostasis, cardiomyocytes endogenously express free-radical scavenging enzymes such as superoxide dismutase, catalase, or glutathione peroxidase [54]. After ischemia and reperfusion, however, anti-oxidative detoxification mechanisms are overwhelmed due to significantly increased rates of ROS generation. Mitochondria become damaged during the ischemic episode and exhibit thereafter an impaired electron transfer system of the respiratory chain [55]. After restoration of the oxygen support, disturbed electron transport leads to massive ROS generation, and these highly reactive radicals are able to inactivate mitochondrial enzymes or initiate lipid peroxidation [55]. Severe mitochondrial damage, in turn, can lead to the release of proapoptotic mediators such as cytochrome C, Smac, Diablo, or the apoptosis-inducing factor [54].

Aside from mitochondria-derived ROS, two enzymes were identified as important sources for ROS in terms of ischemia/reperfusion—namely, xanthine oxidase (XO) and nicotinamide adenine dinucleotide phosphate (NADPH) oxidase [56]. During the ischemic period, ATP-derived hypoxanthine is further metabolized by XO producing superoxide. Predominantly expressed in endothelial cells, the vasculature is primarily affected by XO-mediated superoxide generation [56]. In contrast, NADPH oxidase is highly expressed in neutrophils that numerously infiltrate the infarcted myocardium [56,57]. Neutrophils generate high amounts of ROS in a process called respiratory burst and contribute significantly to the oxidative damage post-MI [58].

2.2.3.2 *Role of Oxidative Stress for Cardiac Necrosis and Inflammation*

The contribution of ROS to cardiac inflammation and remodeling is highly pleiotropic. An unbearable burden of oxidative stress can lead to uncontrolled cell death due to a strong disturbance of the intracellular $Ca2^+$ levels [54]. Oxidative stress restrains the sequestration of $Ca2^+$ from the cytosol because ROS depress the sarcoplasmic reticulum $Ca2^+$ pump ATPase, resulting in increased cytosolic $Ca2^+$ levels [59]. Additionally, ROS cause release from intracellular $Ca2^+$ stores and promote the leakage from the extracellular space into the cell due to lipid membrane peroxidation impairing membrane barrier function [60]. Cytosolic $Ca2^+$ overload, in turn, triggers the opening of the mitochondrial permeability transition (MPT) pore in the outer mitochondrial membrane, resulting in a breakdown of the transmembrane potential at the mitochondrial inner membrane. As a consequence, ATP synthesis stops and mitochondria swell and ultimately rupture [61]. However, although the MPT pathway may be involved in apoptosis, MPT activation has been identified as a major cause of necrotic cell death during oxidative stress [62]. In turn, ROS-mediated aggravation of cardiomyocyte necrosis influences infarct size and therefore the extent of cardiac remodeling. Furthermore, an aggravation of necrosis potentiates DAMP release and complement activation, which, again, fuels cardiomyocyte death constituting a vicious cycle. Additionally, a direct ROS-mediated initiation of the complement cascade may be relevant *in vivo* [63].

Aside from the roles in cardiomyocyte death and complement activation, ROS are involved in inflammatory signaling pathways after an activation of stress-responsive protein kinases. The apoptosis signal-regulating kinase-1 (ASK-1), for instance, is ubiquitously expressed in most mammalian cells [60]. In the steady state, ASK-1 function is efficiently inhibited by thioredoxin 1. After ROS-mediated oxidation, however, the inhibitory function of thioredoxin 1 is repealed, capacitating AKS-1 to trigger inflammatory signaling pathways that end up in an activation of NFκB, JNK, p38, or AP-1 [60,64–66]. In this regard, H_2O_2 has been shown to directly induce myocardial TNFα synthesis driven by p38 MAPK signaling [67]. In endothelial cells, ROS signaling effectuates an upregulation of selectins and cell adhesion molecules that parallel an NFκB-mediated expression of leukocyte-attracting chemokines. ROS, thus, facilitate leukocyte recruitment and are involved in modulating the transcriptional profile of infiltrating leukocytes by activating proinflammatory pathways [68,69] (Figure 2.1).

2.2.4 Mechanical Stimuli

In addition to ROS and DAMPs, mechanical stress such as hemodynamic cardiac stretch has also been demonstrated to induce the activation of inflammatory transcription factors that end up in the synthesis of inflammatory cytokines [70]. Potential mechanosensors include integrins, the cytoskeleton, and sarcolemmal proteins that transduce the mechanical stimulus into several cross-talking signaling pathways—namely, mitogen-activated protein kinase (MAPK) signaling, JAK-signal transducer and activator of transcription (STAT) signaling, and calcineurin-dependent pathways [71].

2.3 EFFECTORS OF INNATE IMMUNITY

Upon activation, innate immunity harnesses a broad arsenal of humoral and cellular effectors to comply with its biological obligations. Effector functions are thereby highly diverse, and the body of evidence underscoring their significance for cardiac remodeling continues to grow. However, only a selection of central effectors will be presented here to outline the mechanistic concepts regarding how immunity modulates healing and remodeling.

2.3.1 Cytokines

Most cytokines are not expressed in the heart under physiological conditions but are significantly upregulated after cardiac injury [72,73]. The induced cytokines influence gene expression in endothelial cells, fibroblasts, or infiltrating leukocytes and modulate the physiology of surviving cardiomyocytes [74]. In the case of small MI, the upregulated cytokine synthesis returns to baseline levels after completion of the healing process. After large MI, however, cytokine levels remain elevated over a longer period of time, correlating with chronic cardiac remodeling [75,76].

Cytokine expression is predominantly driven by the transcription factors NFκB and AP-1 in both the infarct and noninfarct zone [74,77,78]. The proinflammatory cytokines TNFα, IL-6, and IL-1β are rapidly induced after MI and have highly pleiotropic and partially redundant effects on various cell types [74]. Generation of active IL-1β requires processing of the inactive precursor pro-IL-1β by the converting enzyme caspase-1. The activity of caspase-1 is thereby regulated by a high molecular weight complex called "inflammasome" becoming activated, for instance, during oxidative stress [79]. Early cytokine producers in the injured myocardium are local cardiac-resident cells, that is, cardiomyocytes, the vasculature or pre-existing cardiac fibroblasts [15,74,77].

Generally, proinflammatory cytokines have the potential to amplify their physiological impact in a positive feedback loop involving NFκB that drives the cytokines' own expression [80]. The amplification mechanism safeguards efficacious cytokine effects and, moreover, accounts for cytokine expression in the noninjured remote zone. Additionally, cytokine effects are amplified by infiltrating leukocytes secreting high amounts of soluble mediators post-activation [4,81].

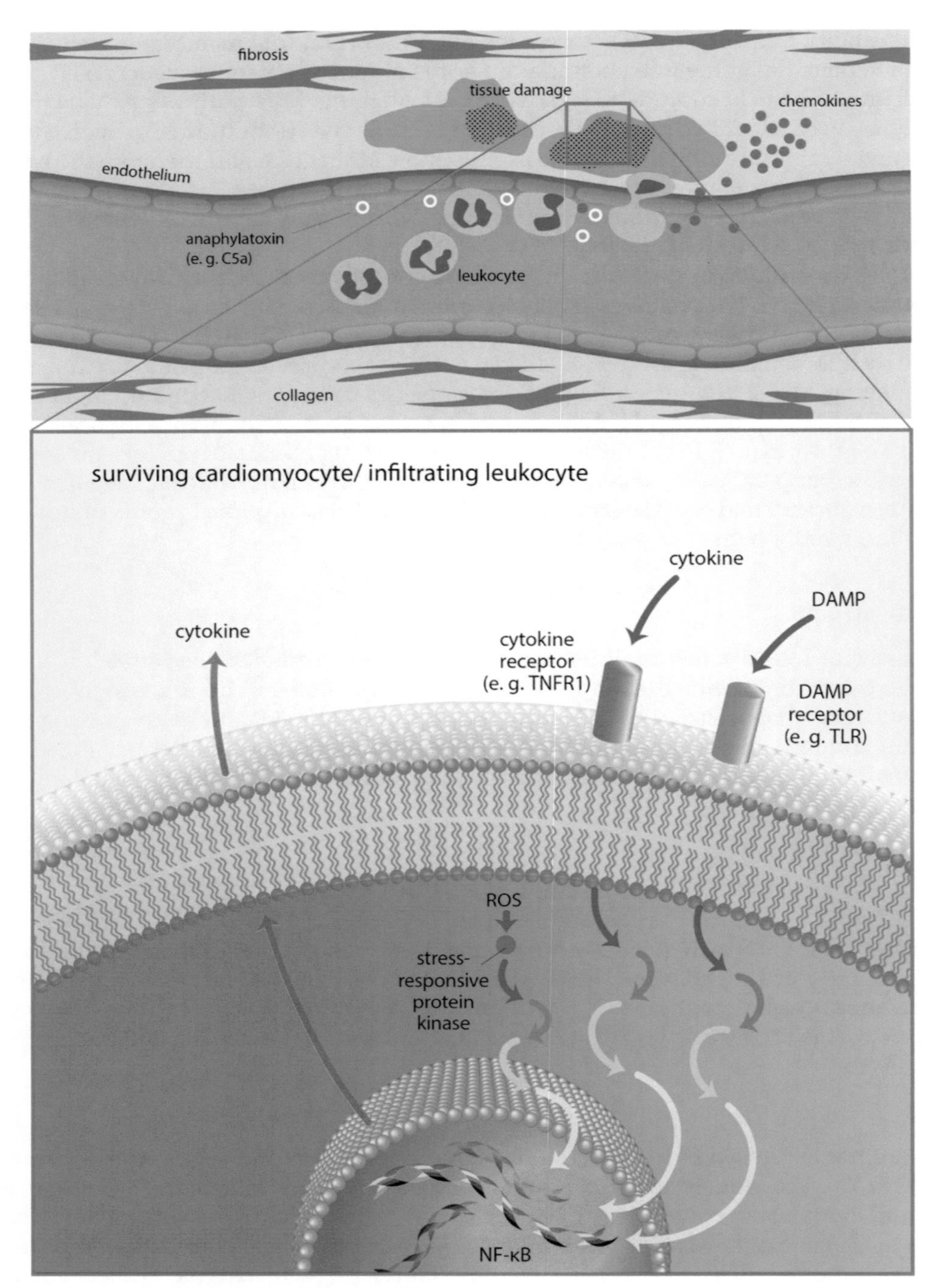

FIGURE 2.1 Cardiomyocyte and leukocyte activation post-MI. After ischemic tissue injury, surviving cardiomyocytes and tissue-infiltrating leukocytes become activated in response to inflammatory mediators, resulting in an expression of various cytokines.

After MI, an early cytokine-mediated influence of cardiac remodeling comprises effects on cardiomyocyte survival. During wound healing and chronic cardiac remodeling, cytokines are involved in inflammation resolution, scar tissue formation, cardiomyocyte hypertrophy, and vascular remodeling [74].

2.3.1.1 Cytokine Effects on Cardiomyocyte Survival

Cytokine effects are highly pleiotropic, depending on the dose, the cytokine combination, and the cell type they engage [82]. In terms of cardiac injury, the effects of TNFα have been extensively investigated. Generally, TNFα can engage two distinct receptors—TNF receptor 1 and TNF receptor 2—the former belonging to the so-called "death receptor" family [83]. After receptor ligation, members of this family trigger signaling pathways that stimulate caspase-8 to initiate the executive phase of apoptosis [84]. Activation of anti-apoptotic Bcl-2 family members, however, may countervail caspase-8 effector functions and, consequently, the net balance of pro- and anti-apoptotic signals determines whether apoptosis will proceed. Additionally to caspase-dependent apoptosis induction, TNFα is known to induce cell death by an activation of FAN, a protein associated with neutral sphingomyelinase activation and ceramide production [85]. Ceramides, in turn, initiate cell apoptosis in a mechanism involving various interconnected downstream effectors [86]. Consistently, noninfarcted mice transgenically expressing TNFα in a cardiac-specific manner show an induction of multiple cell death pathways associated with an early hypertrophic response and development of a dilative cardiomyopathy [87]. Moreover, systemic infusion of TNFα at concentrations observed in heart failure patients elicits pronounced LV dilatation and an impaired LV function [81].

Paradoxically, TNFα is also able to exert cytoprotective effects by activating NFκB and the stress-activated protein kinase/c-Jun N-terminal kinase (JNK), resulting in expression of genes transmitting prosurvival signals [88,89]. The cardioprotective properties of TNFα are best illustrated in a murine model of combined genetic ablation of both TNF receptor 1 and 2. Infarcted mice have a larger infarct size and exhibit increased frequencies of cardiomyocyte apoptosis, indicating that TNFα favors cell preservation in this setting [90].

Aside from TNFα, IL-1β has similar effects on infarct size and myocyte survival post-MI. ATP release from dying cardiomyocytes triggers the purinergic P2X7 receptor that results in inflammasome formation in both cardiomyocytes at the infarct border and within the granulation tissue. Experimental interference with inflammasome activation limits the infarct size and prevents LV enlargement, indicating that inflammasome formation in terms of MI causes loss of functional myocardium [91].

Another cytokine upregulated in the infarct zone is IL-6. Signaling cascades triggered by IL-6 are redundantly activated by various endogenous ligands and global IL-6 KO mice have apparently no obvious phenotype regarding infarct healing and remodeling [92]. However, infusion of IL-6/soluble IL-6 receptor complexes amplifying IL-6 signaling has been demonstrated to prevent apoptosis in a model of ischemia and reperfusion [93].

2.3.1.2 Cytokines Influence Granulation Tissue Formation and Vascular Remodeling Post-injury

After MI, cytokines substantially influence the expression profile in the cardiac vasculature. Early after ischemic injury, cytokines support the formation of a granulation tissue by amplifying the endothelial upregulation of selectins, cell adhesion molecules, and chemokines facilitating leukocyte recruitment (see Section 2.3.2.1).

A hallmark of the granulation tissue is a vigorous neoangiogenesis [94]. Formation of new blood vessels is a critical step in the healing process to safeguard oxygen and nutrient supply to the healing infarct and the peri-infarct zone. Angiogenesis is a complex process involving the interaction between the ECM, endothelial cells (ECs), and pericytes wrapping the endothelium [94,95]. Angiogenesis is initiated by a local imbalance of angiostatic and angiogenic factors with a preponderance of the latter [96–98]. In a murine model of MI, expression of the angiogenic isoform vascular endothelial growth factor 120 ($VEGF_{120}$) shows a biphasic upregulation whereas isoforms $VEGF_{164}$ and $VEGF_{188}$ exhibit a sustained elevated expression [99]. Potent stimulators as well as regulators of angiogenesis are so-called "CXC chemokines" containing two amino terminal cysteines separated by a single nonconserved amino acid. CXC chemokines that contain an ELR motif such as CXCL-8 have strong angiogenic capacity, whereas CXC chemokines without the ELR motif are potent angiostatic factors, for example, the interferon-inducible protein 10 (IP-10) [100]. After ischemia and reperfusion, local TNFα induces IP-10 synthesis in the vascular endothelium, suppressing angiogenesis until the injured myocardium is cleared from debris and until a provisional matrix is formed [101,102]. After 24 h of reperfusion, however, the emergence of transforming growth factor beta (TGFβ) suppresses IP-10 expression and shifts the balance toward neovessel formation [101]. Additionally, TGFβ stimulates expression of both b-FGF and VEGF in endothelial and smooth muscle cells further supporting angiogenesis [74].

Targeting the cytokine network in order to modulate angiogenesis has been demonstrated to improve cardiac remodeling post-MI. VEGF gene delivery by adeno-associated viral vectors into the infarct core and peri-infarct zone, for instance, attenuates cardiomyocyte apoptosis and mitigates LV enlargement [103]. Also, enhancing reparative angiogenesis in the peri-infarct zone by specific inhibition of angiogenesis-suppressing micro-RNAs results in reduced infarct size along with improved remodeling [104].

2.3.1.3 Cytokines Modulate Scar Tissue Formation After Injury

The ECM is a network of cross-linked scaffold proteins comprising collagens, fibronectin, and elastin that interact with integrins and focal adhesion kinases at the cell-ECM junction. After MI, an early ECM breakdown and the loss of vital tissue provoke the formation of a collagenous scar replacing the irreversibly injured myocardium [3]. Cytokines play an important role in stimulating and regulating collagen deposition by (myo-) fibroblasts during scar tissue formation. Furthermore, cytokines drive the synthesis of both matrix metalloproteinases (MMPs) and tissue inhibitors of metalloproteinases (TIMPs) and, therefore, crucially influence ECM turnover.

Members of the TGFβ family are one of the most pleiotropic and multifunctional mediators known. In terms of wound healing and scar tissue formation, the TGFβ isoforms stimulate collagen synthesis in fibroblasts [105]. In addition, TGFβ is able to provoke fibroblast proliferation and myofibroblast transdifferentiation [106]. Further profibrotic effects of TGFβ can be attributed to its capacity to induce protease inhibitors such as TIMPs and plasminogen activator inhibitor-1 (PAI-1), resulting in an attenuated ECM degradation [107,108].

Given the powerful effect of TGFβ on ECM formation, experimental studies have been carried out targeting TGFβ-mediated fibrosis. Blunting TGFβ signaling by adenoviral expression of a soluble TGFβ decoy receptor during the subacute remodeling phase, for instance, prevents LV enlargement by reducing fibrous tissue formation [109].

Comparable to TGFβ, IL-10 has the capacity to stimulate TIMP-1 expression in mononuclear cells that multitudinously infiltrate the infarcted myocardium. With respect to cardiac remodeling, exogenous administration of IL-10 leads to an attenuated LV enlargement, which may partially result from impeded MMP activity [110].

In contrast to TGFβ and IL-10, the proinflammatory cytokines TNFα, IL-1β, and IL-6 efficiently suppress the formation of a solid scar tissue by restraining collagen synthesis in fibroblasts while simultaneously stimulating MMP expression [111]. Transgenic mice constitutively expressing TNFα in the heart develop a dilated cardiomyopathy that comes along with increased cardiac MMP activity. Counterbalancing the effects of TNFα in these mice by adenoviral expression of the soluble TNFα receptor p55 attenuates MMP activity, in line with a correction of LV dysfunction [112].

Another proinflammatory cytokine that exerts significant impact on scar tissue quality is the matricellular protein osteopontin (OPN). Global OPN KO mice exhibit reduced collagen content and disarrayed collagen fiber alignment in the healing infarct region. In line with these scarring disturbances, OPN KO mice show an increase in LV end-systolic and diastolic diameters, suggesting a protective role of OPN in terms of remodeling after MI [113,114].

2.3.1.4 Cytokines and ROS

Proinflammatory cytokines have the ability to induce ROS production in a variety of cells. TNFα is a strong inducer of mitochondrial ROS in cardiomyocytes and stimulates NADPH oxidase activity in infiltrating leukocytes [115,116]. Furthermore, both IL-6 and IL-1β have also been implicated in ROS production in other animal models of cardiovascular disease [117,118]. The physiological relevance of ROS for cardiac remodeling has already been highlighted in detail earlier in this chapter.

2.3.1.5 Cytokines in Inflammation Resolution

After an initial inflammatory reaction in response to cardiac injury, inflammation resolution is a critical step to pave the way for elaborate healing and scarring [11]. Moreover, anti-inflammatory cytokines shut down immunity preventing collateral damage of surviving cardiomyocytes [11].

In addition to the capacity to induce wound-stabilizing TIMP-1, IL-10 is a prominent suppressor of the inflammatory transcriptional program in mononuclear cells [119,120]. After cardiac injury, a distinct macrophage subset is the most abundant source of IL-10 within the healing myocardium [121]. However, in terms of wound healing and remodeling, contradictory results regarding the role of IL-10 have been proposed. Yang *et al.* employed global IL-10 KO mice undergoing ischemia and reperfusion. Compared to WT animals, IL-10-deficient mice showed an aggravated inflammatory response in line with deteriorated survival [122]. In contrast, Zymek *et al.* were not able to find significant differences between WT mice and genetically IL-10-ablated animals. Genotypes showed a similar phenotype regarding leukocyte recruitment, scar tissue formation, and cardiac remodeling, suggesting that endogenous IL-10 is not a key player in inflammation resolution [123]. However, a surplus of IL-10 by exogenous IL-10 administration leads to a reduction of cardiac inflammation and attenuates remodeling after MI [110,124].

The members of the TGFβ family are not only important in the reparative response, but also are crucially involved in the suppression of immunity [125]. Depending on the dose, TGFβ has potent pro- or anti-inflammatory properties. Femtomolar concentrations of TGFβ provoke a strong chemotactic stimulus for monocytes, implying a role in monocyte recruitment into the injured myocardium [126]. Also regarding monocytes, picomolar TGFβ concentrations elicit the synthesis of inflammatory cytokines and an upregulation of integrins [126,127]. However, after tissue infiltration, monocytes can differentiate into mature macrophages that experience a markedly deactivation of the inflammatory program in the presence of TGFβ [128,129]. Furthermore, TGFβ signaling blunts the cytokine-driven induction of adhesion molecules in endothelial cells, attenuating leukocyte recruitment [130].

Targeting TGFβ in experimental settings has been a successful approach to modulate remodeling after MI. TGFβ administration during the inflammatory phase of healing reduces infarct size [131]. In accordance with this observation, an early inhibition of TGFβ function by adenoviral expression of a soluble TGFβ decoy receptor leads to an increased mortality along with exacerbated LV remodeling [132]. However, expression of the decoy receptor during late-stage healing mitigates adverse remodeling due to an attenuated fibrotic response [109].

In summary, the cytokine network constitutes a central pillar of immunity that influences virtually all steps of healing and remodeling. Given the redundant and pleiotropic nature of individual cytokines, a modulation of this intricate cytokine system is challenging. However, especially due to their broad biological properties, a modulation of distinct cytokine pathways seems appealing to beneficially influence remodeling after cardiac injury.

2.3.2 Cellular Effectors

2.3.2.1 Leukocyte Recruitment

Leukocyte extravasation occurs predominantly in post-capillary venules where hemodynamic shear stress is low. At first, flowing leukocytes decelerate by rolling on the activated endothelium. The adhesive interaction between blood-borne cells and the endothelium is thereby mediated by mutual binding of specific cell adhesion receptors—namely, the selectins comprising L-selectin (CD62L), E-selectin (CD62E), and P-selectin (CD62P) [133–135]. L-selectin is constitutively expressed on all leukocytes fractions. E-selectin, in contrast, is synthesized *de novo* in endothelial cells in response to inflammatory cytokines as TNFα or IL-1β. P-selectin exists in preformed endothelial granules that are mobilized to the surface by inflammatory stimuli. Moreover, P-selectin expression in endothelial cells is induced in response to TNFα [136,137].

After selectin-mediated leukocyte capturing, cell arrest occurs by firm adhesion between integrins on leukocytes and intercellular adhesion molecules that coat the endothelium. Firm cell adhesion requires interaction with β2 integrins that share a common β chain (CD18), which pairs with CD11a, CD11b, or CD11c on the leukocyte surface [138]. The endothelial counterparts include the vascular cell adhesion molecule-1, intercellular adhesion molecule (ICAM)-1/ICAM-2, as well as the members of the junctional adhesion molecules family [139–142]. Constitutive ICAM-1/ICAM-2 expression on endothelial cells may be sufficient to mediate leukocyte recruitment into the tissue. However, ICAM-1, for instance, can additionally be upregulated in response to cytokines such as IL-6 or TNFα [143].

Integrin activation is induced by chemokines that are localized at glycosaminoglycan moieties on both the endothelium and ECM components [144]. Generally, CC chemokines function as potent chemoattractants for mononuclear cells and ELR motif-containing CXC chemokines efficiently mediate neutrophil recruitment [145]. Chemokines are induced in response to DAMPs or oxidative stress. Moreover, proinflammatory cytokines such as TNFα are able to induce chemokine expression, for example, CXCL8 [4]. Also, IL-6 seems to exert a potent function regarding CXC chemokine upregulation [146].

Ultimately, transendothelial migration occurs by paracellular trafficking between the endothelial cells or intracellular trafficking through endothelial cytoplasmatic pores. The emergence of distinct leukocyte populations in the injured myocardium occurs in a temporally defined sequence to meet the requirements for proper tissue replacement post-MI (Figure 2.2).

2.3.2.2 Neutrophils

Neutrophils are the most abundant fraction of leukocytes and account for 35-75% of all circulating cells in the steady state [147]. Neutrophils are rapidly recruited to sites of tissue injury and constitute the first line of defense against potentially wound-invading pathogens. The microbicidal function is mediated by pathogen phagocytosis, release of antimicrobial granules, or formation of extracellular traps (NETs) preventing further pathogen spreading [148]. However, aside of vital roles in innate immune reactions against pathogens, neutrophils crucially modulate tissue remodeling after sterile injury [149].

Neutrophil development takes place in the bone marrow where pluripotent hematopoietic cells differentiate into myeloblasts, the neutrophil progenitor cells. After neutrophil egress from the bone marrow, the cells circulate freely in the blood as terminally differentiated cells that have lost their proliferative capacity [150]. The life span of circulating neutrophils is rather short; studies have suggested a maximum life expectancy of approximately 5 days [151]. After cardiac damage, however, neutrophils become rapidly deployed from the bone marrow and constitute the first leukocyte fraction that infiltrates the myocardium [57,152]. Potent neutrophil chemoattractants comprise chemokines and anaphylatoxins such as CXCL2, leukotriene B4, CXCL1, CXCL8, or complement 5a [58]. Neutrophils constitute the first leukocyte fraction that enters the infarct zone. In the nonreperfused murine myocardium, infiltration can be observed as early as several hours after MI, peaking between days 1-3 and subsequently declining [57]. Neutrophil turnover in the tissue is rapid. Having executed their biological function, the cells undergo apoptosis *in situ*.

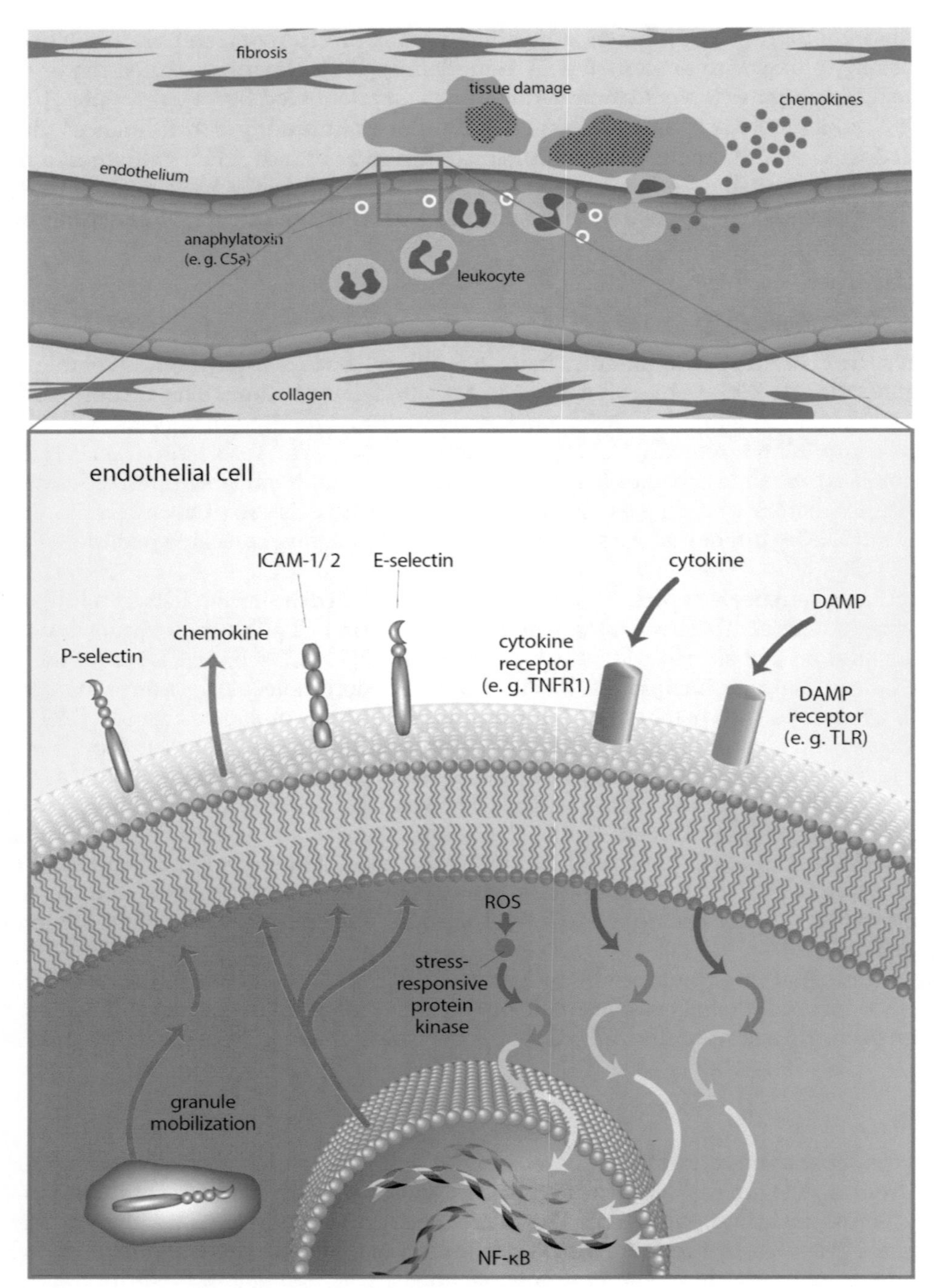

FIGURE 2.2 Endothelium activation after MI. In response to stimuli such as DAMPs, ROS, or inflammatory cytokines, endothelial cells upregulate ICAM-1/-2 expression and initiate chemokine secretion. Moreover, granules containing P-selectin become mobilized transporting P-selectin to the cell surface. Cell adhesion molecules along with anaphylatoxins and chemokine gradients facilitate leukocyte extravasation into the injured myocardium.

2.3.2.2.1 NEUTROPHIL-MEDIATED CARDIAC INJURY

2.3.2.2.1.1 REACTIVE OXYGEN SPECIES After recruitment to the healing myocardium, neutrophils become activated in response to DAMPs, anaphylatoxins, or inflammatory cytokines, resulting in the release of large amounts of ROS and toxic granules [153]. In resting cells, the ROS-producing multicomponent NADPH oxidase is not assembled. After cell activation, however, NADPH subunits assemble into an active enzyme complex leading to a significant increase of ROS production [153,154].

The role of ROS for cardiomyocyte death as well as for initiation and perpetuation of the inflammatory response has been highlighted earlier in this chapter. However, after cardiac injury, the activation of matrix metalloproteinases (MMPs) is of paramount importance for the progression of remodeling [3]. Inflammatory stimuli provoke the synthesis and release of collagenases, gelatinases, stromelysins, and membrane-type MMPs from fibroblasts or infiltrating leukocytes, mediating ECM degradation within both the infarct and noninfarct zone [155]. The activity of individual MMPs is tightly regulated by tissue inhibitors of matrix metalloproteinases (TIMPs) and ECM turnover therefore influenced by the balance of these antagonistic protein families. A key process in the pathophysiology of LV enlargement is an exuberant activation of MMPs within the extracellular space, resulting in a "slippage" of the myofibrils [3,155]. In terms of oxidative stress, ROS influence the overall MMP activity in the heart. ROS have the capacity to stimulate MMP gene expression while simultaneously suppressing TIMP synthesis [155,156]. Furthermore, MMPs are initially synthesized as enzymatically inactive zymogens (pro-MMPs), and ROS have the capacity to directly impose enzymatic activity by zymogen oxidation [155,157,158]. Thus, the presence of ROS favors an elevated MMP activity accounting for an increased ECM degradation and LV enlargement post-MI.

In consideration of the fact that oxidative stress can contribute to cardiac injury and remodeling, therapeutic strategies have been developed to attenuate ROS-mediated pathologies. Rats treated with superoxide dismutase, for instance, exhibit a decreased infarct size following ischemia and reperfusion [159]. In line with this observation, over-expression of the oxidant scavenger glutathione peroxidase prevented LV dilatation in mice after MI [160]. However, in the clinical arena, trials have mostly provided discouraging or contradictory results. Clinical treatments included administration of the thiolic antioxidant N-acetylcysteine, the XO inhibitor allopurinol, antioxidant agents such as vitamin E or selenium, and iron chelators as deferoxamine [161]. A more recent study, however, demonstrated a beneficial effect of the radical scavenger edaravone [162]. Patients treated with this drug presented reduced oxidative stress, decreased reperfusion arrhythmias, and smaller infarct size implying an improvement of long-term cardiac remodeling. The long-term clinical benefit of edaravone, however, needs further evaluation.

In summary, ROS directly damage cardiomyocytes during reperfusion after MI resulting in an accentuation of the inflammatory response. ROS activate inflammatory pathways that facilitate the recruitment of innate immune cells and reinforce the inflammatory profile in these heart-infiltrating leukocytes. Oxidative stress significantly stimulates MMP activity within the infarct zone and promotes thereby ECM and global cardiac remodeling. Although antioxidant treatment modalities may constitute a promising approach for preventing remodeling, clinical studies targeting ROS have been widely unsuccessful so far.

2.3.2.2.1.2 GRANULE TOXICITY Neutrophil function is considerably based on the exocytosis of granule components. Neutrophil-derived granules are grouped into four distinct granule types—azurophilic, specific, gelatinase, and secretory granules containing approximately 300 different proteins [149]. *In vivo*, however, release of toxic granules can only be observed from adherent neutrophils, suggesting a ligand-specific interaction between neutrophils and parenchymal cells after tissue infiltration [4,163]. In terms of cardiac injury, several studies have identified an interaction between ICAM-1 and the CD11b/CD18 hetero-dimer mediating neutrophil adhesion-dependent cytotoxicity [164,165]. Accordingly, ICAM-1 is not only expressed by endothelial cells, but also is induced in several cell types including cardiomyocytes in the presence of IL-1, TNFα, or IL-6 [165–167].

The role of neutrophil granule components for cardiac injury and remodeling has been studied over decades. One of the cardinal mediators of neutrophil function is the enzyme myeloperoxidase (MPO). MPO catalyzes the oxidation of halide ions to hypohalous acids and tyrosine to tyrosyl radical under consumption of hydrogen peroxide. Both hypohalous acids and tyrosyl radical are highly cytotoxic and are therefore important for pathogen clearance in terms of infections [168,169]. However, MPO products are also crucially involved in the immunological misfire that collaterally harms surviving cardiomyocytes after ischemic tissue damage. MPO serum levels are elevated in patients with MI and extreme high levels are associated with impaired survival [170,171]. Cytotoxic MPO products include the tyrosyl radical and derivatives of glycine and threonine, that is, formaldehyde and acrolein [169]. Moreover, the MPO-catalyzed synthesis of chlorinating species accounts for the generation of toxic 2-chlorohexadecanal, which further reduces ventricular performance [172]. Consistently, genetic ablation of MPO in mice attenuates leukocyte recruitment into the infarct zone and attenuates LV dilatation [173].

In addition to MPO, MMP-9 constitutes another important neutrophil-derived enzyme involved in cardiac remodeling [155]. MMP-9 is stored in gelatinase granules and released in the presence of a chemotactic stimulus. After MI, MMP-9 is not exclusively released from infiltrating neutrophils, but also is secreted from macrophages, myocytes, fibroblasts, or vascular cells [58]. However, neutrophils are an important early source of MMP-9 within the infarct zone mediating an initial degradation of the injured ECM. In terms of cardiac remodeling, however, the detrimental effects of MMP-9 seem to outweigh the enzyme's beneficial value. Global MMP-9-deficient mice show an attenuated LV dysfunction and diminished cardiac fibrosis, in line with an improved angiogenesis [174,175].

More recently, neutrophil-mediated injury has been addressed to the formation of so-called neutrophil extracellular traps (NETs), a network of extracellular fibers containing DNA, citrullinated histones, and cytotoxic enzymes [176]. NETs have been demonstrated to play a fundamental role in killing microbes in terms of an infection. However, NETs also significantly contribute to cardiac injury after MI and the resolution of NETs by DNase treatment results in infarct size reduction after ischemia/reperfusion, implying a role of NET formation in terms of cardiac remodeling [177].

Given the broad arsenal of proinflammatory and destructive mediators, efforts have been taken to restrain the function of the neutrophil compartment in its entity. Interference with neutrophil infiltration by CD18 blockade or genetic ablation of ICAM-1, but also specific neutrophil depletion by the use of leukocyte filters, leads to a reduction of infarct size after ischemia/reperfusion [178–180]. The encouraging results from these experimental models spurred a clinical trial targeting neutrophil recruitment to the heart. However, impairing neutrophil infiltration by administration of an inhibitory anti-CD18 monoclonal antibody did not improve the clinical outcome in patients with MI [181].

Conclusively, neutrophils are early cellular effectors that numerously infiltrate the infarct zone. The cells become activated by cytokines and chemokines and modulate healing and remodeling by releasing their high payload of inflammatory mediators, such as proteases and ROS. Albeit hitherto discouraging clinical trials, manipulating neutrophil function may constitute an eligible modality to prevent adverse remodeling.

2.3.2.3 *Mononuclear Cells*

Monocytes and macrophages are protagonists in wound healing and remodeling after MI [182]. The classic, though over-simplified model proposes that monocytes circulate freely after development from bone marrow precursors, transdifferentiating into macrophages upon tissue infiltration [183]. However, the spatial restriction of monocytes and macrophages to distinct tissues and the linear developmental trajectory of this model have been revisited. It has become clear that monocytes have the potential to develop extramedullarily, that is, outside of the bone marrow, especially during inflammatory conditions [184–186]. Also, many tissue macrophages do not derive from monocytes, but from progenitors that have seeded the tissue from the yolk sac before the onset of hematopoiesis [187,188]. In the steady state, macrophages reside in the heart at low numbers that significantly increase post-MI. In regard to the latter observation, monocyte recruitment and subsequent macrophage transdifferentiation, but also local proliferation of pre-existing macrophages, may contribute to the rise of cardiac macrophage numbers [57,189].

The compartment of monocytic cells does not constitute a homogenous population, but comprises distinct monocyte subsets that fulfill various complementary tasks [190]. Similarly, macrophages exhibit marked plasticity and execute a broad spectrum of physiological functions depending on the context of activation [191,192]. The following section highlights the role of the different monocyte and macrophage subsets in cardiac remodeling after MI (Figure 2.3).

2.3.2.3.1 MONOCYTES

In mice, monocyte subsets are discriminated on the basis of Ly-6C expression, a GPI-anchored surface protein of so far unknown physiological function. In the steady state, up to 60% of circulating monocytes belong to the "inflammatory" Ly-6C^{high} CCR2^{high} CX3CR1^{low} CD62L^{+} subset. During inflammatory conditions, the proportion of this

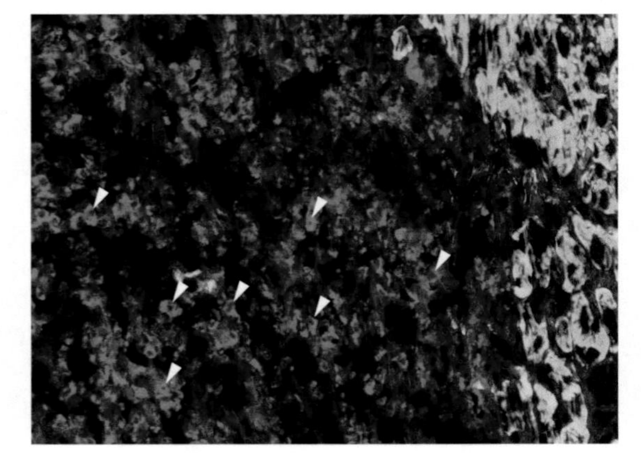

FIGURE 2.3 Monocytes and macrophages in the murine heart 5 days post-MI. Monocytes and macrophages were stained using an antibody against CD68 that is highly expressed on both leukocyte populations. White arrow heads indicate exemplary CD68-positive cells depicted red. Surviving cardiomyocytes adjacent to the infarct region were stained by using Alexa Fluor 488 phalloidin. The cells are depicted green. Nuclei are indicated blue (DAPI, 200-fold magnification).

subset rises due to increased monocytopoiesis in both bone marrow and spleen. The Ly-6C^{high} subset accumulates preferentially at inflammatory sites where these monocytes can mature into macrophages [182,193].

The remaining Ly-6C^{low} CCR2^{low} CX3CR1^{high} CD62L^{-} monocyte subset primarily patrols the vasculature and accumulates at low numbers in solid tissues. From the developmental view, Ly-6C^{low} monocytes arise from Ly-6C^{high} monocytes and not likely from a separate progenitor [182,193]. In humans, however, monocyte subsets are distinguished on the basis of CD14 and CD16 expression [194]. The functional homology between mouse and human monocyte subsets is controversially discussed and a third monocyte subset with individual functions has additionally been described in men [195,196].

Generally, two sequential phases of monocyte infiltration can be identified after MI. During the first days, predominantly Ly-6C^{high} monocytes infiltrate the infarct zone via the chemokine MCP-1 peaking 2 days post-MI [57]. A few days later, as the resolution of inflammation begins, Ly-6C^{high} monocytes wane and Ly-6C^{low} monocytes accumulate in the myocardium, outnumbering Ly-6C^{high} monocytes 7 days after MI. Gradually, monocytes disappear owing to frequent apoptosis or transdifferentiation into mature macrophages [57,186].

Ly-6C^{high} monocytes have a high payload of inflammatory mediators such as TNFα or proteases. These inflammatory monocytes are crucially involved in the phagocytosis and digestion of ECM and cell debris to pave the road for granulation tissue formation [190]. In contrast, Ly-6C^{low} monocytes have less content of inflammatory mediators but exhibit pronounced expression of VEGF and TGFβ stimulating angiogenesis and collagen deposition [190]. However, ablation of either monocyte subset results in disturbed healing and, consistently, a crude depletion of the entire monocyte/macrophage compartment results in aggravated LV remodeling [57,197].

2.3.2.3.2 MACROPHAGES

Traditionally, macrophages are grouped into two major subsets—namely, classically activated M1 cells and alternatively activated M2 cells [198]. The classification was introduced by the observation that macrophages generated from bone marrow cells exhibit a completely different transcriptome in response to different stimuli. Macrophages generated in the presence of interferon-γ or lipopolysaccharide develop into M1 cells that have potent inflammatory properties. In contrast, the presence of IL-4, IL-13, or a combination of IL-10 and TGFβ drives the differentiation into M2 cells that have an anti-inflammatory and wound-healing profile [191,198]. *In vivo*, however, macrophage polarization is rather likely a continuum between the two M1/M2 extremes.

In terms of wound healing and remodeling, the biphasic emergence of the monocyte subsets reflects the kinetics of M1 and M2 marker gene expression in the infarct zone [121]. Therefore, it is tempting to speculate that M1 cells derive from Ly-6C^{high} monocytes and M2 cells later from Ly-6C^{low} monocytes, although evidence for this developmental relationship is missing. However, the features of macrophage subsets qualitatively mirror the functions of monocyte subsets. After MI, inflammatory M1 cells gain the initial inflammatory reaction by secreting factors such as TNFα, IL-6, or IL-1β [121]. In contrast, M2 cells subsequently release factors such as IL-10, TGFβ, or wound-stabilizing OPN [121,199].

In terms of MI, interfering with the transition from an M1-prevalence to an M2-prevalence has been proved to have significant impact on the outcome of remodeling. An impaired or delayed polarization toward M2 cells correlates with deteriorated healing, LV dilatation, and impaired LV function [200,201]. In accordance with this observation, experimental modulation of the macrophage polarization toward an M2 state has been demonstrated to attenuate LV remodeling by improving healing [202].

In summary, monocytes and macrophages show a biphasic response post-MI. Respective subsets emerge sequentially in the heart and fulfill a broad spectrum of complementary tasks. Given their functional plasticity and multifaceted roles in terms of healing, monocytes and macrophages are highly promising targets to influence remodeling.

2.4 REVERSE REMODELING

Reverse remodeling is characterized by an improvement of myocardial function in patients with seemingly end-stage disease. Most data regarding this phenomenon stem from clinical studies describing physiological changes after left-ventricular assist device (LVAD) implantation in heart failure patients [203,204]. However, although a causal relationship between alterations in inflammation and reverse remodeling is hitherto not evident, the incidence of both phenomena correlates. Myocardial recovery in patients after LVAD implantation is associated with a decline of inflammatory mediators such as IL-1, TNFα, IL-8, or C3a in both the myocardium and the serum [204]. Furthermore, certain drugs with the capacity to reverse LV remodeling have anti-inflammatory side effects implying a potential impact of immunosuppression on LV recovery. Angiotensin converting enzyme (ACE) inhibitors, for instance, have been demonstrated to potentially moderate or even reverse remodeling [205,206]. Since angiotensin II receptor type 1 signaling provokes a profound inflammatory response, ACE inhibitors may, at least partially, support reverse remodeling by restraining the inflammatory reaction [207,208].

In summary, although a mechanistic link between the inflammatory response and reverse remodeling is missing, a decline in inflammation may have a supporting role.

2.5 CLINICAL IMPLICATIONS: IS THERE A CAUSAL LINK BETWEEN DYSEQUILIBRATED INFLAMMATION AND REMODELING?

The concept that an exuberant inflammatory immune response deteriorates remodeling is widely accepted. A broad array of experimental studies underscores the importance of inflammation resolution for cardiac healing and prevention of remodeling. On the other hand, an initial inflammatory reaction is crucial to pave the way for healing, and a premature suppression of immunity is therefore considered likewise disadvantageous. The balance of the reaction but also the timing of shifting from inflammation to resolution is of paramount importance to prevent adverse remodeling.

To support the therapeutic significance of this concept in the clinical arena, the central question concerns the relevance of inflammation for remodeling in patients with MI. In support of the concept, heart failure patients show increased clinical signs of inflammation such as elevated C-reactive protein and cytokine levels along with increased leukocyte counts, underscoring the likelihood of an inflammatory component in cardiac remodeling [209–212].

The approach of targeting the inflammatory response to improve healing and remodeling in patients is not new. Decades ago, experiments in large animals led to clinical trials aiming to improve outcome by restraining inflammation. Broad and unspecific immunosuppression by administration of methylprednisolone, however, yielded catastrophic results. Also, more specific approaches preventing complement activation or leukocyte recruitment by inhibitory monoclonal antibodies were unable to improve clinical outcome. Given the extensive experimental efficacy of anti-inflammatory strategies, why, thus, were these clinical trials targeting inflammatory pathways unable to mitigate injury and remodeling?

First, although providing important insight into physiological and pathological mechanisms, the value of animal models for predicting the success of a treatment modality in human patients is limited. An animal model constitutes a defined system where single parameters can be changed according to the experimenter's desire. Reality in the clinic, however, is far more complicated, making it difficult to predict the efficacy of a therapeutic approach. Potential effects on the outcome include factors such as age, genetic polymorphisms, gender, and comorbid conditions such as diabetes, hypertension, or the timing of reperfusion. Studies focusing on the correlation between age and the quality of healing and remodeling have clearly demonstrated the difficulties in translating concepts from animals to humans. In (immuno-) senescent mice, aggravated remodeling and wound healing are associated with a suppressed and delayed inflammatory response as compared to young mice [213]. Since most experimental data are derived from young animals exhibiting vigorous inflammatory reactions, the potentially deleterious effect of inflammation on healing and remodeling might have been overestimated in the elderly.

Second, proper healing requires a fine-tuning of pro- and anti-inflammatory reactions, and the timing of shifting from debris clearance (inflammation) to healing (inflammation resolution) is of paramount importance to prevent adverse remodeling. Strategies suppressing inflammation and immunity may therefore have baneful effects by interfering with debris clearance from the wound. In fact, clinical trials targeting an early disruption of inflammation may have *de facto prolonged* cardiac inflammation by suppressing infarct clearance, resulting in a delayed initiation of healing.

However, modulating immunity is still a promising approach to improving outcome in patients that bear a high risk to undergo remodeling. Moreover, manipulating immunity may become an eligible modality to support or even induce reverse remodeling. Novel therapeutic concepts should specifically target selected components of immunity and modulate the inflammatory reaction in an optimal temporal context.

References

[1] Cohn JN, Ferrari R, Sharpe N. Cardiac remodeling—concepts and clinical implications: a consensus paper from an international forum on cardiac remodeling. Behalf of an International Forum on Cardiac Remodeling. J Am Coll Cardiol 2000;35(3):569–82.
[2] McMurray JJ, Pfeffer MA. Heart failure. Lancet 2005;365(9474):1877–89.
[3] Jugdutt BI. Ventricular remodeling after infarction and the extracellular collagen matrix: when is enough enough? Circulation 2003;108(11):1395–403.

[4] Frangogiannis NG, Smith CW, Entman ML. The inflammatory response in myocardial infarction. Cardiovasc Res 2002;53(1):31–47.
[5] Ertl G, Frantz S. Healing after myocardial infarction. Cardiovasc Res 2005;66(1):22–32.
[6] Sutton MG, Sharpe N. Left ventricular remodeling after myocardial infarction: pathophysiology and therapy. Circulation 2000;101(25):2981–8.
[7] Pfeffer MA. Left ventricular remodeling after acute myocardial infarction. Annu Rev Med 1995;46:455–66.
[8] Czubryt MP. Common threads in cardiac fibrosis, infarct scar formation, and wound healing. Fibrogenesis Tissue Repair 2012;5(1):19.
[9] Frantz S, Bauersachs J, Ertl G. Post-infarct remodelling: contribution of wound healing and inflammation. Cardiovasc Res 2009;81(3):474–81.
[10] McKay RG, Pfeffer MA, Pasternak RC, Markis JE, Come PC, Nakao S, et al. Left ventricular remodeling after myocardial infarction: a corollary to infarct expansion. Circulation 1986;74(4):693–702.
[11] Frangogiannis NG. Regulation of the inflammatory response in cardiac repair. Circ Res 2012;110(1):159–73.
[12] de Haan JJ, Smeets MB, Pasterkamp G, Arslan F. Danger signals in the initiation of the inflammatory response after myocardial infarction. Mediat Inflamm 2013;2013:206039.
[13] Delvaeye M, Conway EM. Coagulation and innate immune responses: can we view them separately? Blood 2009;114(12):2367–74.
[14] Opal SM, Esmon CT. Bench-to-bedside review: functional relationships between coagulation and the innate immune response and their respective roles in the pathogenesis of sepsis. Crit Care 2003;7(1):23–38.
[15] Frangogiannis NG. Chemokines in ischemia and reperfusion. Thromb Haemost 2007;97(5):738–47.
[16] Srikrishna G, Freeze HH. Endogenous damage-associated molecular pattern molecules at the crossroads of inflammation and cancer. Neoplasia 2009;11(7):615–28.
[17] Mogensen TH. Pathogen recognition and inflammatory signaling in innate immune defenses. Clin Microbiol Rev 2009;22(2):240–73, Table of Contents.
[18] Boller T, Felix G. A renaissance of elicitors: perception of microbe-associated molecular patterns and danger signals by pattern-recognition receptors. Annu Rev Plant Biol 2009;60:379–406.
[19] Ionita MG, Arslan F, de Kleijn DP, Pasterkamp G. Endogenous inflammatory molecules engage Toll-like receptors in cardiovascular disease. J innate immun 2010;2(4):307–15.
[20] Feng Y, Chao W. Toll-like receptors and myocardial inflammation. Int J Inflamm 2011;2011:170352.
[21] Akira S, Uematsu S, Takeuchi O. Pathogen recognition and innate immunity. Cell 2006;124(4):783–801.
[22] Kawai T, Akira S. The role of pattern-recognition receptors in innate immunity: update on toll-like receptors. Nat Immunol 2010;11(5):373–84.
[23] Blasius AL, Beutler B. Intracellular toll-like receptors. Immunity 2010;32(3):305–15.
[24] Mann DL. The emerging role of innate immunity in the heart and vascular system: for whom the cell tolls. Circ Res 2011;108(9):1133–45.
[25] Favre J, Musette P, Douin-Echinard V, Laude K, Henry JP, Arnal JF, et al. Toll-like receptors 2-deficient mice are protected against postischemic coronary endothelial dysfunction. Arterioscler Thromb Vasc Biol 2007;27(5):1064–71.
[26] Arslan F, Smeets MB, O'Neill LA, Keogh B, McGuirk P, Timmers L, et al. Myocardial ischemia/reperfusion injury is mediated by leukocytic toll-like receptor-2 and reduced by systemic administration of a novel anti-toll-like receptor-2 antibody. Circulation 2010;121(1):80–90.
[27] Mersmann J, Habeck K, Latsch K, Zimmermann R, Jacoby C, Fischer JW, et al. Left ventricular dilation in toll-like receptor 2 deficient mice after myocardial ischemia/reperfusion through defective scar formation. Basic Res Cardiol 2011;106(1):89–98.
[28] Oyama J, Blais Jr. C, Liu X, Pu M, Kobzik L, Kelly RA, et al. Reduced myocardial ischemia-reperfusion injury in toll-like receptor 4-deficient mice. Circulation 2004;109(6):784–9.
[29] Chong AJ, Shimamoto A, Hampton CR, Takayama H, Spring DJ, Rothnie CL, et al. Toll-like receptor 4 mediates ischemia/reperfusion injury of the heart. J Thorac Cardiovasc Surg 2004;128(2):170–9.
[30] Riad A, Jager S, Sobirey M, Escher F, Yaulema-Riss A, Westermann D, et al. Toll-like receptor-4 modulates survival by induction of left ventricular remodeling after myocardial infarction in mice. J Immunol 2008;180(10):6954–61.
[31] Shimamoto A, Chong AJ, Yada M, Shomura S, Takayama H, Fleisig AJ, et al. Inhibition of toll-like receptor 4 with eritoran attenuates myocardial ischemia-reperfusion injury. Circulation 2006;114(1 Suppl):I270–4.
[32] Sarma JV, Ward PA. The complement system. Cell Tissue Res 2011;343(1):227–35.
[33] Ricklin D, Hajishengallis G, Yang K, Lambris JD. Complement: a key system for immune surveillance and homeostasis. Nat Immunol 2010;11(9):785–97.
[34] Jiang H, Cooper B, Robey FA, Gewurz H. DNA binds and activates complement via residues 14-26 of the human C1q A chain. J Biol Chem 1992;267(35):25597–601.
[35] Hill JH, Ward PA. The phlogistic role of C3 leukotactic fragments in myocardial infarcts of rats. J Exp Med 1971;133(4):885–900.
[36] Rossen RD, Michael LH, Hawkins HK, Youker K, Dreyer WJ, Baughn RE, et al. Cardiolipin-protein complexes and initiation of complement activation after coronary artery occlusion. Circ Res 1994;75(3):546–55.
[37] Williams JP, Pechet TT, Weiser MR, Reid R, Kobzik L, Moore Jr. FD, et al. Intestinal reperfusion injury is mediated by IgM and complement. J Appl Physiol 1999;86(3):938–42.
[38] Weiser MR, Williams JP, Moore Jr. FD, Kobzik L, Ma M, Hechtman HB, et al. Reperfusion injury of ischemic skeletal muscle is mediated by natural antibody and complement. J Exp Med 1996;183(5):2343–8.
[39] Zhang M, Alicot EM, Chiu I, Li J, Verna N, Vorup-Jensen T, et al. Identification of the target self-antigens in reperfusion injury. J Exp Med 2006;203(1):141–52.
[40] Zhang M, Michael LH, Grosjean SA, Kelly RA, Carroll MC, Entman ML. The role of natural IgM in myocardial ischemia-reperfusion injury. J Mol Cell Cardiol 2006;41(1):62–7.
[41] Nijmeijer R, Lagrand WK, Lubbers YT, Visser CA, Meijer CJ, Niessen HW, et al. C-reactive protein activates complement in infarcted human myocardium. Am J Pathol 2003;163(1):269–75.
[42] Mihlan M, Blom AM, Kupreishvili K, Lauer N, Stelzner K, Bergstrom F, et al. Monomeric C-reactive protein modulates classic complement activation on necrotic cells. FASEB J 2011;25(12):4198–210.

[43] Schoos MM, Munthe-Fog L, Skjoedt MO, Ripa RS, Lonborg J, Kastrup J, et al. Association between lectin complement pathway initiators, C-reactive protein and left ventricular remodeling in myocardial infarction-a magnetic resonance study. Mol Immunol 2013;54(3–4):408–14.
[44] Dreyer WJ, Michael LH, Nguyen T, Smith CW, Anderson DC, Entman ML, et al. Kinetics of C5a release in cardiac lymph of dogs experiencing coronary artery ischemia-reperfusion injury. Circ Res 1992;71(6):1518–24.
[45] Birdsall HH, Green DM, Trial J, Youker KA, Burns AR, MacKay CR, et al. Complement C5a, TGF-beta 1, and MCP-1, in sequence, induce migration of monocytes into ischemic canine myocardium within the first one to five hours after reperfusion. Circulation 1997;95(3):684–92.
[46] Vakeva AP, Agah A, Rollins SA, Matis LA, Li L, Stahl GL. Myocardial infarction and apoptosis after myocardial ischemia and reperfusion: role of the terminal complement components and inhibition by anti-C5 therapy. Circulation 1998;97(22):2259–67.
[47] van der Pals J, Koul S, Andersson P, Gotberg M, Ubachs JF, Kanski M, et al. Treatment with the C5a receptor antagonist ADC-1004 reduces myocardial infarction in a porcine ischemia-reperfusion model. BMC Cardiovasc Disord 2010;10:45.
[48] Mahaffey KW, Granger CB, Nicolau JC, Ruzyllo W, Weaver WD, Theroux P, et al. Effect of pexelizumab, an anti-C5 complement antibody, as adjunctive therapy to fibrinolysis in acute myocardial infarction: the COMPlement inhibition in myocardial infarction treated with thromboLYtics (COMPLY) trial. Circulation 2003;108(10):1176–83.
[49] Granger CB, Mahaffey KW, Weaver WD, Theroux P, Hochman JS, Filloon TG, et al. Pexelizumab, an anti-C5 complement antibody, as adjunctive therapy to primary percutaneous coronary intervention in acute myocardial infarction: the COMplement inhibition in Myocardial infarction treated with Angioplasty (COMMA) trial. Circulation 2003;108(10):1184–90.
[50] Cadenas E. Biochemistry of oxygen toxicity. Annu Rev Biochem 1989;58:79–110.
[51] Novo E, Parola M. Redox mechanisms in hepatic chronic wound healing and fibrogenesis. Fibrogenesis Tissue Repair 2008;1(1):5.
[52] Li X, Fang P, Mai J, Choi ET, Wang H, Yang XF. Targeting mitochondrial reactive oxygen species as novel therapy for inflammatory diseases and cancers. J Hematol Oncol 2013;6:19.
[53] McCord JM, Fridovich I. Superoxide dismutase: the first twenty years (1968-1988). Free Radic Biol Med 1988;5(5–6):363–9.
[54] Gustafsson AB, Gottlieb RA. Heart mitochondria: gates of life and death. Cardiovasc Res 2008;77(2):334–43.
[55] Murphy E, Steenbergen C. Mechanisms underlying acute protection from cardiac ischemia-reperfusion injury. Physiol Rev 2008;88(2):581–609.
[56] Zweier JL, Talukder MA. The role of oxidants and free radicals in reperfusion injury. Cardiovasc Res 2006;70(2):181–90.
[57] Nahrendorf M, Swirski FK, Aikawa E, Stangenberg L, Wurdinger T, Figueiredo JL, et al. The healing myocardium sequentially mobilizes two monocyte subsets with divergent and complementary functions. J Exp Med 2007;204(12):3037–47.
[58] Ma Y, Yabluchanskiy A, Lindsey ML. Neutrophil roles in left ventricular remodeling following myocardial infarction. Fibrogenesis Tissue Repair 2013;6(1):11.
[59] Dixon IM, Hata T, Dhalla NS. Sarcolemmal Na(+)-K(+)-ATPase activity in congestive heart failure due to myocardial infarction. Am J Physiol 1992;262(3 Pt 1):C664–71.
[60] Hori M, Nishida K. Oxidative stress and left ventricular remodelling after myocardial infarction. Cardiovasc Res 2009;81(3):457–64.
[61] Halestrap AP, McStay GP, Clarke SJ. The permeability transition pore complex: another view. Biochimie 2002;84(2–3):153–66.
[62] Leung AW, Halestrap AP. Recent progress in elucidating the molecular mechanism of the mitochondrial permeability transition pore. Biochim Biophys Acta 2008;1777(7–8):946–52.
[63] Shingu M, Nobunaga M. Chemotactic activity generated in human serum from the fifth component of complement by hydrogen peroxide. Am J Pathol 1984;117(2):201–6.
[64] Hayakawa Y, Hirata Y, Nakagawa H, Sakamoto K, Hikiba Y, Kinoshita H, et al. Apoptosis signal-regulating kinase 1 and cyclin D1 compose a positive feedback loop contributing to tumor growth in gastric cancer. Proc Natl Acad Sci U S A 2011;108(2):780–5.
[65] Karin M, Shaulian E. AP-1: linking hydrogen peroxide and oxidative stress to the control of cell proliferation and death. IUBMB Life 2001;52(1–2):17–24.
[66] Soga M, Matsuzawa A, Ichijo H. Oxidative stress-induced diseases via the ASK1 signaling pathway. Int J Cell Biol 2012;2012:439587.
[67] Meldrum DR, Dinarello CA, Cleveland Jr JC, Cain BS, Shames BD, Meng X, et al. Hydrogen peroxide induces tumor necrosis factor alpha-mediated cardiac injury by a P38 mitogen-activated protein kinase-dependent mechanism. Surgery 1998;124(2):291–6, discussion 7.
[68] Asehnoune K, Strassheim D, Mitra S, Kim JY, Abraham E. Involvement of reactive oxygen species in toll-like receptor 4-dependent activation of NF-kappa B. J Immunol 2004;172(4):2522–9.
[69] Chandel NS, Trzyna WC, McClintock DS, Schumacker PT. Role of oxidants in NF-kappa B activation and TNF-alpha gene transcription induced by hypoxia and endotoxin. J Immunol 2000;165(2):1013–21.
[70] Kapadia SR, Oral H, Lee J, Nakano M, Taffet GE, Mann DL. Hemodynamic regulation of tumor necrosis factor-alpha gene and protein expression in adult feline myocardium. Circ Res 1997;81(2):187–95.
[71] Beg AA, Baltimore D. An essential role for NF-kappaB in preventing TNF-alpha-induced cell death. Science 1996;274(5288):782–4.
[72] Yue P, Massie BM, Simpson PC, Long CS. Cytokine expression increases in nonmyocytes from rats with postinfarction heart failure. Am J Physiol 1998;275(1 Pt 2):H250–8.
[73] Deten A, Volz HC, Briest W, Zimmer HG. Cardiac cytokine expression is upregulated in the acute phase after myocardial infarction. Experimental studies in rats. Cardiovasc Res 2002;55(2):329–40.
[74] Nian M, Lee P, Khaper N, Liu P. Inflammatory cytokines and postmyocardial infarction remodeling. Circ Res 2004;94(12):1543–53.
[75] Irwin MW, Mak S, Mann DL, Qu R, Penninger JM, Yan A, et al. Tissue expression and immunolocalization of tumor necrosis factor-alpha in postinfarction dysfunctional myocardium. Circulation 1999;99(11):1492–8.
[76] Ono K, Matsumori A, Shioi T, Furukawa Y, Sasayama S. Cytokine gene expression after myocardial infarction in rat hearts: possible implication in left ventricular remodeling. Circulation 1998;98(2):149–56.
[77] Gordon JW, Shaw JA, Kirshenbaum LA. Multiple facets of NF-kappaB in the heart: to be or not to NF-kappaB. Circ Res 2011;108(9):1122–32.
[78] Shimizu N, Yoshiyama M, Omura T, Hanatani A, Kim S, Takeuchi K, et al. Activation of mitogen-activated protein kinases and activator protein-1 in myocardial infarction in rats. Cardiovasc Res 1998;38(1):116–24.

[79] Schroder K, Tschopp J. The inflammasomes. Cell 2010;140(6):821–32.
[80] Nakamura H, Umemoto S, Naik G, Moe G, Takata S, Liu P, et al. Induction of left ventricular remodeling and dysfunction in the recipient heart after donor heart myocardial infarction: new insights into the pathologic role of tumor necrosis factor-alpha from a novel heterotopic transplant-coronary ligation rat model. J Am Coll Cardiol 2003;42(1):173–81.
[81] Kakio T, Matsumori A, Ono K, Ito H, Matsushima K, Sasayama S. Roles and relationship of macrophages and monocyte chemotactic and activating factor/monocyte chemoattractant protein-1 in the ischemic and reperfused rat heart. Lab Invest 2000;80(7):1127–36.
[82] Ozaki K, Leonard WJ. Cytokine and cytokine receptor pleiotropy and redundancy. J Biol Chem 2002;277(33):29355–8.
[83] MacEwan DJ. TNF receptor subtype signalling: differences and cellular consequences. Cell Signal 2002;14(6):477–92.
[84] MacEwan DJ. TNF ligands and receptors—a matter of life and death. Br J Pharmacol 2002;135(4):855–75.
[85] Dbaibo GS, El-Assaad W, Krikorian A, Liu B, Diab K, Idriss NZ, et al. Ceramide generation by two distinct pathways in tumor necrosis factor alpha-induced cell death. FEBS Lett 2001;503(1):7–12.
[86] Mullen TD, Obeid LM. Ceramide and apoptosis: exploring the enigmatic connections between sphingolipid metabolism and programmed cell death. Anti Cancer Agents Med Chem 2012;12(4):340–63.
[87] Haudek SB, Taffet GE, Schneider MD, Mann DL. TNF provokes cardiomyocyte apoptosis and cardiac remodeling through activation of multiple cell death pathways. J Clin Invest 2007;117(9):2692–701.
[88] Malinin NL, Boldin MP, Kovalenko AV, Wallach D. MAP3K-related kinase involved in NF-kappaB induction by TNF, CD95 and IL-1. Nature 1997;385(6616):540–4.
[89] Liu ZG, Hsu H, Goeddel DV, Karin M. Dissection of TNF receptor 1 effector functions: JNK activation is not linked to apoptosis while NF-kappaB activation prevents cell death. Cell 1996;87(3):565–76.
[90] Kurrelmeyer KM, Michael LH, Baumgarten G, Taffet GE, Peschon JJ, Sivasubramanian N, et al. Endogenous tumor necrosis factor protects the adult cardiac myocyte against ischemic-induced apoptosis in a murine model of acute myocardial infarction. Proc Natl Acad Sci U S A 2000;97(10):5456–61.
[91] Mezzaroma E, Toldo S, Farkas D, Seropian IM, Van Tassell BW, Salloum FN, et al. The inflammasome promotes adverse cardiac remodeling following acute myocardial infarction in the mouse. Proc Natl Acad Sci U S A 2011;108(49):19725–30.
[92] Fuchs M, Hilfiker A, Kaminski K, Hilfiker-Kleiner D, Guener Z, Klein G, et al. Role of interleukin-6 for LV remodeling and survival after experimental myocardial infarction. FASEB J 2003;17(14):2118–20.
[93] Matsushita K, Iwanaga S, Oda T, Kimura K, Shimada M, Sano M, et al. Interleukin-6/soluble interleukin-6 receptor complex reduces infarct size via inhibiting myocardial apoptosis. Lab Invest 2005;85(10):1210–23.
[94] Tonnesen MG, Feng X, Clark RA. Angiogenesis in wound healing. J Investig Dermatol Symp Proc 2000;5(1):40–6.
[95] Ren G, Michael LH, Entman ML, Frangogiannis NG. Morphological characteristics of the microvasculature in healing myocardial infarcts. J Histochem Cytochem 2002;50(1):71–9.
[96] Ferrara N, Alitalo K. Clinical applications of angiogenic growth factors and their inhibitors. Nat Med 1999;5(12):1359–64.
[97] Folkman J. Angiogenesis and angiogenesis inhibition: an overview. EXS 1997;79:1–8.
[98] Carmeliet P. Mechanisms of angiogenesis and arteriogenesis. Nat Med 2000;6(4):389–95.
[99] Heba G, Krzeminski T, Porc M, Grzyb J, Dembinska-Kiec A. Relation between expression of TNF alpha, iNOS, VEGF mRNA and development of heart failure after experimental myocardial infarction in rats. J Physiol Pharmacol 2001;52(1):39–52.
[100] Belperio JA, Keane MP, Arenberg DA, Addison CL, Ehlert JE, Burdick MD, et al. CXC chemokines in angiogenesis. J Leukoc Biol 2000;68(1):1–8.
[101] Frangogiannis NG, Mendoza LH, Lewallen M, Michael LH, Smith CW, Entman ML. Induction and suppression of interferon-inducible protein 10 in reperfused myocardial infarcts may regulate angiogenesis. FASEB J 2001;15(8):1428–30.
[102] Angiolillo AL, Sgadari C, Taub DD, Liao F, Farber JM, Maheshwari S, et al. Human interferon-inducible protein 10 is a potent inhibitor of angiogenesis in vivo. J Exp Med 1995;182(1):155–62.
[103] Tao Z, Chen B, Tan X, Zhao Y, Wang L, Zhu T, et al. Coexpression of VEGF and angiopoietin-1 promotes angiogenesis and cardiomyocyte proliferation reduces apoptosis in porcine myocardial infarction (MI) heart. Proc Natl Acad Sci U S A 2011;108(5):2064–9.
[104] Meloni M, Marchetti M, Garner K, Littlejohns B, Sala-Newby G, Xenophontos N, et al. Local inhibition of microRNA-24 improves reparative angiogenesis and left ventricle remodeling and function in mice with myocardial infarction. Mol Ther 2013;21(7):1390–402.
[105] Lijnen P, Petrov V. Transforming growth factor-beta 1-induced collagen production in cultures of cardiac fibroblasts is the result of the appearance of myofibroblasts. Methods Find Exp Clin Pharmacol 2002;24(6):333–44.
[106] Desmouliere A, Geinoz A, Gabbiani F, Gabbiani G. Transforming growth factor-beta 1 induces alpha-smooth muscle actin expression in granulation tissue myofibroblasts and in quiescent and growing cultured fibroblasts. J Cell Biol 1993;122(1):103–11.
[107] Mauviel A. Transforming growth factor-beta: a key mediator of fibrosis. Methods Mol Med 2005;117:69–80.
[108] Schiller M, Javelaud D, Mauviel A. TGF-beta-induced SMAD signaling and gene regulation: consequences for extracellular matrix remodeling and wound healing. J Dermatol Sci 2004;35(2):83–92.
[109] Okada H, Takemura G, Kosai K, Li Y, Takahashi T, Esaki M, et al. Postinfarction gene therapy against transforming growth factor-beta signal modulates infarct tissue dynamics and attenuates left ventricular remodeling and heart failure. Circulation 2005;111(19):2430–7.
[110] Krishnamurthy P, Rajasingh J, Lambers E, Qin G, Losordo DW, Kishore R. IL-10 inhibits inflammation and attenuates left ventricular remodeling after myocardial infarction via activation of STAT3 and suppression of HuR. Circ Res 2009;104(2):e9–e18.
[111] Siwik DA, Chang DL, Colucci WS. Interleukin-1beta and tumor necrosis factor-alpha decrease collagen synthesis and increase matrix metalloproteinase activity in cardiac fibroblasts in vitro. Circ Res 2000;86(12):1259–65.
[112] Li YY, Feng YQ, Kadokami T, McTiernan CF, Draviam R, Watkins SC, et al. Myocardial extracellular matrix remodeling in transgenic mice overexpressing tumor necrosis factor alpha can be modulated by anti-tumor necrosis factor alpha therapy. Proc Natl Acad Sci U S A 2000;97(23):12746–51.
[113] Singh M, Foster CR, Dalal S, Singh K. Osteopontin: role in extracellular matrix deposition and myocardial remodeling post-MI. J Mol Cell Cardiol 2010;48(3):538–43.

[114] Trueblood NA, Xie Z, Communal C, Sam F, Ngoy S, Liaw L, et al. Exaggerated left ventricular dilation and reduced collagen deposition after myocardial infarction in mice lacking osteopontin. Circ Res 2001;88(10):1080–7.
[115] Suematsu N, Tsutsui H, Wen J, Kang D, Ikeuchi M, Ide T, et al. Oxidative stress mediates tumor necrosis factor-alpha-induced mitochondrial DNA damage and dysfunction in cardiac myocytes. Circulation 2003;107(10):1418–23.
[116] Mollapour E, Linch DC, Roberts PJ. Activation and priming of neutrophil nicotinamide adenine dinucleotide phosphate oxidase and phospholipase A(2) are dissociated by inhibitors of the kinases p42(ERK2) and p38(SAPK) and by methyl arachidonyl fluorophosphonate, the dual inhibitor of cytosolic and calcium-independent phospholipase A(2). Blood 2001;97(8):2469–77.
[117] Wassmann S, Stumpf M, Strehlow K, Schmid A, Schieffer B, Bohm M, et al. Interleukin-6 induces oxidative stress and endothelial dysfunction by overexpression of the angiotensin II type 1 receptor. Circ Res 2004;94(4):534–41.
[118] Ferdinandy P, Danial H, Ambrus I, Rothery RA, Schulz R. Peroxynitrite is a major contributor to cytokine-induced myocardial contractile failure. Circ Res 2000;87(3):241–7.
[119] Deng B, Wehling-Henricks M, Villalta SA, Wang Y, Tidball JG. IL-10 triggers changes in macrophage phenotype that promote muscle growth and regeneration. J Immunol 2012;189(7):3669–80.
[120] Biswas SK, Mantovani A. Macrophage plasticity and interaction with lymphocyte subsets: cancer as a paradigm. Nat Immunol 2010;11(10):889–96.
[121] Troidl C, Mollmann H, Nef H, Masseli F, Voss S, Szardien S, et al. Classically and alternatively activated macrophages contribute to tissue remodelling after myocardial infarction. J Cell Mol Med 2009;13(9B):3485–96.
[122] Yang Z, Zingarelli B, Szabo C. Crucial role of endogenous interleukin-10 production in myocardial ischemia/reperfusion injury. Circulation 2000;101(9):1019–26.
[123] Zymek P, Nah DY, Bujak M, Ren G, Koerting A, Leucker T, et al. Interleukin-10 is not a critical regulator of infarct healing and left ventricular remodeling. Cardiovasc Res 2007;74(2):313–22.
[124] Stumpf C, Seybold K, Petzi S, Wasmeier G, Raaz D, Yilmaz A, et al. Interleukin-10 improves left ventricular function in rats with heart failure subsequent to myocardial infarction. Eur J Heart Fail 2008;10(8):733–9.
[125] Yoshimura A, Wakabayashi Y, Mori T. Cellular and molecular basis for the regulation of inflammation by TGF-beta. J Biochem 2010;147(6):781–92.
[126] Wahl SM, Hunt DA, Wakefield LM, McCartney-Francis N, Wahl LM, Roberts AB, et al. Transforming growth factor type beta induces monocyte chemotaxis and growth factor production. Proc Natl Acad Sci U S A 1987;84(16):5788–92.
[127] Letterio JJ, Roberts AB. Regulation of immune responses by TGF-beta. Annu Rev Immunol 1998;16:137–61.
[128] Kitamura M. Identification of an inhibitor targeting macrophage production of monocyte chemoattractant protein-1 as TGF-beta 1. J Immunol 1997;159(3):1404–11.
[129] Werner F, Jain MK, Feinberg MW, Sibinga NE, Pellacani A, Wiesel P, et al. Transforming growth factor-beta 1 inhibition of macrophage activation is mediated via Smad3. J Biol Chem 2000;275(47):36653–8.
[130] Gamble JR, Khew-Goodall Y, Vadas MA. Transforming growth factor-beta inhibits E-selectin expression on human endothelial cells. J Immunol 1993;150(10):4494–503.
[131] Lefer AM, Tsao P, Aoki N, Palladino Jr. MA. Mediation of cardioprotection by transforming growth factor-beta. Science 1990;249(4964):61–4.
[132] Ikeuchi M, Tsutsui H, Shiomi T, Matsusaka H, Matsushima S, Wen J, et al. Inhibition of TGF-beta signaling exacerbates early cardiac dysfunction but prevents late remodeling after infarction. Cardiovasc Res 2004;64(3):526–35.
[133] Lasky LA. Selectins: interpreters of cell-specific carbohydrate information during inflammation. Science 1992;258(5084):964–9.
[134] Ebnet K, Vestweber D. Molecular mechanisms that control leukocyte extravasation: the selectins and the chemokines. Histochem Cell Biol 1999;112(1):1–23.
[135] McEver RP. Selectin-carbohydrate interactions during inflammation and metastasis. Glycoconj J 1997;14(5):585–91.
[136] Sanders WE, Wilson RW, Ballantyne CM, Beaudet AL. Molecular cloning and analysis of in vivo expression of murine P-selectin. Blood 1992;80(3):795–800.
[137] Hahne M, Jager U, Isenmann S, Hallmann R, Vestweber D. Five tumor necrosis factor-inducible cell adhesion mechanisms on the surface of mouse endothelioma cells mediate the binding of leukocytes. J Cell Biol 1993;121(3):655–64.
[138] Luscinskas FW, Lawler J. Integrins as dynamic regulators of vascular function. FASEB J 1994;8(12):929–38.
[139] Rahman A, Fazal F. Hug tightly and say goodbye: role of endothelial ICAM-1 in leukocyte transmigration. Antioxid Redox Signal 2009;11(4):823–39.
[140] Gorina R, Lyck R, Vestweber D, Engelhardt B. beta2 integrin-mediated crawling on endothelial ICAM-1 and ICAM-2 is a prerequisite for transcellular neutrophil diapedesis across the inflamed blood-brain barrier. J Immunol 2014;192(1):324–37.
[141] Staunton DE, Dustin ML, Springer TA. Functional cloning of ICAM-2, a cell adhesion ligand for LFA-1 homologous to ICAM-1. Nature 1989;339(6219):61–4.
[142] Keiper T, Santoso S, Nawroth PP, Orlova V, Chavakis T. The role of junctional adhesion molecules in cell-cell interactions. Histol Histopathol 2005;20(1):197–203.
[143] Wung BS, Ni CW, Wang DL. ICAM-1 induction by TNFalpha and IL-6 is mediated by distinct pathways via Rac in endothelial cells. J Biomed Sci 2005;12(1):91–101.
[144] Laguri C, Arenzana-Seisdedos F, Lortat-Jacob H. Relationships between glycosaminoglycan and receptor binding sites in chemokines-the CXCL12 example. Carbohydr Res 2008;343(12):2018–23.
[145] Allen SJ, Crown SE, Handel TM. Chemokine: receptor structure, interactions, and antagonism. Annu Rev Immunol 2007;25:787–820.
[146] McLoughlin RM, Witowski J, Robson RL, Wilkinson TS, Hurst SM, Williams AS, et al. Interplay between IFN-gamma and IL-6 signaling governs neutrophil trafficking and apoptosis during acute inflammation. J Clin Invest 2003;112(4):598–607.
[147] Borregaard N, Theilgaard-Monch K, Cowland JB, Stahle M, Sorensen OE. Neutrophils and keratinocytes in innate immunity—cooperative actions to provide antimicrobial defense at the right time and place. J Leukoc Biol 2005;77(4):439–43.
[148] Barletta KE, Ley K, Mehrad B. Regulation of neutrophil function by adenosine. Arterioscler Thromb Vasc Biol 2012;32(4):856–64.

[149] Amulic B, Cazalet C, Hayes GL, Metzler KD, Zychlinsky A. Neutrophil function: from mechanisms to disease. Annu Rev Immunol 2012;30:459–89.
[150] Borregaard N. Neutrophils, from marrow to microbes. Immunity 2010;33(5):657–70.
[151] Pillay J, den Braber I, Vrisekoop N, Kwast LM, de Boer RJ, Borghans JA, et al. In vivo labeling with 2H2O reveals a human neutrophil lifespan of 5.4 days. Blood 2010;116(4):625–7.
[152] Day RB, Link DC. Regulation of neutrophil trafficking from the bone marrow. Cell Mol Life Sci 2012;69(9):1415–23.
[153] Ciz M, Denev P, Kratchanova M, Vasicek O, Ambrozova G, Lojek A. Flavonoids inhibit the respiratory burst of neutrophils in mammals. Oxidative Med Cell Longev 2012;2012:181295.
[154] Kleniewska P, Piechota A, Skibska B, Goraca A. The NADPH oxidase family and its inhibitors. Arch Immunol Ther Exp 2012;60(4):277–94.
[155] Spinale FG. Myocardial matrix remodeling and the matrix metalloproteinases: influence on cardiac form and function. Physiol Rev 2007;87(4):1285–342.
[156] Siwik DA, Pagano PJ, Colucci WS. Oxidative stress regulates collagen synthesis and matrix metalloproteinase activity in cardiac fibroblasts. Am J Physiol Cell Physiol 2001;280(1):C53–60.
[157] Nelson KK, Melendez JA. Mitochondrial redox control of matrix metalloproteinases. Free Radic Biol Med 2004;37(6):768–84.
[158] Smith NJ, Chan HW, Osborne JE, Thomas WG, Hannan RD. Hijacking epidermal growth factor receptors by angiotensin II: new possibilities for understanding and treating cardiac hypertrophy. Cell Mol Life Sci 2004;61(21):2695–703.
[159] Hangaishi M, Nakajima H, Taguchi J, Igarashi R, Hoshino J, Kurokawa K, et al. Cu, Zn-superoxide dismutase limits the infarct size following ischemia-reperfusion injury in rat hearts in vivo. Biochem Biophys Res Commun 2001;285(5):1220–5.
[160] Shiomi T, Tsutsui H, Matsusaka H, Murakami K, Hayashidani S, Ikeuchi M, et al. Overexpression of glutathione peroxidase prevents left ventricular remodeling and failure after myocardial infarction in mice. Circulation 2004;109(4):544–9.
[161] Braunersreuther V, Jaquet V. Reactive oxygen species in myocardial reperfusion injury: from physiopathology to therapeutic approaches. Curr Pharm Biotechnol 2012;13(1):97–114.
[162] Tsujita K, Shimomura H, Kaikita K, Kawano H, Hokamaki J, Nagayoshi Y, et al. Long-term efficacy of edaravone in patients with acute myocardial infarction. Circ J 2006;70(7):832–7.
[163] Jaeschke H, Smith CW. Mechanisms of neutrophil-induced parenchymal cell injury. J Leukoc Biol 1997;61(6):647–53.
[164] Albelda SM, Smith CW, Ward PA. Adhesion molecules and inflammatory injury. FASEB J 1994;8(8):504–12.
[165] Entman ML, Youker K, Shoji T, Kukielka G, Shappell SB, Taylor AA, et al. Neutrophil induced oxidative injury of cardiac myocytes. A compartmented system requiring CD11b/CD18-ICAM-1 adherence. J Clin Invest 1992;90(4):1335–45.
[166] Smith CW, Entman ML, Lane CL, Beaudet AL, Ty TI, Youker K, et al. Adherence of neutrophils to canine cardiac myocytes in vitro is dependent on intercellular adhesion molecule-1. J Clin Invest 1991;88(4):1216–23.
[167] Youker K, Smith CW, Anderson DC, Miller D, Michael LH, Rossen RD, et al. Neutrophil adherence to isolated adult cardiac myocytes. Induction by cardiac lymph collected during ischemia and reperfusion. J Clin Invest 1992;89(2):602–9.
[168] Prokopowicz Z, Marcinkiewicz J, Katz DR, Chain BM. Neutrophil myeloperoxidase: soldier and statesman. Arch Immunol Ther Exp 2012;60(1):43–54.
[169] Vasilyev N, Williams T, Brennan ML, Unzek S, Zhou X, Heinecke JW, et al. Myeloperoxidase-generated oxidants modulate left ventricular remodeling but not infarct size after myocardial infarction. Circulation 2005;112(18):2812–20.
[170] Rudolph V, Goldmann BU, Bos C, Rudolph TK, Klinke A, Friedrichs K, et al. Diagnostic value of MPO plasma levels in patients admitted for suspected myocardial infarction. Int J Cardiol 2011;153(3):267–71.
[171] Mocatta TJ, Pilbrow AP, Cameron VA, Senthilmohan R, Frampton CM, Richards AM, et al. Plasma concentrations of myeloperoxidase predict mortality after myocardial infarction. J Am Coll Cardiol 2007;49(20):1993–2000.
[172] Thukkani AK, Martinson BD, Albert CJ, Vogler GA, Ford DA. Neutrophil-mediated accumulation of 2-ClHDA during myocardial infarction: 2-ClHDA-mediated myocardial injury. Am J Physiol Heart Circ Physiol 2005;288(6):H2955–64.
[173] Askari AT, Brennan ML, Zhou X, Drinko J, Morehead A, Thomas JD, et al. Myeloperoxidase and plasminogen activator inhibitor 1 play a central role in ventricular remodeling after myocardial infarction. J Exp Med 2003;197(5):615–24.
[174] Ducharme A, Frantz S, Aikawa M, Rabkin E, Lindsey M, Rohde LE, et al. Targeted deletion of matrix metalloproteinase-9 attenuates left ventricular enlargement and collagen accumulation after experimental myocardial infarction. J Clin Invest 2000;106(1):55–62.
[175] Lindsey ML, Escobar GP, Dobrucki LW, Goshorn DK, Bouges S, Mingoia JT, et al. Matrix metalloproteinase-9 gene deletion facilitates angiogenesis after myocardial infarction. Am J Physiol Heart Circ Physiol 2006;290(1):H232–9.
[176] Brinkmann V, Zychlinsky A. Neutrophil extracellular traps: is immunity the second function of chromatin? J Cell Biol 2012;198(5):773–83.
[177] Savchenko AS, Borissoff JI, Martinod K, De Meyer SF, Gallant M, Erpenbeck L, et al. VWF-mediated leukocyte recruitment with chromatin decondensation by PAD4 increases myocardial ischemia/reperfusion injury in mice. Blood 2014;123(1):141–8.
[178] Arai M, Lefer DJ, So T, DiPaula A, Aversano T, Becker LC. An anti-CD18 antibody limits infarct size and preserves left ventricular function in dogs with ischemia and 48-hour reperfusion. J Am Coll Cardiol 1996;27(5):1278–85.
[179] Palazzo AJ, Jones SP, Girod WG, Anderson DC, Granger DN, Lefer DJ. Myocardial ischemia-reperfusion injury in CD18- and ICAM-1-deficient mice. Am J Physiol 1998;275(6 Pt 2):H2300–7.
[180] Litt MR, Jeremy RW, Weisman HF, Winkelstein JA, Becker LC. Neutrophil depletion limited to reperfusion reduces myocardial infarct size after 90 minutes of ischemia. Evidence for neutrophil-mediated reperfusion injury. Circulation 1989;80(6):1816–27.
[181] Faxon DP, Gibbons RJ, Chronos NA, Gurbel PA, Sheehan F. The effect of blockade of the CD11/CD18 integrin receptor on infarct size in patients with acute myocardial infarction treated with direct angioplasty: the results of the HALT-MI study. J Am Coll Cardiol 2002;40(7):1199–204.
[182] Nahrendorf M, Swirski FK. Monocyte and macrophage heterogeneity in the heart. Circ Res 2013;112(12):1624–33.
[183] van Furth R, Cohn ZA. The origin and kinetics of mononuclear phagocytes. J Exp Med 1968;128(3):415–35.
[184] Goodman JW, Hodgson GS. Evidence for stem cells in the peripheral blood of mice. Blood 1962;19:702–14.
[185] van Furth R, Diesselhoff-den Dulk MM. Dual origin of mouse spleen macrophages. J Exp Med 1984;160(5):1273–83.
[186] Leuschner F, Rauch PJ, Ueno T, Gorbatov R, Marinelli B, Lee WW, et al. Rapid monocyte kinetics in acute myocardial infarction are sustained by extramedullary monocytopoiesis. J Exp Med 2012;209(1):123–37.

[187] Schulz C, Gomez Perdiguero E, Chorro L, Szabo-Rogers H, Cagnard N, Kierdorf K, et al. A lineage of myeloid cells independent of Myb and hematopoietic stem cells. Science 2012;336(6077):86–90.
[188] Ginhoux F, Greter M, Leboeuf M, Nandi S, See P, Gokhan S, et al. Fate mapping analysis reveals that adult microglia derive from primitive macrophages. Science 2010;330(6005):841–5.
[189] Swirski FK, Nahrendorf M. Leukocyte behavior in atherosclerosis, myocardial infarction, and heart failure. Science 2013;339(6116):161–6.
[190] Nahrendorf M, Pittet MJ, Swirski FK. Monocytes: protagonists of infarct inflammation and repair after myocardial infarction. Circulation 2010;121(22):2437–45.
[191] Mosser DM, Edwards JP. Exploring the full spectrum of macrophage activation. Nat Rev Immunol 2008;8(12):958–69.
[192] Wynn TA, Chawla A, Pollard JW. Macrophage biology in development, homeostasis and disease. Nature 2013;496(7446):445–55.
[193] Gordon S, Taylor PR. Monocyte and macrophage heterogeneity. Nat Rev Immunol 2005;5(12):953–64.
[194] Strauss-Ayali D, Conrad SM, Mosser DM. Monocyte subpopulations and their differentiation patterns during infection. J Leukoc Biol 2007;82(2):244–52.
[195] Ingersoll MA, Spanbroek R, Lottaz C, Gautier EL, Frankenberger M, Hoffmann R, et al. Comparison of gene expression profiles between human and mouse monocyte subsets. Blood 2010;115(3):e10–9.
[196] Zawada AM, Rogacev KS, Rotter B, Winter P, Marell RR, Fliser D, et al. SuperSAGE evidence for CD14++CD16+ monocytes as a third monocyte subset. Blood 2011;118(12):e50–61.
[197] van Amerongen MJ, Harmsen MC, van Rooijen N, Petersen AH, van Luyn MJ. Macrophage depletion impairs wound healing and increases left ventricular remodeling after myocardial injury in mice. Am J Pathol 2007;170(3):818–29.
[198] Murray PJ, Wynn TA. Protective and pathogenic functions of macrophage subsets. Nat Rev Immunol 2011;11(11):723–37.
[199] Murry CE, Giachelli CM, Schwartz SM, Vracko R. Macrophages express osteopontin during repair of myocardial necrosis. Am J Pathol 1994;145(6):1450–62.
[200] Zamilpa R, Kanakia R, Cigarroa J, Dai Q, Escobar GP, Martinez H, et al. CC chemokine receptor 5 deletion impairs macrophage activation and induces adverse remodeling following myocardial infarction. Am J Physiol Heart Circ Physiol 2011;300(4):H1418–26.
[201] Ma Y, Halade GV, Zhang J, Ramirez TA, Levin D, Voorhees A, et al. Matrix metalloproteinase-28 deletion exacerbates cardiac dysfunction and rupture after myocardial infarction in mice by inhibiting M2 macrophage activation. Circ Res 2013;112(4):675–88.
[202] Courties G, Heidt T, Sebas M, Iwamoto Y, Jeon D, Truelove J, et al. In vivo silencing of the transcription factor IRF5 reprograms the macrophage phenotype and improves infarct healing. J Am Coll Cardiol 2014;63:1556–66.
[203] Ambardekar AV, Buttrick PM. Reverse remodeling with left ventricular assist devices: a review of clinical, cellular, and molecular effects. Circ Heart Fail 2011;4(2):224–33.
[204] Burkhoff D, Klotz S, Mancini DM. LVAD-induced reverse remodeling: basic and clinical implications for myocardial recovery. J Card Fail 2006;12(3):227–39.
[205] Solomon SD, Skali H, Anavekar NS, Bourgoun M, Barvik S, Ghali JK, et al. Changes in ventricular size and function in patients treated with valsartan, captopril, or both after myocardial infarction. Circulation 2005;111(25):3411–9.
[206] Sharpe N, Smith H, Murphy J, Greaves S, Hart H, Gamble G. Early prevention of left ventricular dysfunction after myocardial infarction with angiotensin-converting-enzyme inhibition. Lancet 1991;337(8746):872–6.
[207] Saavedra JM, Angiotensin II. AT(1) receptor blockers as treatments for inflammatory brain disorders. Clin Sci 2012;123(10):567–90.
[208] Di Raimondo D, Tuttolomondo A, Butta C, Miceli S, Licata G, Pinto A. Effects of ACE-inhibitors and angiotensin receptor blockers on inflammation. Curr Pharm Des 2012;18(28):4385–413.
[209] Schulze PC, Biolo A, Gopal D, Shahzad K, Balog J, Fish M, et al. Dynamics in insulin resistance and plasma levels of adipokines in patients with acute decompensated and chronic stable heart failure. J Card Fail 2011;17(12):1004–11.
[210] Dixon DL, Griggs KM, Bersten AD, De Pasquale CG. Systemic inflammation and cell activation reflects morbidity in chronic heart failure. Cytokine 2011;56(3):593–9.
[211] Maekawa Y, Anzai T, Yoshikawa T, Asakura Y, Takahashi T, Ishikawa S, et al. Prognostic significance of peripheral monocytosis after reperfused acute myocardial infarction:a possible role for left ventricular remodeling. J Am Coll Cardiol 2002;39(2):241–6.
[212] Engstrom G, Melander O, Hedblad B. Leukocyte count and incidence of hospitalizations due to heart failure. Circulation Heart failure 2009;2(3):217–22.
[213] Bujak M, Kweon HJ, Chatila K, Li N, Taffet G, Frangogiannis NG. Aging-related defects are associated with adverse cardiac remodeling in a mouse model of reperfused myocardial infarction. J Am Coll Cardiol 2008;51(14):1384–92.

CHAPTER

3

The Role of Inflammation in Myocardial Infarction

Evangelos P. Daskalopoulos, Kevin C.M. Hermans, Lieke van Delft, Raffaele Altara, W. Matthijs Blankesteijn

Department of Pharmacology, Cardiovascular Research Institute Maastricht (CARIM), Maastricht University, Maastricht, The Netherlands

3.1 INTRODUCTION

The inflammatory response (originating from the latin word *inflammare*, meaning "to set something on fire") is a defense mechanism triggered by tissue damage that acts to protect the organism from further damage. It incorporates machinery that localizes the injury, removes damaged/necrotic tissue, and promotes healing. Nevertheless, its effects might range from beneficial to deleterious and so it has been the focus of research for decades [1]. It is implicated in numerous pathological conditions [2], also affecting the cardiovascular system, and comprises a major contributing factor implicated in the development and acute phase of wound healing following myocardial infarction (MI).

MI is a common cardiovascular pathology, affecting a sizable proportion of the population, especially in Western societies. The processes leading to MI (namely the development of atherosclerosis, establishment of atheromatous plaque, and the transition of the later from stable to unstable) [3], acute MI itself, and the wound healing that follows it [4] are extremely complex, dynamic, and multifactorial. Inflammation constitutes a primary element of the process before (atherosclerosis development) [5] and after (wound healing) MI [6] and involves several pivotal players both on the cellular (cytokines, chemokines) and the molecular (macrophages, monocytes, neutrophils, lymphocytes, etc.) level. The inflammatory response is thus indispensable for a normal wound healing. Nevertheless, excessive inflammatory response that persists over the normal time period, might lead to an expansion of the immune response to the "healthy" un-infarcted areas of the myocardium and contribute to the adverse remodeling with deleterious effects contributing to heart failure (HF) [7]. The exact interplay of the aforementioned factors in the pathology of MI is still a conundrum that is only starting to reveal its secrets.

The purpose of this chapter is to provide an overview of the pivotal relationship between the inflammatory response and MI. We will cover the basic concepts of atherosclerosis development, the factors leading to MI, and mediators of the wound-healing process that follow it. Furthermore, we will discuss the fundamentals of the inflammatory cascade in the cardiovascular system, and we will attempt to provide a synopsis of the general principles and characteristics of the inflammatory response (at cellular and molecular level) before and after myocardial ischemia. Lastly, we will dedicate the main body of this chapter to describe the current pharmacological tools as well as the most interesting novel strategies aiming to target inflammation and prevent or provide treatment for MI.

Inflammation in Heart Failure
http://dx.doi.org/10.1016/B978-0-12-800039-7.00003-7

© 2015 Elsevier Inc. All rights reserved.

3.2 ROLE OF THE INFLAMMATORY RESPONSE BEFORE MI

3.2.1 Development of the Atherosclerotic Plaque

In the past, atherosclerosis has been regarded as a process of lipid accumulation in the vessel wall. The current view on this is far more complicated. The process of atherosclerosis development involves a myriad of mechanisms that are still not completely understood. Behind the simple process of plaque formation in large and mid-sized arteries, shelters a complex chronic inflammatory process in which cells of the innate as well as the adaptive immune system play an essential role.

Atheroma development starts off with activation, dysfunction, and structural modifications of the endothelial lining of the vessel wall, which can be caused by dyslipidemia. Infiltrated lipid components such as low-density lipoprotein (LDL) in the intima are prone to modification through oxidation by radicals (such as reactive oxygen species [ROS]) and enzymatic attack (e.g., lipoxygenase and myeloperoxidase) and can thereby activate the endothelium [8,9]. Endothelial activation preferentially occurs at sites with low average shear but high oscillatory shear stress [10]. This causes augmented expression of chemokines and adhesion molecules (e.g., vascular cell adhesion protein 1 [VCAM-1], E-selectin) and consequently initiates the inflammatory process. In addition, activated platelets interact with leukocytes and endothelial cells and also induce chemokine and cell adhesion molecule expression. These platelets also secrete chemokines (like CXCL4, CXCL7, and CCL5) and cytokines such as interleukin (IL)-1β and transforming growth factor β (TGF-β), which in turn will attract other cell types and enhance the activation of leukocytes [11].

3.2.2 Immune Cells Involved

Many different immune cells are involved in the pathogenesis of atherosclerosis. Monocytes are the first immune cells adhering to the activated endothelium, and once attached, chemokines produced in the intima trigger their migration to the subendothelial space; this is called diapedesis. Once in the intima, monocytes differentiate into macrophages under the influence of macrophage colony-stimulating factor (M-CSF) [12]. Macrophages start to ingest and process LDL, and when the balance of LDL influx and efflux is disturbed they become so-called "foam cells." This type of cell is distinctive of early atherogenesis. Macrophages express a plethora of receptors such as cytokine receptors (IL receptors, tumor necrosis factor [TNF] receptors) and pattern recognition receptors (e.g., Toll-like receptors, scavenger receptors), and upon activation they produce chemokines as well as proinflammatory (IL-1, -6, -12, -15 -18, TNF family members) and anti-inflammatory (such as IL-10 and TGF-β) cytokines [13–16]. Neutrophils have not received much attention in studies of the pathogenesis of atherosclerosis since these cells are not as abundant in plaques as are macrophages or T-cells. However, recent studies have shown that these cells play an important role as well. For example, depletion of circulating neutrophils in mice impairs plaque formation and inhibits development of aortic aneurysms [17,18]. The proinflammatory activity of neutrophils can mainly be credited to the production of ROS and release of granule proteins such as myeloperoxidase and azurocidin. In addition, they also affect the advanced atherosclerotic plaque by secretion and activation of several matrix metalloproteinases (MMPs) and thereby decrease plaque stability [19,20].

Although the presence of dendritic cells (DCs) in aortas was described in 1995, precise details on their mechanisms of action are just emerging but still are far from completely understood [21]. The total amount of DCs present in the aorta before atherosclerosis starts to develop is very modest whereas it dramatically expands upon development of the disease [22]. Thus far, it has been suggested that DCs are involved in cholesterol accumulation and homeostasis [23,24], antigen presentation [25], and cytokine/chemokine production [26], thereby contributing to increased inflammation and plaque growth.

T-cell recruitment to the forming atheroma takes place in a manner comparable to monocyte recruitment. Although monocytes and macrophages are more abundant in the plaque compared to T-cells, they are regarded as similarly important in the inflammatory response. Their activation is dependent on encountering an antigen-presenting cell. This can lead to several responses of which the T helper-1 (Th1) response is the most prevalent in atherosclerosis. Other T-cell subsets that participate in the inflammatory response are T helper-2 (Th2), $CD4^+$, $CD8^+$, natural killer- (NK), and regulatory T-cells (Treg) with all of them having distinct effects on atherogenesis. The Th1, NK, $CD4^+$, and $CD8^+$ subsets are proatherogenic and thereby promote lesion formation and plaque vulnerability, whereas Treg cells have an anti-atherogenic effect. The net effect of Th2 cells is debatable since several studies show controversial outcomes [27,28].

B-cells have a low abundance in the lesion; nevertheless, they play an essential role in atherosclerosis suggested by increased susceptibility to this disease in splenectomized mice and humans. Transfer of B-cells from the spleen can reverse this phenotype in mice [29,30]. This anti-atherogenic effect is possibly caused by antibodies produced by these cells, which are able to recognize epitopes in oxidized LDL or other plaque antigens and thereby contribute to the elimination of these components. In contrast, other studies have reported adverse effects of specific subsets of B-cells, implying that they encompass both pro- and anti-atherogenic effects [31–33].

3.2.3 Maturation and Rupture of the Atherosclerotic Plaque

Fatty streaks appear when the presence of foam cells at the site of plaque formation expands. At this stage, a lipid core has been formed that will progress into a mature atherosclerotic plaque following additional influx of different inflammatory cell types and extracellular lipids. This continued influx of cells results in a core region that is separated from the arterial lumen by a formation of a fibrous cap, which consists of recruited smooth-muscle cells and extracellular matrix (ECM) deposition and can cause reduction of the luminal area. At this stage, the center of the core can become necrotic as a result of apoptotic macrophages as well as other succumbing cells. In addition, neovascularization of capillaries can arise from the vasa vasorum and may allow leakage of other blood components and cause hemorrhages in the lesion. As the atherosclerotic plaque further matures, MMPs, and other matrix-degrading proteases secreted by mainly inflammatory cells, will decrease the thickness of the fibrotic cap, thereby increasing its vulnerability. Ultimately, when the impaired plaque can no longer cope with the hemodynamic strain, it ruptures and exposes its debris into the arterial lumen, leading to thrombus formation. This thrombus can occlude the artery or it can travel to areas further downstream where it causes ischemia [27,28,34,35], a process that in the heart is known as MI.

3.3 THE ROLE OF THE INFLAMMATORY RESPONSE IN MI

3.3.1 MI and Wound Healing

The wound healing that follows an acute MI is a dynamic and complex process that aims to confine the myocardial injury and sustain the functional capacity of the myocardium. The inflammatory cascade is activated at a very early point following MI, and it is generally regarded as the cornerstone of the wound-healing process. Although the wound-healing process after MI has been under the spotlight of research for decades, its mysteries are still poorly understood. In the following section, we will give a synopsis of MI, address the stages of wound healing, and discuss the major effectors (humoral, cellular, and other) of the inflammatory response following MI.

Acute MI is a leading cause of mortality and morbidity worldwide, and according to the WHO, coronary heart disease (CHD) was the culprit for the loss of 7.2 million lives worldwide in 2004 alone [36]. Furthermore, although survival after MI has improved in the last decades, the prognosis is still poor, with 12% of MI sufferers dying within the first 6 months [37] and more than half of patients over age 65 dying within 5 years of first MI [38]. MI is a major contributor of HF development, which is a chronic, debilitating, and eventually deadly condition. HF has been described as an emerging epidemic syndrome, and its prevalence is approximately 23 million around the globe [39]. About 20-30% of patients over age 65 who suffer a first MI, eventually progress into HF, while mortality is extremely high—50% of patients diagnosed with HF die within 5 years [38]. In addition, the economic burden on the healthcare systems due to MI and HF is colossal [38,40,41].

According to the latest report by the Global MI Task Force, MI is characterized by myocardial (cardiomyocyte) necrosis in a clinical setting, associated with acute and prolonged ischemia (2-4 h are usually adequate to lead to complete death of cardiomyocytes in the area at risk) that prevents blood flow to parts of the myocardium. The magnitude of the damage to the cardiomyocytes depends on various aspects, such as ischemia followed by reperfusion (ischemia/reperfusion [I/R] models when referring to animal work), ischemia without reperfusion (permanent coronary occlusion models), the number of alternative collateral coronaries perfusing the affected myocardium, the requirements of blood supply (which are variable for each organism), and other variables [42].

Wound healing following ischemia is a dynamic process that is dependent on specific spatial and temporal characteristics of various cellular and molecular factors. Its main aim is to produce a strong scar in the areas where cardiomyocytes have perished and hence maintain (at the best possible level) the beating function of the heart muscle [43]. Normally, the wound-healing process is completed within 5-6 weeks in humans [42] and is characterized by four distinct phases.

The *first phase* is defined by the death of cardiomyocytes due to ischemia. This occurs as early as 6 h after the initiation of the ischemic injury and might involve both apoptosis and necrosis [44]. The latter appears to be the stimulating factor for the initiation of the inflammatory phase (*second phase* of wound healing) [45], although factors like the complement C5, TGF-β, and monocyte chemoattractant protein 1 (MCP-1) might be involved a lot earlier to prepare the inflammatory response [46]. The first immune cells attracted to the border zone of the injured area are a subpopulation of leukocytes called polymorphonuclear neutrophils (PMNs), which remove various debris via phagocytosis and coordinate the synthesis of MMPs and their counterbalancing tissue inhibitors of metalloproteinases (TIMPs) [44,46]. Neutrophils, monocytes, and macrophages play crucial roles during this stage with the aid of various cytokines, such as TNF-α, several ILs like IL-1β, IL-6, and IL-10, chemokines and their receptors (namely, CCR2, CXCR2, etc.) [47–49]. It has to be noted here that an excessive or prolonged inflammatory stage might lead to an increased ECM degradation (which leads to a disturbed collagen production/degradation balance and the formation of a weak and sensitive-to-stretching scar) and an extended release of apoptotic stimulators (causing extra cardiomyocyte death), leading to deleterious effects on the cardiac remodeling [49]. The major factors leading to the suppression of the inflammatory response are signals transmitted following the apoptotic death of neutrophils and the recruitment of $Ly6C^{lo}/CX3CR1^{hi}$ monocytes [49]. This leads to the *third phase*, which is the deposition of granulation tissue [44]. The first unique attribute of this phase is the increased presence of myofibroblasts. This cell type is regarded as the activated form of cardiac fibroblasts (CFs), although the source of the myofibroblasts is a controversial subject and other cell types (fibrocytes, pericytes, endothelial, and epithelial cells) might serve as precursors [43]. During the granulation tissue stage, the myofibroblast, which possesses smooth-muscle cell-like characteristics, secretes various ECM components (including collagen fibers) that offer support in the areas where cardiomyocytes have died [50]. The myofibroblast is a crucial component of the wound-healing process and previous work by van den Borne *et al.* has shown that higher myofibroblast counts are associated with beneficial effects on cardiac function and architecture [51]. Furthermore, the second important component of the granulation tissue phase is the production of new blood vessels (neovascularization), which leads to the improved perfusion of the injured area and has been shown to improve wound healing following permanent MI as well as I/R [52]. Lastly, the *fourth phase* is characterized by the maturation of the ECM and the production of a robust and stable scar. The scar is a dynamic and metabolically active tissue, and an adequately healed scar is of colossal importance for the continuation of the myocardial performance. After the maturation of the scar, myofibroblasts slowly degrade; however, as shown by Willems *et al.*, myofibroblasts remain resident in the well-healed infarcts even years after MI [53]. It should be noted here that the four phases of wound healing after MI are overlapping and might also show deviations from person to person and between species; hence, one should be careful when extrapolating findings from animal models to humans.

As mentioned earlier, wound healing and the cellular and molecular remodeling following it are adaptive processes to maintain an adequate cardiac function. Nevertheless, a defective wound-healing process can have devastating effects to the injured myocardium. The major consequence of inadequate wound healing is the development of adverse ventricular remodeling, which is characterized by changes in the ventricular architecture (dilatation, wall thinning), stiffness due to collagen deposition, as well as changes in the molecular and cellular level that occur both in the infarct and in the remote area [54]. All of these phenomena, incorporating cardiomyocyte hypertrophy, fibrosis of the infarcted area and also of the remote—uninjured—areas, dilatation of the left ventricle (LV) and so forth, eventually lead to HF [55].

The *type of MI*—either in the clinic or in the laboratory—can play a major role in the extent by which the inflammatory response is activated. Models of myocardial ischemia in laboratory animals are of two kinds: permanent ligation of the left anterior descending (LAD) coronary artery or ischemia-reperfusion (I/R). The first is a quite straightforward procedure that follows the series of phenomena described above (see four stages of wound-healing post-MI above). Without prompt reperfusion, the whole area-at-risk supplied by the blocked coronary will of course become necrotic. In the case of the I/R model, which is actually closer to what currently occurs in the clinic with thrombolysis and percutaneous coronary intervention (PCI), reperfusion reestablishes blood flow through the coronaries and provides the ischemic tissue with blood (i.e., O_2 and nutrients). This leads to a substantial reduction of necrosis and a restriction of the damage caused by the ischemia in the first place. On the other hand, reperfusion leads to further cardiomyocyte injury (lethal reperfusion injury), with effects ranging from activation of ROS to increases in Ca^{2+} overload or stimulation of the inflammatory cascade and can have deleterious effects on the already injured myocardium [56,57]. The main culprits of this action are believed to be the leukocytes [58], as well as endothelial cells that are activated by various proinflammatory cytokines (IL-6, IL-8, TNF-α) and the complement system (C5a/C5b mainly) [59]. Lastly, various adaptations of the I/R protocol have been investigated (involving pre- and post-conditioning) and have shown to protect from the I/R injury, with anti-inflammatory effects forming the basis of this protective action. For a more in-depth review of the I/R and the role of inflammation, refer to a recent review paper by Vander Heide and Steenbergen [60]. Similar effects are observed following reperfusion of patients who have suffered an acute MI. The mechanical effect of PCI stimulates monocytes, leukocytes, and neutrophils, leading not only to inflammatory response damage but also to restenosis and thrombosis in some cases [61], although it has been claimed that the modern PCI methods are less prone to induce an extensive inflammatory response [62].

Infarct rupture is a complication of MI that can lead to sudden death due to the LV free wall rupturing at the site of a transmural infarction. The causative factor of infarct rupture is an excessive early inflammatory response, which induces the ECM degradation, leading to thinning of the LV and, consequently, to rupture of the free wall. This usually occurs within the first week after MI and almost invariably causes instant death due to cardiac tamponade [63]. Infarct rupture is closely linked to inflammation, and it has been shown that neutrophils and macrophages (expressing MMP-9 and MMP-2, respectively) are major contributors [64]. Anzai *et al.* showed that serum C-reactive protein (CRP) values are associated with rupture risk [65]. In addition, elevated mRNA levels of TNF-α, IL-6, and MCP-1 in the infarct are also associated with higher risk for rupture, as observed in studies comparing 129SV mice (well-known for their cardiac rupture sensitivity) and C57Bl/6 mice [66]. Of course, other factors (apart from inflammation) can play decisive roles whether rupture will take place—namely, blood pressure (with low blood pressure reducing the risk) [51], smoking, and other comorbidities [67]. Thankfully, in the last 10-20 years, incidence of cardiac rupture has decreased substantially due to thrombolysis and PCI, which salvage the myocardium and prevent transmural infarct development [68].

Lastly, a recent paper by Dutta *et al.* showed that MI initiates a vicious circle of inflammatory response that activates inflammation, which in turn stimulates atherosclerotic plaques for several months, and this may form the basis of a *reinfarction* [69]. Hence, one should consider not only the acute but also the more long-term effects of the inflammatory response following MI.

3.3.2 Humoral Immune Response Post-MI

3.3.2.1 *Cytokines*

Cytokines are mediator molecules fundamental for the inflammatory process. They are, in general, not expressed at a basal level in the heart, but gene expression is only activated following injury, or ischemia. Cytokines such as TGF-β, TNF-α, IL-1, and IL-6 can have profound effects on cardiomyocyte survival, the trafficking of immune cells, myocardial contractility, and the scar formation; hence, they are the molecules that actually drive not only the inflammatory process, but the whole wound-healing process following MI [70,71].

TNF-α is a key proinflammatory cytokine that exerts its effects via two types of receptors—TNFR1 and TNFR2. The effects of TNF-α are tightly connected to nuclear factor kappa-light-chain-enhancer of activated B cells (NF-κB) activation, as well as other signaling pathways, such as JNK, p38MAPK, and so forth [72]. Following myocardial injury, TNF-α is released by a variety of cell types, such as macrophages, lymphocytes, CFs, endothelial cells, and mast cells [73,74]. TNF-α has effects on a wide range of functions and cell types that are involved in wound healing following ischemia; it depresses cardiac contractility [75], induces myocyte apoptosis [76], and mediates the deposition of ECM [77]. TNF-α confers its effects during wound healing directly, but also indirectly via an upregulation of other proinflammatory cytokines such as IL-1α/β and IL-6 in CFs, which could imply that CFs are also contributing to the immune response following MI [78]. Furthermore, TNF-α can have profound effects on various MMPs, via which it contributes to the development of adverse remodeling and eventually to HF [70,79]. A recent study by Monden *et al.* has proposed an interesting theory about the bimodal effect of TNF-α, depending on which receptor is activated: Effects that are mediated via TNFR1 appear to be deleterious for the myocardium post-MI, while activation of TNFR2 leads to cardioprotection and amelioration of adverse remodeling [80]. Furthermore, these striking differences between the effects of myocardial TNFR1 and TNFR2 have also been shown in nonischemic models of disease [81].

ILs are also key players of the inflammatory response and several members of the family have been implicated in the post-MI inflammatory response—namely, IL-1α/β, IL-6, IL-10, IL-23, IL-33, and others. IL-1 has been described as a proinflammatory cytokine coming in two forms—IL-1α and IL-1β. IL-1α is mainly membrane-bound or is released after cell death, while Il-1β is circulating in the blood stream [82]. IL-1 regulates macrophage activation, the infiltration of leukocytes, as well as ROS formation and apoptotic death of cardiomyocytes [83]. Recent studies have also indicated other functions of IL-1, such as mediating the phenotype of cardiac (myo)fibroblasts, which can have profound effects on the ECM production and the scar formation [84]. IL-6 is also an important proinflammatory cytokine that is closely associated with CRP levels. Following MI, its levels are increased within 1-2 days and it can remain elevated even 3 months afterward [85], stimulating various effects on a wide range of cell types [86]. In addition, IL-10 is an anti-inflammatory cytokine, which along with TGF-β plays an important role in the inhibition of the inflammatory phase and in the initiation of the granulation phase. It suppresses the release of both cytokines (e.g., IL-1α/β, IL-6, TNF-α) and chemokines (several CC and CXC) [87], although it is not clear whether IL-10 is indispensable for the wound-healing process [88]. Furthermore, as shown by a study with transgenic (Tg) mice, ablation of IL-23 leads to increased inflammation and suppressed CF activation, which eventually lead to deleterious effects for the ischemic myocardium [89]. Lastly, IL-33 has recently been proposed as another player in the post-MI wound-healing process, however, its actions are still poorly understood since it can act as a pro- or anti-inflammatory cytokine, depending on the disease state [90].

TGF-β is a pleiotropic cytokine that controls crucial functions regarding inflammation and cellular proliferation and orchestrates wound healing following ischemia [91] by acting on a wide range of cells, including macrophages, T-cells and other immune cells, cardiomyocytes, (myo)fibroblasts, and endothelial cells [92]. TGB-β can associate with two types of receptors—type I (TβRI) and type II (TβII)—and activates the TGF/Smad axis (Smad proteins being the downstream effectors) as well as other signaling pathways (including JNK, ERK, p38MAPK, and others) [93]. The inflammatory effects of TGF-β are very diverse and can range from inductive effects on monocyte trafficking (and so promoting cytokine/chemokine production) [94] to suppressive effects on neutrophils and macrophages, which eventually inhibit cytokine and chemokine release [91,95]. A decade ago, Ikeuchi *et al.* demonstrated that inhibition of TGF-β signaling following MI can have beneficial effects at first (in the early phase of wound healing); however, when the suppression is sustained it leads to adverse LV remodeling and HF development. The cardioprotective effect of this anti-TGF-β treatment in the early post-MI period and the adverse effects (including increased mortality) observed later were attributed to enhanced TNF-α, IL-1β, and MCP-1 levels in the infarct [96].

3.3.2.2 *Chemokines*

The inflammatory process implicates the mobilization of various immune cells that are strongly dependent on stimuli regulated by chemotactic chemokines. Several types of chemokines exist and each can bind to several receptors, increasing the complexity of the system. The importance of chemokines is enormous for the pre-MI phenomena [97], as well as for the post-MI wound-healing processes [98], and it is characteristic that different chemokines regulate each of the distinct four phases of the wound-healing process. Chemokines are divided into large families (CC, CXC, and CX3C—named according to the amino acid numbers present between the first two cysteines) and are of two kinds, either "homeostatic," or involved in basal activity of leukocytes, or "inducible," meaning that they play major roles in the inflammatory response after injury, or ischemia, leading to leukocyte activation. Furthermore, major roles in the activation of chemokines are played by the Toll-like receptors (TLRs) and the NF-κB system [99].

As explained in the following section (Cellular Immune Response Post-MI), the first cells that are recruited after MI (within minutes to hours) are PMNs, which directly lead to tissue injury via proteolytic enzymes or ROS activation. The two major families of chemokines orchestrating the triggering of leukocyte infiltration are CC and CXC. IL-8 (also known as CXCL8) is of paramount importance in the trafficking of neutrophils in dog and rabbit MI models, while it does not appear to be expressed in mice. Another important chemokine is interferon γ-induced protein 10 (IP-10 or CXCL10), which has anti-angiogenic and anti-fibrotic effects, while playing a role in the activation of T-cells. Furthermore, MCP (also known as CCL2) is implicated in the infiltration of monocytes/macrophages, as well as in angiogenesis and (myo)fibroblast migration and activation [6]. Several other chemokines with important functions shortly following MI or I/R are CXCL1, CXCL2, CXCL6, CCL3, and CCL5 [98,100]. When the inflammatory phase is completed, TGF-β (possibly together with IL-10) inhibits IP-10 and mediates the release of proangiogenic molecules such as CXCL1, CXCL2, CXCL12, CCL2, and macrophage inhibitory factor (MIF), which initiate the *de novo* development of blood vessels and capillaries [98]. It has to be noted that MIF mediates its proinflammatory effects via CXCR and CCR receptors [101] and leads to the recruitment of monocytes and T-cells and the regulation of smooth-muscle cells [102]. Its role in the development of atherosclerosis and in the plaque destabilization is well-defined [101], while the latest research proposes a direct effect of MIF in the inflammatory response following MI [102] or I/R [103]. It is beyond the scope of this chapter to provide an extensive review on the various cytokines and chemokines that characterize the inflammatory response before or after MI. For a more in-depth discussion regarding the different humoral effectors of the inflammatory response following MI, refer to Chapter 2 by Weirather and Frantz, as well as some excellent review papers by Frangogiannis [49] or Frangogiannis and Entman [104].

3.3.3 Cellular Immune Response Post-MI

The complex phenomena taking place during the post-MI immune response are characterized by the activation of a wide range of cells in the injured myocardium, which are activated in an extremely sophisticated and timely manner. As for the humoral effectors, a more exhaustive overview of the characteristics of leukocytes, neutrophils, monocytes, and macrophages can be found in Chapter 2. In this chapter, we will only provide a short resume of the aforementioned cell types and also give a synopsis of other cell types (e.g., platelets, mast cells, DCs, CFs) that might also play important roles during the immune response following MI.

3.3.3.1 *Leukocytes*

The first inflammatory cell type that infiltrates the myocardium following ischemia is the *PMN*. PMNs are the most abundant white blood cell subtype in the circulation. They are crucial regulators of the immune response because they are important for the recruitment of monocytes and their activation into macrophages as well as for the activity of DCs and natural killer cells [105]. PMNs are the first cell type to be recruited into the infarct (within hours post-MI) and various chemoattractants, such as CXCL1, CXCL2, and IL-8 (CXCL8) as well as Complement 5a, mobilize PMNs from the blood stream into the infarct [106]. Before moving to the injury area, blood circulation PMNs first have to be attached to the endothelium (via cell adhesion receptors called selectins) [107] and then be directed to the injured tissue mainly under the influence of the intercellular adhesion molecule -1 and -2 (ICAM-1/2) [108]. Furthermore, the transmigration of the PMNs is regulated by various chemokines [109]. Once infiltrating leukocytes reach the injured area, they release potent cytokines such as IL-1β and TNF-α, which in turn can stimulate the release of other cytokines like IL-1α/β, IL-6, and IL-8 [110] and drive the inflammatory response. Furthermore, the leukocyte-derived MIF is also another important proinflammatory factor mediating macrophage infiltration and the expression of several inflammatory markers [111], playing crucial roles in a later stage.

3.3.3.2 *Monocytes*

Monocytes are white blood cells that have a fundamental role in the inflammatory process [112] and are the circulating precursors of macrophages. Monocytes express several chemokines—namely, CCL2, CCL7, CX3CL1, and various chemokine receptors such as CCR1, CCR2, CCR5, CCR6, CCR7, CCR8, and CXCR2. These chemokines (and receptors) are the determinant factors in the regulation of monocyte trafficking [112]. Monocytes are important for the development of atherosclerosis [113] as well as the post-MI period [48]. The main source of monocytes following MI is the spleen, via IL-1β-mediated activation [114]. Monocytes infiltrate the myocardium within the first 12h following the ischemic injury under the influence of neutrophil-derived chemokines [105]; two major types of monocytes exist in the mouse: the Ly-6C^{high} monocytes, which are associated with inflammation and appear first, and the Ly-6C^{low} monocytes (less involved in inflammation) that start accumulating a bit later [115]. Following accumulation, monocytes start expressing several proinflammatory cytokines, such as TNF-α, IL-1β, as well as MMPs, which stimulate the digestion of the necrotic tissue. Afterward (after the 5th day post-MI), another subset of monocytes express factors that promote healing—namely, IL-10, TGF-β, vascular endothelial growth factor (VEGF), and so forth [115]. According to Nahrendorf *et al.*, this coordinated activation of the two monocyte subsets is of paramount importance for a balanced and successful wound healing following MI. It is characteristic that in patients who have suffered an acute MI, there is an inverse relationship between EF and high levels of circulating monocyte levels [116].

3.3.3.3 *Macrophages*

Macrophages are multifunctional cells that are major components of the innate immune system. Several phenotypes exist and one designation is between M1 (inflammation-related) and M2 (noninflammatory), playing pivotal roles in atherosclerosis development (via their differentiation to foam cells) as well as following ischemia [117]. Various stimuli play a decisive role in the polarization of macrophages toward the M1 direction (e.g., Toll-like receptors, IFN-γ, TNF-α, etc.) or M2 (such as IL-4/10/13) [118]. The source of macrophages in the inflammatory phase following MI is mainly from the bone marrow and the spleen, where macrophages originate from blood monocytes, with CXCL12 and IL-1β playing important roles. Following recruitment in the border zone and then in the infarct area, macrophages start removing debris (via phagocytosis) in order for the scar formation to be possible [117]. M1 macrophages secrete proinflammatory cytokines (IL-12, IL-23) and chemokines (CXCL9, CXCL10). The M2 phenotype is associated with anti-inflammatory molecules (such as IL-10, CCL17/22/24) [119], and they can release a wide range of other molecules that play indispensable roles during wound healing, such as MCP-1, TNF-α, MMPs, angiotensin converting enzyme and angiotensinogen, various growth factors, and so forth [120]. This orchestrated biphasic function of M1/M2 macrophages appears to be crucial for the normal wound healing; as it has been demonstrated by the group of Lindsey, disturbances in the normal M1/M2 balance following MI can have devastating effects for the remodeling of the injured myocardium [121].

3.3.3.4 Nonimmune Cells

Platelets are mostly known for their crucial importance in hemostasis and thrombosis and have been targeted pharmacologically. Furthermore, their involvement in the inflammatory response that leads to the development of the atherosclerotic plaque is also well established [122], with platelets expressing a vast variety of cytokines and chemokines (CCL2, CCL5 or RANTES, CXCL1/2/4/8, to name a few) and growth factors as well as activating the complement system [123]. Moreover, their ability to interact with and activate lymphocytes and DCs (see below) and the fact that they express TLRs just like some immune cells make it easy to associate them as important players in the inflammatory response during atherogenesis [124]. In addition, it appears that their importance is not limited to the phenomena leading to MI, but also during wound healing after an ischemic episode by accumulating (even within 6h post-MI) in the infarct area and maintaining local inflammation [125]. MPC-1, ICAM-1, and IL-1β originating from platelets have been shown to stimulate the endothelium and to promote the infiltration of neutrophils and monocytes to the endothelial surface, thus promoting the inflammatory response [126,127]. Furthermore, platelets appear to be associated with an induction of IL-1β and TNF-α levels in the infarct, contributing to LV remodeling and even infarct rupture in a mouse MI model [125].

Mast cells are best known for the production, storage, and release of histamine and are implicated in a variety of pathological conditions; however, their role in the heart and more importantly following MI is not very clear [128]. Frangogiannis *et al.* have shown not only that mast cells increase dramatically in number following MI [129], but also that mast cells are the major producers of TNF-α in the heart of dogs that have been subjected to I/R [130]. Furthermore, mast cells can release a wide range of other factors such as histamine, chemokines, and cytokines (e.g., IL-6/8/13), growth factors (e.g., vascular endothelial growth factor [VEGF]), prostaglandins, and so forth, all contributing to the inflammatory response [131].

In addition, *DCs* are antigen-presenting cells that patrol the blood and stimulate the activation of T-cells to induce an immune response; hence, they play an important role in the inflammatory response of various pathologies. DCs were shown to infiltrate the infarcted myocardium in an early stage and could be major culprits in the development of HF [132]. Furthermore, DCs are the regulators of the fine balance between the monocyte and macrophage phenotype. Mice with defunct DCs show suppression of the anti-inflammatory type of monocytes ($Ly6C^{low}$) and M2 macrophages and an increase in the proinflammatory-related monocytes ($Ly6C^{high}$), M1 macrophages and IL-10, all leading eventually to adverse remodeling and higher mortality after MI [133].

The *CF* is a multifaceted cell type that exerts pleiotropic effects and together with its activated form—the myofibroblast—has been shown to be essential for a successful wound-healing post-MI. The (myo)fibroblast confers its main function by producing and depositing ECM proteins, regulating a fine balance between MMPs and TIMPs, and playing a decisive role in the robustness of the scar [43]. Additionally, the CFs also appears to take part in the inflammatory response by releasing various proinflammatory cytokines and chemokines. Not only do CFs produce such mediators (TGF-β, TNF-α, IL-1β/6/8, CXCL1/8, etc.) but they also are stimulated by a wide range of factors (TGF-β, TNF-α, IL-1α/β, IL-17a/18, etc.) that are released from neighboring cells and affect the (myo) fibroblast ECM producing/degrading action [134,135]. Lastly, the CFs are shown to play a critical role in the termination of the inflammatory response, a milestone that is necessary before the granulation tissue healing stage is initiated [135].

3.3.4 Other Factors Modulating the Immune Response Post-MI

Lately, the *inflammasome*, which is a group of protein complexes that recognize inflammation-related stimuli [136], has been suggested as another key player for the post-MI wound-healing process. It acts as a sensor for the inflammatory response, especially during I/R injury, and amplifies the inflammatory response. The presence of danger signals stimulates a set of proteins residing in the cytosol and this results in the activation of IL-1β and IL-18 [137]. Recently, it was shown that inhibition of the NLRP3 inflammasome following MI can suppress the injury after I/R [138]. Inhibition of the inflammasome in cardiomyocytes has been shown to be cardioprotective [139]: however, the complexity and variance of the different inflammasome types is large [140], so more research is needed to reveal the secrets of the inflammasome in infarct healing.

CRP is a major player in the inflammatory response, and it is a biomarker of inflammation in various pathologies. It has proatherogenic effects on blood vessels by inducing the monocyte-endothelial cell adhesion [141]; however, it is also regarded as a predictor of poor prognosis following MI. Patients with high CRP levels have increased risk of LV remodeling 2 and 24 weeks post-MI, compared to patients with lower CRP levels [142]. The deleterious effects of CRP following MI were also shown in a study by Griselli *et al.* who demonstrated that injection of CRP in rat infarcts leads to a dramatic increase in the infarct size by approximately 40% [143]. Increased plasma CRP levels in the immediate period after MI provide an indication regarding cardiomyocyte necrosis and are connected to poor prognosis (including increased myocardial rupture, LV aneurysm, and overall cardiac death risks) even 1 year after the infarct [65]. Furthermore, increased CRP levels might also be predictive for the failure of thrombolysis in patients who have suffered ST-elevation MI (STEMI) [144], making the measurement of CRP levels an invaluable piece of information that can be utilized by clinicians for optimal management of each patient after an acute episode.

3.4 INFLAMMATION AS A PHARMACOLOGICAL AND BIOCELLULAR TARGET

3.4.1 Therapy Aimed at Inflammation Before MI

As mentioned earlier, the financial and societal burden for societies due to MI is enormous, hence, a reasonable strategy is the primary prevention of atherosclerosis development and MI occurrence. Various agents that are currently available in the market have been shown to modulate the inflammatory response via direct and indirect ways and hence can be used in anti-atherosclerotic therapy. Furthermore, various research groups have been active in the field, trying to utilize knowledge on inflammatory response mediators and formulate novel therapies to target inflammation before MI develops. In the following section, we will provide an overview of the current agents aiming to halt atheroma development and prevent MI from occurring and a synopsis of the latest novel research in the field. Figure 3.1 provides a graphic overview of these current and novel approaches.

3.4.1.1 Current Pharmacotherapy Targeting Inflammation Before MI

As discussed in Sections 3.2.1–3.2.3, the inflammatory response forms the basis of atherosclerosis; hence, any strategy directed toward the modulation of inflammatory mediators can have profound effects on the development of the atherosclerotic plaque and the prognosis for a patient (whether this will lead to CHD or not).

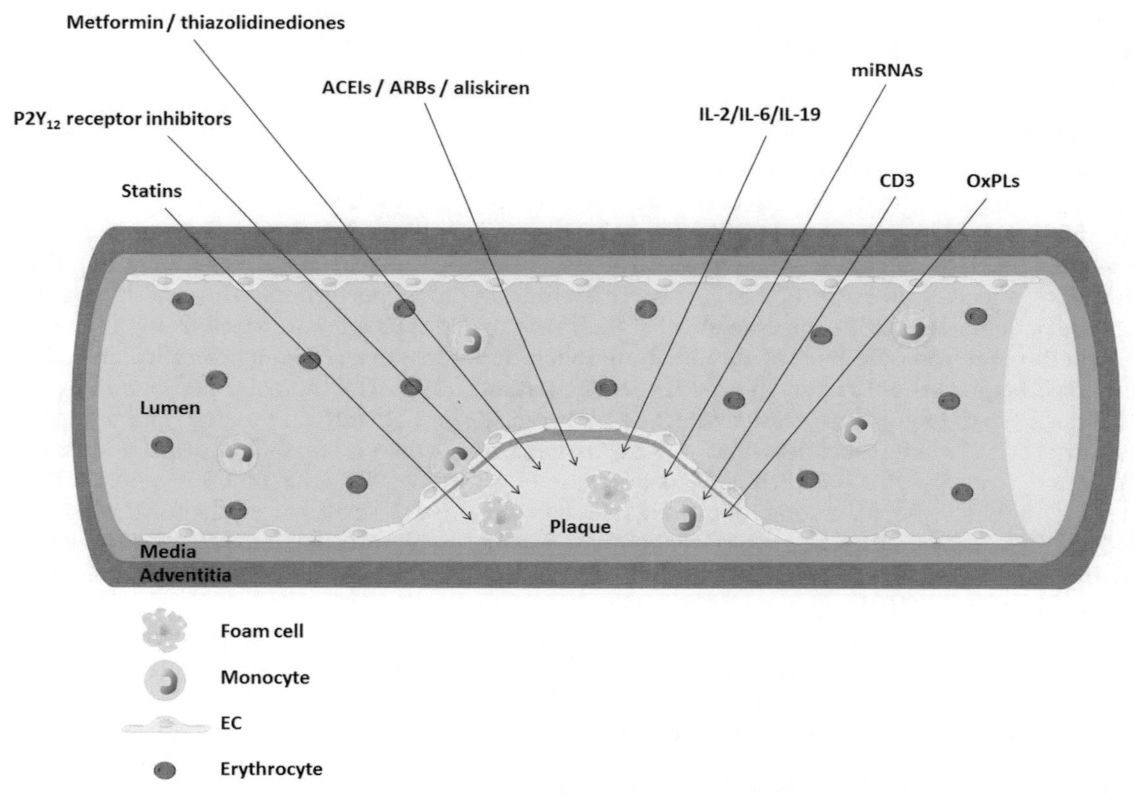

FIGURE 3.1 Current (left-hand side) and novel (right-hand side) therapeutic approaches in manipulating the inflammatory response during atherosclerosis. ACEIs, Angiotensin coverting enzymer inhibitors; ARBs, angiotensin-II receptor blockers; CD3, cluster of differentiation 3; IL-2, interleukin-2; IL-6, interleukin-6; IL-19, interleukin-19: interleukin-2/6/19; miRNAs, microRNAs; OxOLs, (synthetic) oxidized phospholipids.

3.4.1.1.1 STATINS

Statins (HMG CoA reductase inhibitors) are lipid-lowering agents routinely used in cardiology for a wide variety of indications, ranging from primary and familial hypercholesterolemia, hyperlipidemia, and the prevention of cardiovascular events [145]. The administration of such agents is mainly focused on taking advantage of their lipid-lowering effects, thus, preventing atherosclerosis development. However, research is providing evidence for other accessory mechanisms, via which statins mediate their beneficial pleiotropic effects [146]. These include effects on the migration of various cell types (inflammatory and noninflammatory), the thrombogenicity of the atherosclerotic plaque, the physiological function of the endothelium [147], as well as ROS-mediated effects [148].

Statins have been shown to abrogate the inflammatory response within the atherosclerotic plaque, hence, reducing the magnitude of the atheroma buildup as well as improving the plaque stability, thus, lowering the risk of a rupture leading to MI [149]. Statins have been reported to supress the activation of various cell types. Atorvastatin or rosuvastatin can interrupt atherosclerotic development by enhancing the migration of macrophages away from the plaque [150]. On the other hand, atorvastatin inhibits neutrophil infiltration in the early phase of the atherosclerotic buildup via nitric oxide, and this appears to be independent of any lowering effects on the cholesterol levels [151]. Recent studies on leukocytes isolated from dyslipidemic patients are showing that simvastatin reduced the proinflammatory chemokine IL-8 [152]. In addition, combinational use of an Angiotensin-II receptor blockers [ARB] (telmisartan) with a statin (rosuvastatin) in patients with carotid atherosclerosis showed a remarkable effect on suppressing a wide range of inflammatory factors including IL-1β/2/6/10/17/23, TNF-α, CRP, and MCP-1 [153]. Lastly, Duivenvoorden *et al.* recently provided an impressive report utilizing nanoparticles with reconstituted high-density lipoprotein (HDL) that can deliver statins right into the atherosclerotic plaque of apoE KO mice. This therapy was shown to suppress TNF-α and MCP-1 mRNA levels, as well as macrophage presence in the plaque without causing any side effects in the liver [154], opening new horizons in the drug therapy of atherosclerosis.

Furthermore, high-dose statin therapy has been studied by various groups and has been shown to provide evidence for the inflammatory response as a target of statins. As mentioned earlier, high-sensitivity CRP is regarded as an independent predictor for CHD risk [155]; various clinical trials have demonstrated the inverse relationship between statin use and CRP levels, for example, the CARE [156], PRINCE [157], and JUPITER trials [158] with the CANTOS and CINT trials expected to provide more evidence on the role of inflammation in the statin-mediated effects [159]. Still, more research is needed to unveil the whole spectrum of the anti-inflammatory effects of statins that go beyond their lipid-lowering properties. For a more in-depth discussion on the effects of statins on HF, the reader is referred to Chapter 12.

3.4.1.1.2 HYPOGLYCEMIC AGENTS

Diabetes mellitus type II is a common comorbidity in patients with ischemic heart disease. Diabetic patients are treated with hypoglycemic agents, with several drug classes existing, including *metformin* and *thiazolidinediones* (with the latter group discussed in more detail in Chapter 11). These agents have significant effects on the control of blood glucose; nevertheless, they also reduce the overall cardiovascular risk. The reason for the latter is, in part, due to their anti-inflammatory effects. It is obvious that most of the clinical studies performed with oral hypoglycemic drugs involve diabetic patients; hence, the interpretation of their results might be an issue when trying to explain phenomena in nondiabetic conditions. Carter *et al.* demonstrated that the widely-prescribed metformin can reduce the levels of CRP (but not C3 complement factor) in type II diabetic patients [160]. The thiazolidinedione rosiglitazone (4 mg daily) has been shown to have a suppressive effect on high-sensitivity CRP (68% reduction) and to lead to reduction of carotid artery intima-media thickness that is a major marker of atherosclerosis buildup [161]. Interestingly, in both aforementioned studies, the beneficial effects of the oral hypoglycemic agents on CRP were not blood glucose-related; hence, it might also be interesting to investigate whether comparable effects are demonstrated in nondiabetic patients that are high-risk for atherosclerosis development. Furthermore, rosiglitazone (8 mg daily) reduces the presence of inflammatory cells, as well as TNF-α and NF-κB, leading to increased stability of atherosclerotic plaques of a group of diabetic patients [162].

On the other hand, there is data weakening the potential of hypoglycemic agents as potential strategies to reduce inflammation in atherosclerosis, such as the ability of metformin to increase TNF-α levels in lean subjects with CHD [163] and the increased risk of MI when patients are given rosiglitazone therapy [164]. Hence, more research is needed in the field to determine the real anti-inflammatory and cardioprotective benefit of the hypoglycemic agents.

3.4.1.1.3 RENIN ANGIOTENSIN SYSTEM TARGETING

Drugs targeting the renin angiotensin system (RAS)—namely, angiotensin converting enzyme inhibitors (*ACEIs*), *ARBs*, and *renin inhibitors* (e.g., aliskiren)—have also been shown to affect the inflammatory response. Although they are not indicated for primary prevention of CHD, they are prescribed to patients with hypertension to lower the overall risk for CHD [145]. RAS has a multimodal effect on the pathogenesis of atherosclerosis. It acts via several molecular mediators (e.g., Angiotensin II, NADPH oxidase, peroxisome proliferator-activated receptor (PPAR), etc.), hence, the treatment strategy can be directed toward various targets [165]. It is hypothesized that the beneficial effects of the drugs targeting RAS are conferred not only via their effects on blood pressure, but also via mediation of the inflammatory response in the atherosclerotic plaque. In a very early study, Hernandez-Presa *et al.* demonstrated that the ACEI quinapril can suppress IL-8, MCP-1 and attenuate the infiltration of macrophages in the rabbit atheromatous lesions, leading to their stabilization [166]. Enalapril has shown its effectiveness against Ang-II-induced atherosclerosis by suppressing several inflammatory-related factors, including MCP-1 and inducing PPAR-α/γ, which possess anti-inflammatory actions [167]. In addition, a novel ACEI called XJP-1 exhibited suppressive effects on monocyte adhesion and TNF-α and MCP-1 levels in vitro (on HUVECs), increasing our knowledge on the mode of action of ACEI on endothelial cells [168]. Administration of the ARB candesartan to monocytes suppresses IL-1β/6, TNF-α, and MCP-1 levels, an effect that is mediated via an inhibition of TLR2/4 receptors [169] that are related to the inflammatory response as previously mentioned. A very recent study has reported anti-inflammatory effects of the ARB irbesartan in atheromatous plaques of apoE KO mice, via a suppression of macrophage infiltration [170]. Additionally, a meta-analysis of nine clinical trials has proven that telmisartan suppresses IL-6 and TNF-α [171], although this is not confirmed by a recent study measuring inflammatory factors in hypertensive and atherosclerotic patients [172]. The combined use of ACEI and ARB seems to confer additional beneficial anti-inflammatory and anti-atherosclerotic effects in apoE$^{-/-}$ mice, mainly via effects on macrophage numbers [173]. Lastly, aliskiren has been shown to have plaque stabilizing actions [174] and has been lately associated with anti-inflammatory effects in the context of atherosclerosis development. A recent study on a Western-type diet fed mouse model showed that aliskiren exhibits an inhibitory effect on MCP-1 production of the aorta, as well as a suppressive effect on macrophages and adherence of monocytes and T-cells numbers. These effects appear to contribute to the halting of the progression of the atherosclerotic plaque, an effect that can be further improved when aliskiren is coadministered with a statin [175]. For a thorough review paper on the RAS as a target for the prevention of atherosclerosis development with data from both basic and clinical research, the reader is referred to the work of Montecucco *et al.* [176], as well as Chapter 10 (by Carbone & Montecucco).

3.4.1.1.4 P2Y$_{12}$ RECEPTOR INHIBITORS

Lastly, various reports have linked platelets in the modulation of the inflammatory response during the development of the atherosclerotic plaque [177], hence, targeting the platelets could potentially be an interesting strategy. The platelet P2Y$_{12\ ADP}$ receptor is a well-established target of the thienopyridine-class of *antiplatelet drugs* (e.g., clopidogrel). Blockade of the P2Y$_{12}$ receptor leads to suppression of CD40 (which plays major roles in the interaction between platelets and other cell types contributing to atherogenesis), CRP and the prevention of platelet-leukocyte aggregation in various clinical pathologies involving atherosclerosis [122]. This effect that was not confirmed by a clinical study investigating the potential anti-inflammatory effects of ticagrelor or clopidogrel [178]. Furthermore, knocking out P2Y$_{12}$ receptors in mice fed a high-fat diet led to a suppression of platelet factor 4 secretion, which is closely associated with the infiltration of monocytes [179]. Nevertheless, a very recent study that also used P2Y$_{12}$ KO mice advocated that it is the effect of P2Y$_{12}$ receptors of the vessel wall and not of the platelets playing a part in atherogenesis, raising questions about the effectiveness of the drugs like clopidogrel in early development of atherosclerosis. Thus, further studies are needed to gain more insight in this area [180].

3.4.1.2 Novel Strategies Targeting Inflammation Before MI

As mentioned above, the current therapeutic agents against atherosclerosis are mainly focused on alleviating hyperlipidemia, hyperglycemia, and hypertension. However, recent evidence has demonstrated that statin treatment also has favorable effects on inflammation (as discussed in more detail below) [181]. Since inflammation has a primary role in the development of atherosclerosis, a considerable number of novel therapeutic strategies are emerging to intervene in the inflammatory cascade in order to dampen this development. Currently, several clinical trials are ongoing in an attempt to interfere with the inflammatory cascade, including immunosuppressives (methotrexate), PPAR agonists (thiazolidinediones), IL-1 receptor antagonists (Anakinra), and HDL mimetics (e.g., Apoa1-Milano) [182]. Notwithstanding all these novel trials, basic science is still in the search for new targets to reduce the development of atherosclerosis.

MicroRNAs (miRNAs or miRs) have emerged as important regulators of the inflammatory response [183] and might thereby serve as potential therapeutic targets to control atherosclerosis. Recently, it has been demonstrated that systemic delivery of miR-181b in a mouse model for atherosclerosis inhibits the activation of NF-κB and thereby inhibits lesion formation, proinflammatory gene expression, and influx of lesional macrophages and CD4+ T-cells in the vascular wall [184]. In addition, administration of miR-126 enables CXCR4 that in its turn increases the production of CXCL12 and limited atherosclerosis [185]. Many other miRNAs have been considered as potential candidates for targeting atherosclerosis such as miR-155 and -146 (involved in DC functioning) [186] and miR-31 and -17-3p (involved in EC activation) [187]. However, our knowledge of the individual function of many miRNAs is still in an embryonic state, and their role in the setting of atherosclerosis development still has to be evaluated. For an in-depth analysis of miRNAs and their association with inflammation, the reader is reffered to Chapter 13.

Several other studies have attempted to unveil novel therapeutic targets of inflammation to prevent or reduce the development of atherosclerosis. One of these potential targets is *CD3*, which seems to play a major role in recruitment of inflammatory cells to the developing plaque. Administration of anti-CD3 antibody in mice that had already developed atherosclerosis demonstrated a regression of atherosclerosis and reduced accumulation of macrophages and CD4+ T-cells in the plaques, whereas the Treg population was increased [188]. Another possible candidate is *IL-19*. Recent evidence suggests that this is a potent inhibitor of atherosclerosis by acting through several mechanisms, including a decrease in macrophage infiltration and immune cell polarization [189]. Additionally, Wolfs *et al.* demonstrated that administration of antigens produced by helminths is able to reduce plaque size by 44% in a mouse atherosclerosis model. This outcome is mediated by diminishing the inflammatory response as was shown by reduced circulatory neutrophils and inflammatory monocytes as well as increased production of the anti-inflammatory IL-10. Furthermore, the atherosclerotic lesion itself showed reduced inflammation as well, since it incorporated fewer inflammatory cells and had reduced expression of inflammatory markers [190]. The targeting of another IL, *IL-2*, has also given promising results. An Australian group examined the effects of an IL-2/anti-IL-2 antibody complex in apoE$^{-/-}$ mice that were fed a high-fat diet. The therapy was shown to affect CD4$^+$CD25$^+$Foxp3$^+$ regulatory T-cells in atherosclerotic lesions and hence suppress the development of atherosclerosis [191]. Lastly, *IL-6* has been extensively investigated as a potential target for anti-atherosclerotic treatment with the monoclonal antibody tocilizumab being currently in the forefront [192]. Furthermore, the effects of *synthetic oxidized phospholipids (OxPLs)*, which are native regulators of inflammation on monocyte chemotaxis, have been investigated. *In vivo* administration of a specific OxPL—VB-201—in a mouse model for atherosclerosis reduced atheroma development by diminishing macrophage infiltration [193]. On the other hand, hypercholesterolemia-induced activation and priming of hematopoietic stem and progenitor cells, as well as endothelial specific overexpression of the lectin-like oxLDL receptor, aggravates atherosclerotic plaque development [194,195]. Counteracting these mechanisms could also be attractive strategies to reduce atherosclerosis. Lastly, *in vitro* work with macrophages has shown that a potent nutritional supplement, β-D-glucan, is able to inhibit oxLDL-induced proinflammatory effects through regulation of p38 MAPK phosphorylation [196].

Taken all together, there are numerous potential targets to attack the inflammatory response in the atherosclerotic plaque. Nevertheless, more knowledge has to be acquired to further understand the exact role of these therapeutic targets in atherosclerosis before new candidates advance into clinical trials.

3.4.2 Therapy Aimed at Inflammation After MI

3.4.2.1 Current Pharmacotherapy Targeting Inflammation After MI

The current pharmacotherapy for patients suffering an acute MI consists of ACEIs, ARBs, statins, beta-blockers, mineralocorticoid receptor antagonists (like spironolactone), antiplatelets (aspirin and $P2Y_{12}$ inhibitors), glycoprotein (GP) IIb/IIIa receptor antagonists, calcium-channel blockers, nitrates, and others. Several of these drug classes form part of the algorithms that are recommended by the most current guidelines of the American College of Cardiology Foundation (ACCF) and the American Heart Association (AHA) for patients who have suffered STEMI [197], or non-STEMI [198]. Nevertheless, not all of these agents have been implicated in the regulation of the inflammatory response after acute MI. In the next section, we will make an attempt to provide a review of the current literature. Figure 3.2 provides a graphic synopsis of both currently available and contemporary strategies.

3.4.2.1.1 STATINS

As previously mentioned, *statins* are routinely prescribed following cardiovascular events [145]. They have been shown to decrease mortality and morbidity of patients with MI, and their beneficial effects were initially attributed to their ability to inhibit the biosynthesis of cholesterol. Nevertheless, further research is pointing toward additional effects (beyond their lipid-lowering action), such as effects on inflammation, oxidative stress, and others (which are not the focus of the current review) [181].

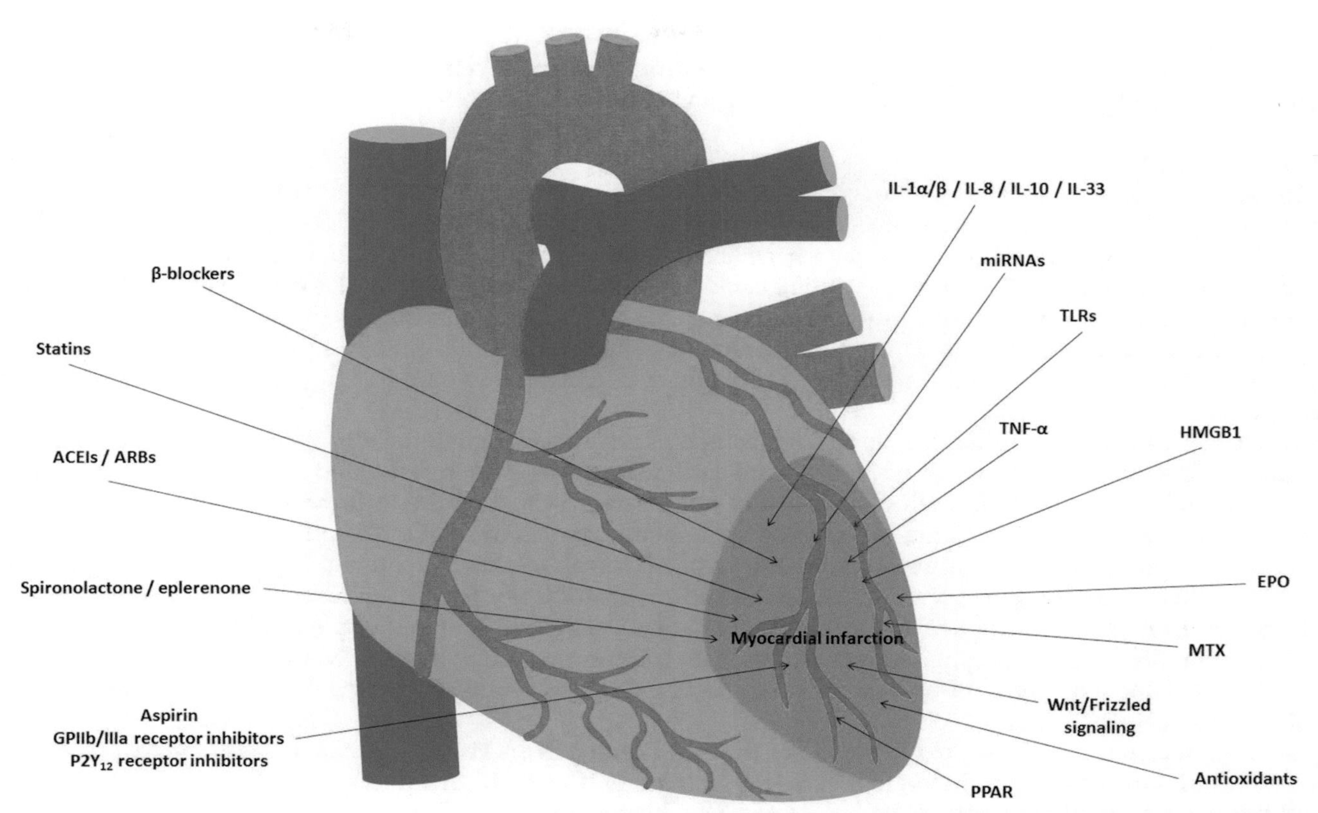

FIGURE 3.2 Current (left-hand side) and novel (right-hand side) therapeutic approaches in manipulating the inflammatory response during post-MI wound healing. ACEIs, angiotensin coverting enzymer Inhibitors; ARBs, angiotensin-II receptor blockers; EPO, erythropoetin; HMGB1, high-mobility group protein B1; IL-1α/β, interleukin-1α/β; IL-8, interleukin-8; IL-10, interleukin-10; IL-33, interleukin-33; MI, myocardial infarction; miRNAs, microRNAs; MTX, methotrexate; PPAR, peroxisome proliferator-activated receptor; TLRs, toll-like receptors; TNF-α, tumor necrosis factor α.

Several animal studies have addressed the role of atorvastatin in the regulation of various critical cytokines following MI. Tawfik *et al.* demonstrated the ability of atorvastatin to suppress CRP and re-establish the balance between pro- (TNF-α) and anti-inflammatory (IL-10) cytokines following MI in the rat serum. [199]. Additionally, atorvastatin appears to target IL-6 and MCP-1 (but the effects on IL-10 are the most prominent) [200]. Similar results were shown by Sun *et al.* [201] in an I/R rat model. Simvastatin has been shown to suppress the proinflammatory TNF-α, and IL-1β/6 and induce the anti-inflammatory IL-10 in the infarct and remote areas, compared to control rats [202]. Analogous results were reported by another study showing beneficial effects of simvastatin on TNF-α and IL-6 levels in MI rats, while it was demonstrated that TLR4 is implicated [203]. Finally, pravastatin (a newer member of the drug class) administration in Zucker rats following MI led to decreased serum levels of TNF-α and IL-1β, and this is speculated to play a role in the improved cardiac function in the treatment group [204].

Of course, the beneficial pleiotropic effects demonstrated by statins in basic research gave the stimulus to investigate whether similar effects are observed in humans. One study (part of the larger PROVE IT—TIMI 22 clinical trial including patients with a first coronary incident) showed clearly that a high dose of atorvastatin (80 mg) reduced the levels of CRP substantially, and this plays a major role in the prognosis of the MI patients [205]. A South Korean group followed up with approximately 100 patients with a first MI for 8 months and showed that a moderate-high dose (40 mg) of atorvastatin decreased TNF-α and IL-6 levels compared to patients receiving a low dose (10 mg) and this was correlated with improved coronary flow [206]. Furthermore, Sposito *et al.* moved a step further to provide evidence regarding the importance of the dose and the duration of simvastatin therapy in the magnitude of the effect in the inflammatory response. They showed that early initiation and high dose was found to be the optimal [207]; while on the other hand, a Greek group demonstrated atorvastatin in a low dose to also be beneficial in decreasing IL-6 [208].

Factors implicated in these effects are mediating endothelial function, inhibition of inflammatory response, suppression of vascular remodeling, stabilization of atheromatous plaques [146,181], and antioxidant (ROS) actions [181]. On the other hand, a wide range of signaling pathways and mediators have been proposed as important for these pleiotropic effects, such as the PPAR/NF-κB axis, Rho/ROCK, Rac, eNOS, PI3K/Akt, heat shock protein 90, and others [148]. Nevertheless, the beneficial cholesterol-independent effects of statins post-MI are still being investigated in greater depth to reveal their exact mechanisms of action.

3.4.2.1.2 RAS TARGETING

The RAS has been established as a major regulator of the cardiovascular physiology and pathophysiology. Two main drug classes that mediate their effects via targeting of the RAS system are the *ACEIs* and the *ARBs*. These drugs are considered the golden standards in the pharmacotherapy following MI and have been shown to reduce morbidity and improve survival after an ischemic attack [209,210]. They have profound Angiotensin-II-related effects (so having reducing effects on blood pressure and beneficial effects on cardiac output and stroke volume), however, research has also been pointing toward anti-inflammatory effects. ACEIs have been implicated in the regulation of the LV remodeling for more than 15 years now [211], however, their immunomodulating actions are only starting to be understood.

In a study conducted about a decade ago focusing on an I/R dog model, the effects of Ang-II inhibition (with ACEIs or ARBs) on the inflammatory response and cardiac function were investigated [212]. The cell type in focus was found to be the leukocyte, with captopril (but not losartan) affecting the infiltration of leukocytes and having substantial effects on contractile function of the ischemic myocardium. On the other hand, a Japanese study suggested the macrophage infiltration as the target of ARB (candesartan) treatment following permanent LAD ligation in Wistar rats, an effect leading to suppressed myocardial fibrosis [213]. Leuschner *et al.* suggested that ischemia of the myocardium activates monocytes that are normally resident in the spleen and their mobilization is dependent on the Ang-II signaling. Indeed, by using an I/R mouse model, they showed that enalapril reduces the migration of monocytes from the spleen, and this has direct effects on EF and infarct size (compared to untreated apoE$^{-/-}$ mice) [214].

Lapointe *et al.* report the suppressing effect of the ACEI captopril on TNF-α levels in a rat MI model [215], while the ARB candesartan has been shown to reverse the changes in IL-6, IL-10, and TNF-α in an I/R canine model [216]. In addition, the cytokine-targeting effects of ACEIs have been demonstrated on patients suffering from HF. Kovacs *et al.* demonstrated that quinapril (10 mg daily) suppressed TNF-α and CRP 2 and 3 months following treatment initiation of MI patients [217]. Additionally, high-dose enalapril (40 mg daily) reduced the IL-6 activity, however, this treatment was unable to completely abrogate the inflammatory activation [211]. In a comparison of ACEI versus ARB treatment in a rat MI model, Sandmann *et al.* showed that olmesartan treatment is substantially more beneficial by suppressing IL-1β, IL-6, and macrophage infiltration (and having overall better effects on cardiac function parameters), compared to ramipril, which suppressed only IL-6 [218]. Lastly, ACEIs and ARBs (and their combination) have been shown to target TGF-β1 and the Smad proteins [219], a signaling pathway that is well-known to regulate (and be regulated by) inflammation [220]. Hence, it is becoming apparent that the beneficial effects of ACEIs and ARBs are—at least partly—mediated via effects on the inflammatory response.

3.4.2.1.3 MINERALOCORTICOID RECEPTOR ANTAGONISTS

As discussed earlier, the RAS targeting has been in the forefront of clinical use as well as research for decades now. *Mineralocorticoid receptor antagonists* (spironolactone and eplerenone) play a central role in the pharmacotherapy post-MI. In contrast to glucocorticoids, which have been shown to impair myocardial function and increase infarct rupture (and so mortality) [221–223], the mineralocorticoid manipulation can have beneficial effects following MI. This is demonstrated by animal model work [224] as well as by large clinical trials on morbidity and mortality in patients with severe HF [225,226]. The implication of aldosterone to the promotion of inflammation is well established [227], with ROS-mediated effects on macrophages, fibroblasts, lymphocytes, and other cell types with proinflammatory actions being to blame; hence, the blockade of aldosterone's effects has beneficial properties. Indeed, recent studies, like the one by Fraccarollo *et al.*, have revealed the involvement of several cytokines (IL-1β/4/6/10 and TNF-α) and the chemokine MCP-1 in the cardioprotective effects of eplerenone in a rat MI model [228].

3.4.2.1.4 BETA-BLOCKERS

Beta-blockers have been routinely used following acute MI for several decades and various clinical trials have proven their beneficial effects on reducing mortality and reinfarction [229], especially in combination with other agents such as ACEI or ARBs [230,231]. Nevertheless, publications connecting beta-adrenoceptor blocking agents with effects on the inflammatory cascade following MI are very scarce; this indicates that the immunomodulatory effects they might have (e.g., on monocytes [232]) are probably not the main mechanism by which they produce their beneficial effects following MI.

Recent research has shown that beta-blockers appear to have only a minimal (nebivolol) or no effect (metoprolol) on proinflammatory genes in endothelial cells [233]. In contrast, in another study carvedilol (a beta-blocker well known for its anti-inflammatory and antioxidant properties) [234] was found to suppress plasma levels of TNF-α and IL-6 in ischemic and nonischemic dilated cardiomyopathy, but with effects more marked in nonischemic patients [235]. It is important to note here that there is evidence against the coadministration of some beta-blockers with statins, as the benefit of the latter might be reduced in acute MI patients [236]. The researchers demonstrated that metoprolol or propranolol might reduce the anti-inflammatory effects of simvastatin (which suppresses CRP), hence their simultaneous use (a very common combination in clinical use) might lead to a reduced statin-mediated beneficial effect.

3.4.2.1.5 ANTIPLATELET AGENTS

Aspirin has been one of the most cost-effective therapies in the management of CHD, including acute MI and primary/secondary CHD prevention. A major study approximately 17 years ago demonstrated that aspirin (162 mg daily) dramatically reduced mortality, reinfarction, and the occurrence of stroke in patients with acute symptoms of MI [237]. Furthermore, it has been suggested that the beneficial cardioprotective effects of aspirin are further expanded when this is coadministered with a statin [238]. Aspirin has been known for decades as an anti-inflammatory agent (and specifically via the NF-κB signaling pathway) [239]. Nevertheless, evidence connecting aspirin with anti-inflammatory effects following MI is very limited. Solheim *et al.* showed that aspirin therapy (160 mg daily) can suppress CRP levels (and TNF-α to a lesser extent) even 3 months after acute MI, however, the research team could not link this reduction to clinical end-points [240]. Similar results were demonstrated by Adamek *et al.* who point out that high-dose aspirin (120 mg/kg of weight) has a suppressive effect on IL-1β and TNF-α in a mouse MI model, however, this has no consequence on functional or volumetric (dilatation) cardiac parameters [241]. Nevertheless, the dose of aspirin used is high and so any extrapolation to the clinical setting should be made with caution.

The *GP IIb/IIIa receptor antagonists* (e.g., abciximab) are potent inhibitors of platelet aggregation utilized routinely after MI and especially in patients that undergo PCI [242] in order to reduce the likelihood of thrombogenesis. Platelets are cells that can drive inflammation by releasing various cytokines (especially IL-1β) once they are activated [243] and by expressing a range of chemokine receptors—namely, CCR1/3/4 and CXCR4 [244]. Furthermore, platelets can form aggregates with white blood cells (leukocytes) and hence play a role in thrombus formation. GP IIb/IIIa receptor antagonists have been found to suppress IL-1β and prevent platelet-leukocyte aggregation and hence prevent thrombosis [245]. The anti-inflammatory actions of GP IIb/IIIa receptor antagonists have also been confirmed by more recent studies (not on ischemia models though) reporting effects on TNF-α, IL-6 [246], as well as CRP [246,247]. Hence, it is possible that GP IIb/IIIa receptor antagonists mediate their beneficial effects following MI, not only via anti-thrombosis but also via inflammatory modulating actions, however, more research should be performed to investigate this in further depth.

Lastly, the *P2Y$_{12}$ receptor inhibitors* (examples including clopidogrel, prasugrel, etc.) have been part of the therapeutic arsenal following acute MI (among other indications), in combination with other agents. Members of this drug class target platelets, inhibit their activation and aggregation, and demonstrate antithrombotic effects. The role of these agents in the regulation of the inflammatory response is well established [248] by affecting neutrophils and macrophages and suppressing IL-1β and TGF-β [249], however, its significance in the post-ischemia period is still not clear. One of the very few studies in which this relationship was investigated is the one by Xiao and Theroux [250]. They showed that clopidogrel (300 mg single dose after an acute coronary syndrome) can suppress the aggregates of platelets with both monocytes and neutrophils; nevertheless, whether this might have any clinical benefit for these patients remains unexplained.

3.4.2.2 *Novel Strategies Targeting Inflammation After MI*

Currently, there is intensive work being performed that focuses on the inflammatory response in order to identify novel targets and strategies for the post-MI treatment. In this section, we will attempt to provide a short overview of several of these novel approaches. For further insight in the contemporary research strategies in the field, the reader is referred to Chapter 14 by Mann *et al.*

The importance of IL-1α/β, IL-8, IL-10, and other ILs, as well as TNF-α during the inflammatory stage post-MI, is well established and has attracted a considerable amount of interest as a target for therapies aiming to prevent adverse remodeling. Several studies have investigated the effects of the manipulation of these factors via various strategies.

There are a number of studies focusing on manipulations of ILs, in order to improve wound-healing post-MI. The neutralization of *IL-1α/β* by the use of a recombinant fusion protein (IL-1 Trap) has been shown to confer beneficial effects in a mouse MI model. Treatment with IL-1 Trap leads to an inhibition of cardiomyocyte apoptosis, suppression of IL-6 levels, and improvement of cardiac function 1 week post-MI [251]. Similar results have been demonstrated in a study that selectively blocked IL-1β in a mouse permanent MI model [252]. The beneficial effects of the IL-1 blockade were confirmed by a clinical study involving the use of the IL-1 antagonist anakinra in patients with STEMI. This trial [253] showed that anakinra can prevent LV remodeling and its demonstrated safety warranted the continuation of the clinical trial on a larger scale. *IL-8* is a proinflammatory chemokine that might also be a potential target following MI. Transfusion of endothelial cells, which overexpressed IL-8 receptors (IL8RA/B) in a rat MI model, demonstrated beneficial effects on infarct size, LV remodeling, neutrophil infiltration, and proinflammatory cytokine levels [254]. Nevertheless, studies involving patients with a first MI and studying the risk for further ischemia have revealed contradictory results. Elevated IL-8 levels in the serum of men are associated with increased risk of MI recurrence, while the opposite was shown for women. Hence, it is not very clear which strategy (induction or inhibition of IL-8) would be optimal for post-MI wound healing and whether this would necessarily apply to both genders [255]. Moreover, the anti-inflammatory *IL-10* has been suggested as a potential target in the last few years. Although early studies with IL-10 KO mice failed to demonstrate any deleterious effect for the wound healing (indicating that IL-10 might not be that important for wound healing) [88], subsequent studies reported controversial results. Two separate studies showed that IL-10 treatment inhibits the inflammatory response and improves LV architecture and function following MI in mice [256] and rats [257], reviving the interest for IL-10 targeting. *IL-33*, which belongs to the IL-1 cytokine superfamily, has shown bimodal anti- and proinflammatory actions [90]. IL-33 and its corresponding receptor ST2 have been implicated in anti-apoptotic beneficial effects after MI in mice [258] leading to improved LV function and architecture as well as reduced mortality. The beneficial effects of IL-33 have been recently confirmed by Yin *et al.* [259] in a mouse MI model. The authors showed that recombinant IL-33 treatment leads to reduced macrophage activation, suppressed cytokine cardiac levels, and improvement in the LV function, making further research on this cytokine of greater interest.

Furthermore, *TNF-α* has been proposed as a potential therapeutic target for a few decades now and both basic research and clinical trials have been performed yielding interesting but conflicting results. Gurevitch *et al.* reported that the administration of an antibody against TNF-α [260] leads to a reduction of the ischemic damage and prevents adverse remodeling in a rat I/R model. Sun *et al.* came to some very important conclusions after studying wild-type (WT) versus $TNF^{-/-}$ mice following permanent coronary artery ligation. The knocking out of TNF was shown to dramatically suppress infarct rupture rate and reduce apoptosis of cardiomyocytes and the overall LV remodeling—effects mediated via a depression on NF-κB, immune cell infiltration and MMP activity [261]. Ramani *et al.* demonstrated that the TNF-α receptor 1 is very important for the observed effects [262]. More recent research with Tg mice in the field of TNF-α targeting reports that a major contributor to the beneficial or detrimental effect on the ischemic myocardium is the type of TNF-α receptor that is stimulated. Activation of TNF-receptor type 1 (TNFR1) leads to adverse cardiac effects, while activation of TNFR2 can act in a cardioprotective way. In this sense, blockade of TNFR1 could confer a beneficial overall effect following ischemia [80,263]. Moreover, a study by Sugano *et al.* confirmed this theory by blocking TNF-α via *in vivo* transfer of soluble TNF receptor 1 to rats in a permanent MI model [264] and showing beneficial effects on LV function and infarct size. Unfortunately, the blockade of TNF-α in the clinical setting—using etanercept, a recombinant human TNF receptor that can render TNF inactive—failed to show beneficial effects in patients with acute MI [265]. Although etanercept showed promising results in animal studies [266], upcoming clinical trials [267] are expected to shed more light in its potential for the prevention and treatment of MI. For a complete outlook in the area of clinical trials focusing on TNF, we refer the reader to Chapter 14.

miRNAs are gaining more and more interest in all kinds of research fields, including cardiology. They are implicated in various processes (physiological and pathological) of the cardiovascular system, and they have been shown to play key roles during wound-healing post-MI [268]. More specifically, several reports have been pointing toward the importance of several miRNAs and effects on the inflammatory response in a wide range of cardiovascular diseases [183]. A study by Zidar *et al.* has shown that miR-146a, miR-150, and miR-155 levels are dramatically increased as early as 24 h after an acute MI and the first two are still elevated after a week [269]. MiR-146a levels are induced by TNF-α [270], while miR-146a has been shown to be induced by NF-κB stimulation of monocytes and be under the control of TLR [271]. Moreover, miR-155 overexpression leads to induction of TNF-α [272]; a more recent study has demonstrated that miRNA-155 is closely associated with the expression of Th17 cells and IL-17a of peripheral mononuclear cell origin in patients with an acute coronary syndrome (including acute MI) [273]. In addition, miR-155 has been implicated in the regulation of inflammation in nonischemic injury involving hypertrophy and HF, with miR-155 KO mice showing depressed monocyte/macrophage activation [274]; thus, it would be extremely interesting to see future studies involving the deletion of miR-155 in MI animal models. Further research on the aforementioned miRNAs involving MI models will illustrate the exact cross-talking mechanisms and regulation pathways with the inflammatory response and whether they could be considered as reliable strategies for post-MI therapies.

TLRs—covered extensively in Chapter 2—belong to a family of receptors that recognize conserved pathogen motifs called pathogen-associated molecular patterns (PAMPs) and thus help in the initiation of the innate immune response after infection. In addition, TLRs can also respond to signals from damage-associated molecular patterns (DAMPs), such as the ones observed following I/R or permanent ischemia. Indeed, TLR stimulation leads to activation of NF-κB, which eventually switch on various immunomodulatory molecules (e.g., TNF-α, IL-1β, interferons, etc.) [275,276]. Hence, the interest for their function in the setting of MI has increased in recent years. Several studies have been performed focusing on TLR2 and TLR4 modulation in mostly I/R models and have demonstrated beneficial effects [276]. Sakata *et al.* demonstrated that knockout of TLR2 causes a suppression of TNF-α and IL-1β levels in Langerdoff-perfused mouse hearts following I/R, and this prevents LV dysfunction [277]; however, Shishido *et al.* could not acknowledge similar effects since their TLR2 KO and WT animals showed the same immune cell infiltration after MI [278]. Lu *et al.* investigated the roles of TLR3 in both permanent MI and transient I/R by utilizing TLR3 KO mice. This group reported a suppression of NF-κB, IL-1β, and TNF-α and inflammatory cell infiltration in the ischemic myocardium [279]. On the other hand, a study on TLR4 KO mice exhibited dramatic effects on the expression of several cytokines (IL-2/6/17, TNF-α, and interferon-γ) and an overall cardioprotective effect following MI in a mouse MI model [280]. For more in-depth information regarding the role of TLR in the control of the immune response following MI, the reader is referred to an excellent review paper by Feng and Chao [276]. *NF-κB* is a major factor playing a role in the inflammatory response. NF-κB activation can lead to the stimulation of transcription of various cytokine modulator genes (such as IL-1/6 and TNF-α), as well as of some chemokines (CXCL10, CCL2, etc.) [281]. Nevertheless, data are conflicting, with studies demonstrating, on the one hand, beneficial effects from the amelioration of NF-κB in a I/R mouse model [282] but also proving that NF-κB might be having an anti-inflammatory effect [283].

As mentioned previously, oxidative stress is one of the key components stimulating the inflammatory response following MI and, hence, *antioxidant agents*—targeting mostly ROS—have been shown to confer beneficial effects after ischemia. The effects of the antioxidant enzymes superoxide dismutase plus catalase were investigated about 30 years ago in an I/R dog model and showed beneficial effects on reduction of the infarct size [284]; nevertheless, a preceding animal study [285] and a clinical trial [286] failed to yield similar results. The group of Sia *et al.* [287] provided evidence that long-term use of the antioxidant (and cholesterol-lowering) probucol can reduce mortality, reduce fibrosis, and improve cardiac function in a rat MI model. The beneficial effects were mediated via a suppressed oxidative stress response and repression of IL-1β/6. Nevertheless, research on probucol has been halted, following the finding that it can lower HDL levels [288]. Lastly, several studies have focused on vitamins (E and C) operating as free-radical scavengers and their potential as antioxidant and anti-inflammatory agents following acute MI. For more information on the matter, the reader is directed to a review publication by Rodrigo *et al.* [289]. To sum up, studies on antioxidant strategies have not yet shown robust results, so more research is needed to ascertain whether they can demonstrate any potential benefits that can be utilized in the clinical setting.

Methotrexate (MTX), a drug agent used in rheumatoid arthritis, psoriasis, and several types of cancer [145] has also been suggested as a novel strategy toward suppressing the deleterious inflammatory response in cardiovascular incidents. MTX is a potent anti-inflammatory agent that has been shown to reduce TNF-α, IL-1β, IL-6, and CRP levels when used in very low doses (10-30 mg once a week) [290,291] and produces effects on various inflammation-related cell types [292]. A meta-analysis on patients suffering from rheumatoid arthritis showed that MTX treatment reduces the risk of cardiovascular events by 21% compared to patients on other disease-modifying anti-rheumatic drugs [293]. Additionally, MTX and its derivative (MX-68) were shown to have beneficial effects on limiting infarct size in a dog I/R model [294]. Two clinical trials, CIRT [295] and TETHYS [296], are aiming to gain insights into the actions of low-dose MTX in reducing cardiovascular inflammation and to investigate whether MTX could be part of a novel treatment strategy to prevent re-occurrence of MI or to treat patients who have suffered an acute ischemic episode.

Erythropoetin (EPO) is a polypeptide hormone that is activated in hypoxic conditions and plays a crucial role in erythropoiesis [297]. Furthermore, EPO expression can, amongst others, be stimulated by TNF-α, IL-1β, and IL-6 [297], making it an attractive target after MI and during the inflammatory response. Notably, various studies have demonstrated beneficial effects of EPO following permanent MI [298], as well as I/R [299] in rodent models, hence, interest grew on EPO. Following I/R, EPO was able to suppress proinflammatory cytokines, such as TNF-α and IL-6, and to induce anti-inflammatory cytokines, like IL-10 [300]. While in a permanent coronary ligation model, EPO had similar effects but also targeted IL-1β and TGF-β [301]. Still, in the latter study, the fact that EPO was administered 4 weeks post-MI might give misleading conclusions, as EPO would not be able to target the early inflammatory response post-MI. Hence, more research needs to be performed before safe conclusions regarding the potential of EPO as an anti-inflammatory strategy can be made.

For a long time, the *Wnt/frizzled signaling* pathway has been in the research focus of several physiological and pathophysiological mechanisms, including the post-MI wound-healing process [302]. The essential immune-regulatory role of *Wnt/frizzled signaling* pathway is also well established. Both divisions of the signaling cascade, the β-catenin "canonical" [303] and the β-catenin independent (noncanonical) [304] have been implicated in the regulation of inflammatory cells and ILs. Wnt5a can alter the expression of macrophage-derived IL-1β and IL-6 [304], while Wnt3a and Wnt5a have been associated with TLR/NF-κB activation (pathways that are known to regulate inflammatory cell recruitment) [305,306]. Hence, a very reasonable approach would be to investigate the potential effects of manipulations of Wnt/frizzled signaling on the inflammatory response post-MI and to verify whether they can be utilized as novel strategies. The direct regulatory role of Wnt signaling factors (such as Dickkopf-1) has been demonstrated in the development of the atherosclerotic plaque [307], however, reports on the Wnt cascade association with the post-MI inflammation are extremely limited. Actually, the only study in this field is the one from Barandon *et al.* The group transplanted bone marrow cells from Tg mice overexpressing secreted frizzled-related protein 1 (sFRP-1) and subjected them to MI. They reported a suppression of the proinflammatory IL-6 and an induction of the anti-inflammatory IL-10 levels in the scar of sFRP-1 transplanted mice 4days post-MI, compared to controls (without any effect observed on TNF-α or IL-1β). These observations were correlated to reduced rupture incidence and scar size while cardiac performance parameters were improved [308]. This study confirms that a fine balance between anti- and proinflammatory factors is crucial for the early wound-healing post-MI, however, more research is needed to investigate the effects of Wnt signaling manipulations in the early inflammatory response post-MI, before any firm conclusions can be drawn.

Recently, it has been suggested that exogenous *high-mobility group protein B1 (HMGB1)* is involved in the inflammatory response and could potentially be part of the therapeutic strategy to modify the inflammatory response after ischemia. HMGB1 is a multifunctioning cytokine that is secreted by inflammatory cells (including monocytes, macrophages, and DC) [309] and targets receptors such as TLR [310]. The actions of HMGB1 on inflammation are quite variable and various studies report both anti-inflammatory [311] and proinflammatory [312] effects. HMGB1 can increase TNF-α in macrophages, TNF-α, and IL-1β/6 in monocytes and TNF-α and IL-1β in neutrophils, while there are reports that it can induce its own release from various inflammatory cell types [312]. Specific research on the cardiovascular system is still inconclusive. An interesting finding was that levels of HMGB1 in serum of MI patients are correlated with CRP and Troponin I levels [313]. Furthermore, Xu *et al.* showed that HMGB1 plays a major role in I/R injury by working via TNF-α and the JNK signaling pathway, implying that the blockade of HMGB1 could yield cardioprotective effects [314]. The injection of HMGB1 in rat hearts improved cardiac function and measurements with a suppressive effect on DCs to be the working mechanism [315], although the administration of HMGB1 at 3weeks post-MI might suggest that the acute inflammatory response after MI was not the target. On the other hand, treatment with an anti-HMGB1 antibody suppressed the upregulation of TNF-α and IL-1β in the infarct and diminished macrophage presence, however, LV adverse remodeling was worsened, an effect due to the HMGB1-inhibitory action on wound healing [316]. It is apparent that the results from the HMGB1 effects on the inflammatory process post-MI or following I/R studies are not conclusive and different models, routes of administration, and different inflammatory cell types targeted might be leading to variable results; thus, more research will shed light in the true potential of HMGB1 as a target.

The nuclear receptor superfamily of *PPAR* (covered extensively in Chapter 11 by Planavila & van Bilsen) and especially the PPAR-α and PPAR-γ families, is a well-established modulator of the inflammatory response in a wide range of conditions [317]. PPAR is expressed in a variety of immune-related cell types, such as neutrophils, monocytes, macrophages, DCs, lymphocytes, and so forth [318], while PPAR ligands have been shown to regulate factors such as TNF-α, IL-1β, and IL-6 [317]. Wayman's group has shown that PPAR-α and -γ ligand administration following I/R in rats can lead to MCP-1 and ICAM-1 suppression, which eventually plays a role in the reduction of the infarct size [319] by modulating immune cell trafficking [109]. In addition, another I/R study demonstrated that a PPAR-α ligand (GW7647) suppresses neutrophils and IL-6 levels, leading to cardioprotective effects [320]. It is interesting to note that both studies have implicated the regulation of NF-κB in the observed effects. Furthermore, PPAR-δ has also been suggested as an important player in I/R models, as its ligand (GW0742) can inhibit IL-6, IL-8, ICAM-1 and MCP-1 in Zucker fatty rats leading to improved cardiac function and reduced infarct size [321]. It should be noted here that the PPAR family has been implicated in the working mechanism of statins following MI [148] and several studies have provided solid evidence for the implication of PPAR-mediated pathways in the statin beneficial effects.

3.5 CONCLUSIONS

In this chapter we have provided an overview of the inflammatory process and its mediators that play a crucial role before the development of MI (atheromatous plaque buildup, transition from stable to unstable), as well as after the ischemic injury occurs (wound-healing post-MI). Targeting of inflammation gained research interest about

20-25 years ago, however, the results from the majority of clinical trials in patients with MI were rather disheartening [6]. New research has been soaring and inflammation is attracting interest again as a potential target to prevent and to heal MI, since the inflammatory response is of critical importance for the evolution of atheroma; hence, targeting of inflammation during the maturation of the atheromatous plaque (and before this ruptures) could be an attractive strategy for the prevention of ischemic episodes. Furthermore, the inflammatory cascade has been shown to serve essential roles in the early stages following MI and thus, more novel strategies are urgently needed to improve wound healing and prevent LV remodeling and progression to HF.

The optimal preventive drug strategy targeting the inflammatory process before MI should lead to a stable atheromatous plaque and prevent MI and other cardiovascular events from occurring. On the other hand, the ideal therapeutic drug approach focusing on inflammation following MI would be directed toward a robust and properly-healed scar, without the establishment of fibrosis in the remote areas or dilatation of the infarcted LV, while it should of course have minimal side effects. The current pharmacotherapeutic arsenal (both in the primary prevention and therapy of MI) is effective in reducing overall mortality and morbidity, however, it cannot avert cardiovascular episodes; following MI it can only delay the development of adverse remodeling, dilatation, and HF. Moreover, all drug classes currently available in this field have a considerable amount of side effects, which can lead to reduced adherence, worse clinical outcomes, and poor quality of life for the patient. From all this, it is obvious that new strategies have to be laid out urgently.

One possible scenario could be to invest on acquiring an in-depth understanding of the exact mechanisms by which current pharmacotherapy achieves its beneficial effects via modulation of the inflammatory response. Surely this approach would save research time and also would reduce the cost, compared to starting research of new targets with new molecules. On the other hand, recent data indicates that there are several novel areas to target the inflammatory response before or after MI; their application could prevent atherosclerosis development (and hence diminish dramatically the occurrence of MI) and also deliver attractive results in the wound healing of the infarcted myocardium. Still, the inflammatory cascade is a remarkably complex network consisting of various steps and determined by a plethora of mediators that depend on spatial and temporal factors. The exact mechanisms by which the inflammatory response leads to atherosclerosis, progresses to MI, and participates in wound healing of the injured myocardium are not entirely understood and further research is required to uncover its secrets. Lastly, an important aspect of the inflammatory cascade that should be highlighted is the lack of sufficient knowledge on its regulation and so its activation or depression should be judged on a specific-case basis. Hence, any novel targets and strategies aiming to affect the inflammatory response in order to target atherosclerosis (MI prevention) or to improve wound-healing post-MI should be tailored-made for each individualized patient (personalized medicine).

References

[1] Scott A, et al. What is "inflammation"? Are we ready to move beyond Celsus? Br J Sports Med 2004;38(3):248–9.
[2] Nathan C. Points of control in inflammation. Nature 2002;420(6917):846–52.
[3] Gutstein DE, Fuster V. Pathophysiology and clinical significance of atherosclerotic plaque rupture. Cardiovasc Res 1999;41(2):323–33.
[4] Daskalopoulos EP, Janssen BJ, Blankesteijn WM. Myofibroblasts in the infarct area: concepts and challenges. Microsc Microanal 2012;18(1):35–49.
[5] Legein B, et al. Inflammation and immune system interactions in atherosclerosis. Cell Mol Life Sci 2013;70(20):3847–69.
[6] Frangogiannis NG. The immune system and cardiac repair. Pharmacol Res 2008;58(2):88–111.
[7] Heymans S, et al. Inflammation as a therapeutic target in heart failure? A scientific statement from the Translational Research Committee of the Heart Failure Association of the European Society of Cardiology. Eur J Heart Fail 2009;11(2):119–29.
[8] Skalen K, et al. Subendothelial retention of atherogenic lipoproteins in early atherosclerosis. Nature 2002;417(6890):750–4.
[9] Leitinger N. Oxidized phospholipids as modulators of inflammation in atherosclerosis. Curr Opin Lipidol 2003;14(5):421–30.
[10] Nakashima Y, et al. Upregulation of VCAM-1 and ICAM-1 at atherosclerosis-prone sites on the endothelium in the ApoE-deficient mouse. Arterioscler Thromb Vasc Biol 1998;18(5):842–51.
[11] Lievens D, von Hundelshausen P. Platelets in atherosclerosis. Thromb Haemost 2011;106(5):827–38.
[12] Moore KJ, Tabas I. Macrophages in the pathogenesis of atherosclerosis. Cell 2011;145(3):341–55.
[13] Seneviratne AN, Sivagurunathan B, Monaco C. Toll-like receptors and macrophage activation in atherosclerosis. Clin Chim Acta 2012;413(1–2):3–14.
[14] Kzhyshkowska J, Neyen C, Gordon S. Role of macrophage scavenger receptors in atherosclerosis. Immunobiology 2012;217(5):492–502.
[15] Koenen RR, Weber C. Chemokines: established and novel targets in atherosclerosis. EMBO Mol Med 2011;3(12):713–25.
[16] Galkina E, Ley K. Immune and inflammatory mechanisms of atherosclerosis (*). Annu Rev Immunol 2009;27:165–97.
[17] Eliason JL, et al. Neutrophil depletion inhibits experimental abdominal aortic aneurysm formation. Circulation 2005;112(2):232–40.
[18] Zernecke A, et al. Protective role of CXC receptor 4/CXC ligand 12 unveils the importance of neutrophils in atherosclerosis. Circ Res 2008;102(2):209–17.
[19] Chevrier I, et al. Myeloperoxidase genetic polymorphisms modulate human neutrophil enzyme activity: genetic determinants for atherosclerosis? Atherosclerosis 2006;188(1):150–4.

[20] Soehnlein O. Multiple roles for neutrophils in atherosclerosis. Circ Res 2012;110(6):875–88.
[21] Bobryshev YV, Lord RS. S-100 positive cells in human arterial intima and in atherosclerotic lesions. Cardiovasc Res 1995;29(5):689–96.
[22] Galkina E, et al. Lymphocyte recruitment into the aortic wall before and during development of atherosclerosis is partially L-selectin dependent. J Exp Med 2006;203(5):1273–82.
[23] Paulson KE, et al. Resident intimal dendritic cells accumulate lipid and contribute to the initiation of atherosclerosis. Circ Res 2010;106(2):383–90.
[24] Gautier EL, et al. Conventional dendritic cells at the crossroads between immunity and cholesterol homeostasis in atherosclerosis. Circulation 2009;119(17):2367–75.
[25] Hermansson A, et al. Inhibition of T cell response to native low-density lipoprotein reduces atherosclerosis. J Exp Med 2010;207(5):1081–93.
[26] Koltsova EK, Ley K. How dendritic cells shape atherosclerosis. Trends Immunol 2011;32(11):540–7.
[27] Hansson GK, Robertson AK, Soderberg-Naucler C. Inflammation and atherosclerosis. Annu Rev Pathol 2006;1:297–329.
[28] Weber C, Zernecke A, Libby P. The multifaceted contributions of leukocyte subsets to atherosclerosis: lessons from mouse models. Nat Rev Immunol 2008;8(10):802–15.
[29] Caligiuri G, et al. Protective immunity against atherosclerosis carried by B cells of hypercholesterolemic mice. J Clin Invest 2002;109(6):745–53.
[30] Witztum JL. Splenic immunity and atherosclerosis: a glimpse into a novel paradigm? J Clin Invest 2002;109(6):721–4.
[31] Kyaw T, et al. Conventional B2 B cell depletion ameliorates whereas its adoptive transfer aggravates atherosclerosis. J Immunol 2010;185(7):4410–9.
[32] Ait-Oufella H, et al. B cell depletion reduces the development of atherosclerosis in mice. J Exp Med 2010;207(8):1579–87.
[33] Kyaw T, et al. Depletion of B2 but not B1a B cells in BAFF receptor-deficient ApoE mice attenuates atherosclerosis by potently ameliorating arterial inflammation. PLoS One 2012;7(1):e29371.
[34] Woollard KJ. Immunological aspects of atherosclerosis. Clin Sci (Lond) 2013;125(5):221–35.
[35] van der Wal AC, Becker AE. Atherosclerotic plaque rupture—pathologic basis of plaque stability and instability. Cardiovasc Res 1999;41(2):334–44.
[36] WHO. WHO Fact Sheet N°317—Cardiovascular diseases (CVDs), 2011. Available from: http://www.who.int/mediacentre/factsheets/fs317/en/; [30.01.13].
[37] Fox KA, et al. Prediction of risk of death and myocardial infarction in the six months after presentation with acute coronary syndrome: prospective multinational observational study (GRACE). BMJ 2006;333(7578):1091.
[38] Go AS, et al. Heart disease and stroke statistics—2013 update: a report from the American Heart Association. Circulation 2013;127(1):e6–e245.
[39] Roger VL. Epidemiology of heart failure. Circ Res 2013;113(6):646–59.
[40] Lacey L, Tabberer M. Economic burden of post-acute myocardial infarction heart failure in the United Kingdom. Eur J Heart Fail 2005;7(4):677–83.
[41] Bui AL, Horwich TB, Fonarow GC. Epidemiology and risk profile of heart failure. Nat Rev Cardiol 2011;8(1):30–41.
[42] Thygesen K, et al. Third universal definition of myocardial infarction. J Am Coll Cardiol 2012;60(16):1581–98.
[43] Daskalopoulos EP, Hermans KC, Blankesteijn WM. Cardiac (myo)fibroblast: novel strategies for its targeting following myocardial infarction. Curr Pharm Des 2014;20(12):1987–2002.
[44] Cleutjens JP, et al. The infarcted myocardium: simply dead tissue, or a lively target for therapeutic interventions. Cardiovasc Res 1999;44(2):232–41.
[45] Frantz S, Bauersachs J, Ertl G. Post-infarct remodelling: contribution of wound healing and inflammation. Cardiovasc Res 2009;81(3):474–81.
[46] Ertl G, Frantz S. Healing after myocardial infarction. Cardiovasc Res 2005;66(1):22–32.
[47] Dewald O, et al. Of mice and dogs: species-specific differences in the inflammatory response following myocardial infarction. Am J Pathol 2004;164(2):665–77.
[48] Nahrendorf M, Pittet MJ, Swirski FK. Monocytes: protagonists of infarct inflammation and repair after myocardial infarction. Circulation 2010;121(22):2437–45.
[49] Frangogiannis NG. Regulation of the inflammatory response in cardiac repair. Circ Res 2012;110(1):159–73.
[50] Dobaczewski M, et al. Extracellular matrix remodeling in canine and mouse myocardial infarcts. Cell Tissue Res 2006;324(3):475–88.
[51] van den Borne SW, et al. Mouse strain determines the outcome of wound healing after myocardial infarction. Cardiovasc Res 2009;84(2):273–82.
[52] Vandervelde S, et al. Increased inflammatory response and neovascularization in reperfused vs. non-reperfused murine myocardial infarction. Cardiovasc Pathol 2006;15(2):83–90.
[53] Willems IE, et al. The alpha-smooth muscle actin-positive cells in healing human myocardial scars. Am J Pathol 1994;145(4):868–75.
[54] Cohn JN, Ferrari R, Sharpe N. Cardiac remodeling—concepts and clinical implications: a consensus paper from an international forum on cardiac remodeling. Behalf of an International Forum on Cardiac Remodeling. J Am Coll Cardiol 2000;35(3):569–82.
[55] Konstam MA, et al. Left ventricular remodeling in heart failure: current concepts in clinical significance and assessment. JACC Cardiovasc Imaging 2011;4(1):98–108.
[56] Cannon 3rd RO. Mechanisms, management and future directions for reperfusion injury after acute myocardial infarction. Nat Clin Pract Cardiovasc Med 2005;2(2):88–94.
[57] Hausenloy DJ, Yellon DM. Myocardial ischemia-reperfusion injury: a neglected therapeutic target. J Clin Invest 2013;123(1):92–100.
[58] Barros LF, et al. Myocardial reperfusion: leukocyte accumulation in the ischemic and remote non-ischemic regions. Shock 2000;13(1):67–71.
[59] Vinten-Johansen J, et al. Inflammation, proinflammatory mediators and myocardial ischemia-reperfusion Injury. Hematol Oncol Clin North Am 2007;21(1):123–45.
[60] Vander Heide RS, Steenbergen C. Cardioprotection and myocardial reperfusion: pitfalls to clinical application. Circ Res 2013;113(4):464–77.
[61] Toutouzas K, Colombo A, Stefanadis C. Inflammation and restenosis after percutaneous coronary interventions. Eur Heart J 2004;25(19):1679–87.
[62] Tiong AY, et al. Lack of widespread inflammation after contemporary PCI. Int J Cardiol 2010;140(1):82–7.

[63] van den Borne SW, et al. Increased matrix metalloproteinase-8 and -9 activity in patients with infarct rupture after myocardial infarction. Cardiovasc Pathol 2009;18(1):37–43.
[64] Tao ZY, et al. Temporal changes in matrix metalloproteinase expression and inflammatory response associated with cardiac rupture after myocardial infarction in mice. Life Sci 2004;74(12):1561–72.
[65] Anzai T, et al. C-reactive protein as a predictor of infarct expansion and cardiac rupture after a first Q-wave acute myocardial infarction. Circulation 1997;96(3):778–84.
[66] Gao XM, et al. Infarct size and post-infarct inflammation determine the risk of cardiac rupture in mice. Int J Cardiol 2010;143(1):20–8.
[67] Lopez-Sendon J, et al. Factors related to heart rupture in acute coronary syndromes in the Global Registry of Acute Coronary Events. Eur Heart J 2010;31(12):1449–56.
[68] Honda S, et al. Changing incidence of cardiac rupture and its determinants in patients with acute myocardial infarction: results from 5,694 patients database over 35 years. J Am Coll Cardiol 2013;61(10_S):E23.
[69] Dutta P, et al. Myocardial infarction accelerates atherosclerosis. Nature 2012;487(7407):325–9.
[70] Siwik DA, Chang DL, Colucci WS. Interleukin-1beta and tumor necrosis factor-alpha decrease collagen synthesis and increase matrix metalloproteinase activity in cardiac fibroblasts in vitro. Circ Res 2000;86(12):1259–65.
[71] Nian M, et al. Inflammatory cytokines and postmyocardial infarction remodeling. Circ Res 2004;94(12):1543–53.
[72] Kleinbongard P, Schulz R, Heusch G. TNFalpha in myocardial ischemia/reperfusion, remodeling and heart failure. Heart Fail Rev 2011;16(1):49–69.
[73] Aggarwal BB. Signalling pathways of the TNF superfamily: a double-edged sword. Nat Rev Immunol 2003;3(9):745–56.
[74] Coggins M, Rosenzweig A. The fire within: cardiac inflammatory signaling in health and disease. Circ Res 2012;110(1):116–25.
[75] Murray DR, Freeman GL. Tumor necrosis factor-alpha induces a biphasic effect on myocardial contractility in conscious dogs. Circ Res 1996;78(1):154–60.
[76] Haudek SB, et al. TNF provokes cardiomyocyte apoptosis and cardiac remodeling through activation of multiple cell death pathways. J Clin Invest 2007;117(9):2692–701.
[77] Li YY, et al. Myocardial extracellular matrix remodeling in transgenic mice overexpressing tumor necrosis factor alpha can be modulated by anti-tumor necrosis factor alpha therapy. Proc Natl Acad Sci U S A 2000;97(23):12746–51.
[78] Turner NA, et al. Mechanism of TNFalpha-induced IL-1alpha, IL-1beta and IL-6 expression in human cardiac fibroblasts: effects of statins and thiazolidinediones. Cardiovasc Res 2007;76(1):81–90.
[79] Bradham WS, et al. Tumor necrosis factor-alpha and myocardial remodeling in progression of heart failure: a current perspective. Cardiovasc Res 2002;53(4):822–30.
[80] Monden Y, et al. Tumor necrosis factor-alpha is toxic via receptor 1 and protective via receptor 2 in a murine model of myocardial infarction. Am J Physiol Heart Circ Physiol 2007;293(1):H743–53.
[81] Al-Lamki RS, et al. TNF receptors differentially signal and are differentially expressed and regulated in the human heart. Am J Transplant 2009;9(12):2679–96.
[82] Dinarello CA. Interleukin-1 in the pathogenesis and treatment of inflammatory diseases. Blood 2011;117(14):3720–32.
[83] Grothusen C, et al. Impact of an interleukin-1 receptor antagonist and erythropoietin on experimental myocardial ischemia/reperfusion injury. ScientificWorldJournal 2012;2012:737585.
[84] Saxena A, et al. IL-1 induces proinflammatory leukocyte infiltration and regulates fibroblast phenotype in the infarcted myocardium. J Immunol 2013;191(9):4838–48.
[85] Gabriel AS, et al. IL-6 levels in acute and post myocardial infarction: their relation to CRP levels, infarction size, left ventricular systolic function, and heart failure. Eur J Intern Med 2004;15(8):523–8.
[86] Huang M, et al. Role of interleukin-6 in regulation of immune responses to remodeling after myocardial infarction. Heart Fail Rev 2014;Apr 23.
[87] Moore KW, et al. Interleukin-10 and the interleukin-10 receptor. Annu Rev Immunol 2001;19:683–765.
[88] Zymek P, et al. Interleukin-10 is not a critical regulator of infarct healing and left ventricular remodeling. Cardiovasc Res 2007;74(2):313–22.
[89] Savvatis K, et al. Interleukin-23 deficiency leads to impaired wound healing and adverse prognosis after myocardial infarction. Circ Heart Fail 2014;7(1):161–71.
[90] Miller AM. Role of IL-33 in inflammation and disease. J Inflamm (Lond) 2011;8(1):22.
[91] Seo D, Hare JM. The transforming growth factor-beta/Smad3 pathway: coming of age as a key participant in cardiac remodeling. Circulation 2007;116(19):2096–8.
[92] Frangogiannis NG. The inflammatory response in myocardial injury, repair, and remodelling. Nat Rev Cardiol 2014;11(5):255–65.
[93] Bujak M, Frangogiannis NG. The role of TGF-beta signaling in myocardial infarction and cardiac remodeling. Cardiovasc Res 2007;74(2):184–95.
[94] Feinberg MW, et al. Essential role for Smad3 in regulating MCP-1 expression and vascular inflammation. Circ Res 2004;94(5):601–8.
[95] Kitamura M. Identification of an inhibitor targeting macrophage production of monocyte chemoattractant protein-1 as TGF-beta 1. J Immunol 1997;159(3):1404–11.
[96] Ikeuchi M, et al. Inhibition of TGF-beta signaling exacerbates early cardiac dysfunction but prevents late remodeling after infarction. Cardiovasc Res 2004;64(3):526–35.
[97] Zernecke A, Weber C. Chemokines in the vascular inflammatory response of atherosclerosis. Cardiovasc Res 2010;86(2):192–201.
[98] Liehn EA, et al. Repair after myocardial infarction, between fantasy and reality: the role of chemokines. J Am Coll Cardiol 2011;58(23):2357–62.
[99] Frangogiannis NG. Chemokines in ischemia and reperfusion. Thromb Haemost 2007;97(5):738–47.
[100] Steffens S, Montecucco F, Mach F. The inflammatory response as a target to reduce myocardial ischaemia and reperfusion injury. Thromb Haemost 2009;102(2):240–7.
[101] Zernecke A, Bernhagen J, Weber C. Macrophage migration inhibitory factor in cardiovascular disease. Circulation 2008;117(12):1594–602.
[102] White DA, et al. Pro-inflammatory action of MIF in acute myocardial infarction via activation of peripheral blood mononuclear cells. PLoS One 2013;8(10):e76206.

[103] Liehn EA, et al. Compartmentalized protective and detrimental effects of endogenous macrophage migration-inhibitory factor mediated by CXCR2 in a mouse model of myocardial ischemia/reperfusion. Arterioscler Thromb Vasc Biol 2013;33(9):2180–6.
[104] Frangogiannis NG, Entman ML. Chemokines in myocardial ischemia. Trends Cardiovasc Med 2005;15(5):163–9.
[105] Amulic B, et al. Neutrophil function: from mechanisms to disease. Annu Rev Immunol 2012;30:459–89.
[106] Ma Y, Yabluchanskiy A, Lindsey ML. Neutrophil roles in left ventricular remodeling following myocardial infarction. Fibrogenesis Tissue Repair 2013;6(1):11.
[107] Patel KD, Cuvelier SL, Wiehler S. Selectins: critical mediators of leukocyte recruitment. Semin Immunol 2002;14(2):73–81.
[108] Rahman A, Fazal F. Hug tightly and say goodbye: role of endothelial ICAM-1 in leukocyte transmigration. Antioxid Redox Signal 2009;11(4):823–39.
[109] Frangogiannis NG, Smith CW, Entman ML. The inflammatory response in myocardial infarction. Cardiovasc Res 2002;53(1):31–47.
[110] Blum A, Sheiman J, Hasin Y. Leukocytes and acute myocardial infarction. Isr Med Assoc J 2002;4(11):1060–5.
[111] White DA, et al. Differential roles of cardiac and leukocyte derived macrophage migration inhibitory factor in inflammatory responses and cardiac remodelling post myocardial infarction. J Mol Cell Cardiol 2014;69:32–42.
[112] Shi C, Pamer EG. Monocyte recruitment during infection and inflammation. Nat Rev Immunol 2011;11(11):762–74.
[113] Woollard KJ, Geissmann F. Monocytes in atherosclerosis: subsets and functions. Nat Rev Cardiol 2010;7(2):77–86.
[114] Leuschner F, et al. Rapid monocyte kinetics in acute myocardial infarction are sustained by extramedullary monocytopoiesis. J Exp Med 2012;209(1):123–37.
[115] Nahrendorf M, et al. The healing myocardium sequentially mobilizes two monocyte subsets with divergent and complementary functions. J Exp Med 2007;204(12):3037–47.
[116] Tsujioka H, et al. Impact of heterogeneity of human peripheral blood monocyte subsets on myocardial salvage in patients with primary acute myocardial infarction. J Am Coll Cardiol 2009;54(2):130–8.
[117] Frantz S, Nahrendorf M. Cardiac macrophages and their role in ischaemic heart disease. Cardiovasc Res 2014;102(2):240–8.
[118] Ma J, et al. Regulation of macrophage activation. Cell Mol Life Sci 2003;60(11):2334–46.
[119] Biswas SK, Mantovani A. Macrophage plasticity and interaction with lymphocyte subsets: cancer as a paradigm. Nat Immunol 2010;11(10):889–96.
[120] Lambert JM, Lopez EF, Lindsey ML. Macrophage roles following myocardial infarction. Int J Cardiol 2008;130(2):147–58.
[121] Ma Y, et al. Matrix metalloproteinase-28 deletion exacerbates cardiac dysfunction and rupture after myocardial infarction in mice by inhibiting M2 macrophage activation. Circ Res 2013;112(4):675–88.
[122] Steinhubl SR, et al. Clinical evidence for anti-inflammatory effects of antiplatelet therapy in patients with atherothrombotic disease. Vasc Med 2007;12(2):113–22.
[123] Habets KL, Huizinga TW, Toes RE. Platelets and autoimmunity. Eur J Clin Invest 2013;43(7):746–57.
[124] Rondina MT, Weyrich AS, Zimmerman GA. Platelets as cellular effectors of inflammation in vascular diseases. Circ Res 2013;112(11):1506–19.
[125] Liu Y, et al. Novel role of platelets in mediating inflammatory responses and ventricular rupture or remodeling following myocardial infarction. Arterioscler Thromb Vasc Biol 2011;31(4):834–41.
[126] Hawrylowicz CM, Howells GL, Feldmann M. Platelet-derived interleukin 1 induces human endothelial adhesion molecule expression and cytokine production. J Exp Med 1991;174(4):785–90.
[127] Gawaz M, et al. Activated platelets induce monocyte chemotactic protein-1 secretion and surface expression of intercellular adhesion molecule-1 on endothelial cells. Circulation 1998;98(12):1164–71.
[128] Levick SP, et al. Cardiac mast cells: the centrepiece in adverse myocardial remodelling. Cardiovasc Res 2011;89(1):12–9.
[129] Frangogiannis NG, et al. Stem cell factor induction is associated with mast cell accumulation after canine myocardial ischemia and reperfusion. Circulation 1998;98(7):687–98.
[130] Frangogiannis NG, et al. Resident cardiac mast cells degranulate and release preformed TNF-alpha, initiating the cytokine cascade in experimental canine myocardial ischemia/reperfusion. Circulation 1998;98(7):699–710.
[131] Theoharides TC, et al. Mast cells and inflammation. Biochim Biophys Acta 2012;1822(1):21–33.
[132] Yilmaz A, et al. Emergence of dendritic cells in the myocardium after acute myocardial infarction—implications for inflammatory myocardial damage. Int J Biomed Sci 2010;6(1):27–36.
[133] Anzai A, et al. Regulatory role of dendritic cells in postinfarction healing and left ventricular remodeling. Circulation 2012;125(10):1234–45.
[134] van Nieuwenhoven FA, Turner NA. The role of cardiac fibroblasts in the transition from inflammation to fibrosis following myocardial infarction. Vascul Pharmacol 2013;58(3):182–8.
[135] Shinde AV, Frangogiannis NG. Fibroblasts in myocardial infarction: a role in inflammation and repair. J Mol Cell Cardiol 2014;70C:74–82.
[136] Strowig T, et al. Inflammasomes in health and disease. Nature 2012;481(7381):278–86.
[137] Takahashi M. Role of the inflammasome in myocardial infarction. Trends Cardiovasc Med 2011;21(2):37–41.
[138] Marchetti C, et al. A novel pharmacologic inhibitor of the nlrp3 inflammasome limits myocardial injury after ischemia-reperfusion in the mouse. J Cardiovasc Pharmacol 2014;63(4):316–22.
[139] Mezzaroma E, et al. The inflammasome promotes adverse cardiac remodeling following acute myocardial infarction in the mouse. Proc Natl Acad Sci U S A 2011;108(49):19725–30.
[140] Stutz A, Golenbock DT, Latz E. Inflammasomes: too big to miss. J Clin Invest 2009;119(12):3502–11.
[141] Huang X, et al. C-reactive protein promotes adhesion of monocytes to endothelial cells via NADPH oxidase-mediated oxidative stress. J Cell Biochem 2012;113(3):857–67.
[142] Takahashi T, et al. Serum C-reactive protein elevation in left ventricular remodeling after acute myocardial infarction—role of neurohormones and cytokines. Int J Cardiol 2003;88(2–3):257–65.
[143] Griselli M, et al. C-reactive protein and complement are important mediators of tissue damage in acute myocardial infarction. J Exp Med 1999;190(12):1733–40.
[144] Zairis MN, et al. C-reactive protein levels on admission are associated with response to thrombolysis and prognosis after ST-segment elevation acute myocardial infarction. Am Heart J 2002;144(5):782–9.

[145] British_National_Formulary_66, Cardiovascular System (Chapter 2). 2013; 83–176.
[146] Zhou Q, Liao JK. Statins and cardiovascular diseases: from cholesterol lowering to pleiotropy. Curr Pharm Des 2009;15(5):467–78.
[147] Vaughan CJ, Gotto Jr. AM, Basson CT. The evolving role of statins in the management of atherosclerosis. J Am Coll Cardiol 2000;35(1):1–10.
[148] Antonopoulos AS, et al. Statins as anti-inflammatory agents in atherogenesis: molecular mechanisms and lessons from the recent clinical trials. Curr Pharm Des 2012;18(11):1519–30.
[149] Koh KK. Effects of statins on vascular wall: vasomotor function, inflammation, and plaque stability. Cardiovasc Res 2000;47(4):648–57.
[150] Feig JE, et al. Statins promote the regression of atherosclerosis via activation of the CCR7-dependent emigration pathway in macrophages. PLoS One 2011;6(12):e28534.
[151] Baetta R, et al. Nitric oxide-donating atorvastatin attenuates neutrophil recruitment during vascular inflammation independent of changes in plasma cholesterol. Cardiovasc Drugs Ther 2013;27(3):211–9.
[152] Marino F, et al. Simvastatin down-regulates the production of Interleukin-8 by neutrophil leukocytes from dyslipidemic patients. BMC Cardiovasc Disord 2014;14:37.
[153] Liu Z, et al. Treatment with telmisartan/rosuvastatin combination has a beneficial synergistic effect on ameliorating Th17/Treg functional imbalance in hypertensive patients with carotid atherosclerosis. Atherosclerosis 2014;233(1):291–9.
[154] Duivenvoorden R, et al. A statin-loaded reconstituted high-density lipoprotein nanoparticle inhibits atherosclerotic plaque inflammation. Nat Commun 2014;5:3065.
[155] Shishehbor MH, Hazen SL. Inflammatory and oxidative markers in atherosclerosis: relationship to outcome. Curr Atheroscler Rep 2004;6(3):243–50.
[156] Ridker PM, et al. Long-term effects of pravastatin on plasma concentration of C-reactive protein. The Cholesterol and Recurrent Events (CARE) Investigators. Circulation 1999;100(3):230–5.
[157] Albert MA, et al. Effect of statin therapy on C-reactive protein levels: the pravastatin inflammation/CRP evaluation (PRINCE): a randomized trial and cohort study. JAMA 2001;286(1):64–70.
[158] Ridker PM, et al. Rosuvastatin to prevent vascular events in men and women with elevated C-reactive protein. N Engl J Med 2008;359(21):2195–207.
[159] Ridker PM. Moving beyond JUPITER: will inhibiting inflammation reduce vascular event rates? Curr Atheroscler Rep 2013;15(1):295.
[160] Carter AM, et al. Metformin reduces C-reactive protein but not complement factor C3 in overweight patients with Type 2 diabetes mellitus. Diabet Med 2005;22(9):1282–4.
[161] Stocker DJ, et al. A randomized trial of the effects of rosiglitazone and metformin on inflammation and subclinical atherosclerosis in patients with type 2 diabetes. Am Heart J 2007;153(3): 445.e1-6.
[162] Marfella R, et al. The ubiquitin-proteasome system contributes to the inflammatory injury in ischemic diabetic myocardium: the role of glycemic control. Cardiovasc Pathol 2009;18(6): 332–45.
[163] Carlsen SM, et al. Metformin increases circulating tumour necrosis factor-alpha levels in non-obese non-diabetic patients with coronary heart disease. Cytokine 1998;10(1):66–9.
[164] Nissen SE, Wolski K. Effect of rosiglitazone on the risk of myocardial infarction and death from cardiovascular causes. N Engl J Med 2007;356(24):2457–71.
[165] Koh KK, et al. Combination therapy for treatment or prevention of atherosclerosis: focus on the lipid-RAAS interaction. Atherosclerosis 2010;209(2):307–13.
[166] Hernandez-Presa MA, et al. ACE inhibitor quinapril reduces the arterial expression of NF-kappaB-dependent proinflammatory factors but not of collagen I in a rabbit model of atherosclerosis. Am J Pathol 1998;153(6):1825–37.
[167] da Cunha V, et al. Enalapril attenuates angiotensin II-induced atherosclerosis and vascular inflammation. Atherosclerosis 2005;178(1):9–17.
[168] Fu R, et al. XJP-1, a novel ACEI, with anti-inflammatory properties in HUVECs. Atherosclerosis 2011;219(1):40–8.
[169] Dasu MR, Riosvelasco AC, Jialal I. Candesartan inhibits Toll-like receptor expression and activity both in vitro and in vivo. Atherosclerosis 2009;202(1):76–83.
[170] Zhao Y, et al. Suppressive effects of irbesartan on inflammation and apoptosis in atherosclerotic plaques of apoE-/- mice: molecular imaging with 14C-FDG and 99mTc-annexin A5. PLoS One 2014;9(2):e89338.
[171] Takagi H, et al. Effects of telmisartan therapy on interleukin-6 and tumor necrosis factor-alpha levels: a meta-analysis of randomized controlled trials. Hypertens Res 2013;36(4):368–73.
[172] Klinghammer L, et al. Impact of telmisartan on the inflammatory state in patients with coronary atherosclerosis—influence on IP-10, TNF-alpha and MCP-1. Cytokine 2013;62(2):290–6.
[173] Fukuda D, et al. Inhibition of renin-angiotensin system attenuates periadventitial inflammation and reduces atherosclerotic lesion formation. Biomed Pharmacother 2009;63(10):754–61.
[174] Nussberger J, et al. Renin inhibition by aliskiren prevents atherosclerosis progression: comparison with irbesartan, atenolol, and amlodipine. Hypertension 2008;51(5):1306–11.
[175] Kuhnast S, et al. Aliskiren inhibits atherosclerosis development and improves plaque stability in APOE*3Leiden.CETP transgenic mice with or without treatment with atorvastatin. J Hypertens 2012;30(1):107–16.
[176] Montecucco F, Pende A, Mach F. The renin-angiotensin system modulates inflammatory processes in atherosclerosis: evidence from basic research and clinical studies. Mediators Inflamm 2009;2009:752406.
[177] Badrnya S, et al. Platelets mediate oxidized low-density lipoprotein-induced monocyte extravasation and foam cell formation. Arterioscler Thromb Vasc Biol 2014;34(3):571–80.
[178] Husted S, et al. Changes in inflammatory biomarkers in patients treated with ticagrelor or clopidogrel. Clin Cardiol 2010;33(4):206–12.
[179] Li D, et al. Roles of purinergic receptor P2Y, G protein-coupled 12 in the development of atherosclerosis in apolipoprotein E-deficient mice. Arterioscler Thromb Vasc Biol 2012;32(8):e81–9.
[180] West LE, et al. Vessel wall, not platelet, P2Y12 potentiates early atherogenesis. Cardiovasc Res 2014;102(3):429–35.
[181] Bonetti PO, et al. Statin effects beyond lipid lowering—are they clinically relevant? Eur Heart J 2003;24(3):225–48.
[182] Zernecke A, Weber C. Improving the treatment of atherosclerosis by linking anti-inflammatory and lipid modulating strategies. Heart 2012;98(21):1600–6.

[183] Schroen B, Heymans S. Small but smart—microRNAs in the centre of inflammatory processes during cardiovascular diseases, the metabolic syndrome, and ageing. Cardiovasc Res 2012;93(4):605–13.
[184] Sun X, et al. Systemic delivery of microRNA-181b inhibits nuclear factor-kappaB activation, vascular inflammation, and atherosclerosis in apolipoprotein E-deficient mice. Circ Res 2014;114(1):32–40.
[185] Zernecke A, et al. Delivery of microRNA-126 by apoptotic bodies induces CXCL12-dependent vascular protection. Sci Signal 2009;2(100):ra81.
[186] Busch M, Zernecke A. microRNAs in the regulation of dendritic cell functions in inflammation and atherosclerosis. J Mol Med (Berl) 2012;90(8):877–85.
[187] Sun X, Belkin N, Feinberg MW. Endothelial microRNAs and atherosclerosis. Curr Atheroscler Rep 2013;15(12):372.
[188] Kita T, et al. Regression of atherosclerosis with anti-CD3 antibody via augmenting a regulatory T-cell response in mice. Cardiovasc Res 2014;102(1):107–17.
[189] Ellison S, et al. Attenuation of experimental atherosclerosis by interleukin-19. Arterioscler Thromb Vasc Biol 2013;33(10):2316–24.
[190] Wolfs IM, et al. Reprogramming macrophages to an anti-inflammatory phenotype by helminth antigens reduces murine atherosclerosis. FASEB J 2014;28(1):288–99.
[191] Dinh TN, et al. Cytokine therapy with interleukin-2/anti-interleukin-2 monoclonal antibody complexes expands CD4+CD25+Foxp3+ regulatory T cells and attenuates development and progression of atherosclerosis. Circulation 2012;126(10):1256–66.
[192] Hartman J, Frishman WH. Inflammation and atherosclerosis: a review of the role of interleukin-6 in the development of atherosclerosis and the potential for targeted drug therapy. Cardiol Rev 2014;22(3):147–51.
[193] Feige E, et al. Inhibition of monocyte chemotaxis by VB-201, a small molecule lecinoxoid, hinders atherosclerosis development in ApoE(-)/(-) mice. Atherosclerosis 2013;229(2):430–9.
[194] Seijkens T, et al. Hypercholesterolemia-induced priming of hematopoietic stem and progenitor cells aggravates atherosclerosis. FASEB J 2014;28(5):2202–13.
[195] Akhmedov A, et al. Endothelial overexpression of LOX-1 increases plaque formation and promotes atherosclerosis in vivo. Eur Heart J 2014. Available from: http://www.ncbi.nlm.nih.gov/pubmed/24419805.
[196] Wang S, et al. beta-Glucan attenuates inflammatory responses in oxidized LDL-induced THP-1 cells via the p38 MAPK pathway. Nutr Metab Cardiovasc Dis 2014;24(3):248–55.
[197] O'Gara PT, et al. 2013 ACCF/AHA guideline for the management of ST-elevation myocardial infarction: a report of the American College of Cardiology Foundation/American Heart Association Task Force on Practice Guidelines. Circulation 2013;127(4):e362–425.
[198] Anderson JL, et al. 2012 ACCF/AHA focused update incorporated into the ACCF/AHA 2007 guidelines for the management of patients with unstable angina/non-ST-elevation myocardial infarction: a report of the American College of Cardiology Foundation/American Heart Association Task Force on Practice Guidelines. Circulation 2013;127(23):e663–828.
[199] Tawfik MK, et al. Atorvastatin restores the balance between pro-inflammatory and anti-inflammatory mediators in rats with acute myocardial infarction. Eur Rev Med Pharmacol Sci 2010;14(6):499–506.
[200] Stumpf C, et al. Atorvastatin enhances interleukin-10 levels and improves cardiac function in rats after acute myocardial infarction. Clin Sci (Lond) 2009;116(1):45–52.
[201] Sun YM, et al. Effect of atorvastatin on expression of IL-10 and TNF-alpha mRNA in myocardial ischemia-reperfusion injury in rats. Biochem Biophys Res Commun 2009;382(2):336–40.
[202] Zhang J, et al. Simvastatin regulates myocardial cytokine expression and improves ventricular remodeling in rats after acute myocardial infarction. Cardiovasc Drugs Ther 2005;19(1):13–21.
[203] Sheng FQ, et al. In rats with myocardial infarction, interference by simvastatin with the TLR4 signal pathway attenuates ventricular remodelling. Acta Cardiol 2009;64(6):779–85.
[204] Li TS, et al. Pravastatin improves remodeling and cardiac function after myocardial infarction by an antiinflammatory mechanism rather than by the induction of angiogenesis. Ann Thorac Surg 2006;81(6):2217–25.
[205] Ridker PM, et al. C-reactive protein levels and outcomes after statin therapy. N Engl J Med 2005;352(1):20–8.
[206] Hong SJ, et al. Low-dose versus moderate-dose atorvastatin after acute myocardial infarction: 8-month effects on coronary flow reserve and angiogenic cell mobilisation. Heart 2010;96(10):756–64.
[207] Sposito AC, et al. Timing and dose of statin therapy define its impact on inflammatory and endothelial responses during myocardial infarction. Arterioscler Thromb Vasc Biol 2011;31(5):1240–6.
[208] Stefanadi E, et al. Early initiation of low-dose atorvastatin treatment after an acute ST-elevated myocardial infarction, decreases inflammatory process and prevents endothelial injury and activation. Int J Cardiol 2009;133(2):266–8.
[209] Abdulla J, et al. A systematic review: effect of angiotensin converting enzyme inhibition on left ventricular volumes and ejection fraction in patients with a myocardial infarction and in patients with left ventricular dysfunction. Eur J Heart Fail 2007;9(2):129–35.
[210] Werner C, et al. RAS blockade with ARB and ACE inhibitors: current perspective on rationale and patient selection. Clin Res Cardiol 2008;97(7):418–31.
[211] Gullestad L, et al. Effect of high- versus low-dose angiotensin converting enzyme inhibition on cytokine levels in chronic heart failure. J Am Coll Cardiol 1999;34(7):2061–7.
[212] de Gusmao FM, et al. Angiotensin II inhibition during myocardial ischemia-reperfusion in dogs: effects on leukocyte infiltration, nitric oxide synthase isoenzymes activity and left ventricular ejection fraction. Int J Cardiol 2005;100(3):363–70.
[213] Kohno T, et al. Angiotensin-receptor blockade reduces border zone myocardial monocyte chemoattractant protein-1 expression and macrophage infiltration in post-infarction ventricular remodeling. Circ J 2008;72(10):1685–92.
[214] Leuschner F, et al. Angiotensin-converting enzyme inhibition prevents the release of monocytes from their splenic reservoir in mice with myocardial infarction. Circ Res 2010;107(11):1364–73.
[215] Lapointe N, et al. Comparison of the effects of an angiotensin-converting enzyme inhibitor and a vasopeptidase inhibitor after myocardial infarction in the rat. J Am Coll Cardiol 2002;39(10):1692–8.
[216] Jugdutt BI, et al. Aging-related early changes in markers of ventricular and matrix remodeling after reperfused ST-segment elevation myocardial infarction in the canine model: effect of early therapy with an angiotensin II type 1 receptor blocker. Circulation 2010;122(4):341–51.

[217] Kovacs I, et al. Correlation of flow mediated dilation with inflammatory markers in patients with impaired cardiac function. Beneficial effects of inhibition of ACE. Eur J Heart Fail 2006;8(5):451–9.
[218] Sandmann S, et al. Differential effects of olmesartan and ramipril on inflammatory response after myocardial infarction in rats. Blood Press 2006;15(2):116–28.
[219] Zhang RY, et al. Effects of angiotensin converting enzyme inhibitor, angiotensin II type I receptor blocker and their combination on postinfarcted ventricular remodeling in rats. Chin Med J (Engl) 2006;119(8):649–55.
[220] Sanjabi S, et al. Anti-inflammatory and pro-inflammatory roles of TGF-beta, IL-10, and IL-22 in immunity and autoimmunity. Curr Opin Pharmacol 2009;9(4):447–53.
[221] Roberts R, DeMello V, Sobel BE. Deleterious effects of methylprednisolone in patients with myocardial infarction. Circulation 1976;53(3 Suppl):I204–6.
[222] Mannisi JA, et al. Steroid administration after myocardial infarction promotes early infarct expansion. A study in the rat. J Clin Invest 1987;79(5):1431–9.
[223] Silverman HS, Pfeifer MP. Relation between use of anti-inflammatory agents and left ventricular free wall rupture during acute myocardial infarction. Am J Cardiol 1987;59(4):363–4.
[224] Delyani JA, Robinson EL, Rudolph AE. Effect of a selective aldosterone receptor antagonist in myocardial infarction. Am J Physiol Heart Circ Physiol 2001;281(2):H647–54.
[225] Pitt B, et al. The effect of spironolactone on morbidity and mortality in patients with severe heart failure. Randomized Aldactone Evaluation Study Investigators. N Engl J Med 1999;341(10):709–17.
[226] Pitt B, et al. Eplerenone, a selective aldosterone blocker, in patients with left ventricular dysfunction after myocardial infarction. N Engl J Med 2003;348(14):1309–21.
[227] Sun Y, et al. Aldosterone-induced inflammation in the rat heart: role of oxidative stress. Am J Pathol 2002;161(5):1773–81.
[228] Fraccarollo D, et al. Immediate mineralocorticoid receptor blockade improves myocardial infarct healing by modulation of the inflammatory response. Hypertension 2008;51(4):905–14.
[229] Gheorghiade M, Goldstein S. Beta-blockers in the post-myocardial infarction patient. Circulation 2002;106(4):394–8.
[230] Vantrimpont P, et al. Additive beneficial effects of beta-blockers to angiotensin-converting enzyme inhibitors in the Survival and Ventricular Enlargement (SAVE) Study. SAVE Investigators. J Am Coll Cardiol 1997;29(2):229–36.
[231] Califf RM, et al. Usefulness of beta blockers in high-risk patients after myocardial infarction in conjunction with captopril and/or valsartan (from the VALsartan In Acute Myocardial Infarction [VALIANT] trial). Am J Cardiol 2009;104(2):151–7.
[232] Kuroki K, et al. beta2-adrenergic receptor stimulation-induced immunosuppressive effects possibly through down-regulation of co-stimulatory molecules, ICAM-1, CD40 and CD14 on monocytes. J Int Med Res 2004;32(5):465–83.
[233] Wolf SC, et al. Influence of nebivolol and metoprolol on inflammatory mediators in human coronary endothelial or smooth muscle cells. Effects on neointima formation after balloon denudation in carotid arteries of rats treated with nebivolol. Cell Physiol Biochem 2007;19(1–4):129–36.
[234] Calo LA, Semplicini A, Davis PA. Antioxidant and antiinflammatory effect of carvedilol in mononuclear cells of hypertensive patients. Am J Med 2005;118(2):201–2.
[235] Kurum T, Tatli E, Yuksel M. Effects of carvedilol on plasma levels of pro-inflammatory cytokines in patients with ischemic and nonischemic dilated cardiomyopathy. Tex Heart Inst J 2007;34(1):52–9.
[236] Quinaglia e Silva JC, et al. Effect of beta blockers (metoprolol or propranolol) on effect of simvastatin in lowering C-reactive protein in acute myocardial infarction. Am J Cardiol 2009;103(4):461–3.
[237] Hennekens CH, Dyken ML, Fuster V. Aspirin as a therapeutic agent in cardiovascular disease: a statement for healthcare professionals from the American Heart Association. Circulation 1997;96(8):2751–3.
[238] Hebert PR, Pfeffer MA, Hennekens CH. Use of statins and aspirin to reduce risks of cardiovascular disease. J Cardiovasc Pharmacol Ther 2002;7(2):77–80.
[239] Yin MJ, Yamamoto Y, Gaynor RB. The anti-inflammatory agents aspirin and salicylate inhibit the activity of I(kappa)B kinase-beta. Nature 1998;396(6706):77–80.
[240] Solheim S, et al. Influence of aspirin on inflammatory markers in patients after acute myocardial infarction. Am J Cardiol 2003;92(7):843–5.
[241] Adamek A, et al. High dose aspirin and left ventricular remodeling after myocardial infarction: aspirin and myocardial infarction. Basic Res Cardiol 2007;102(4):334–40.
[242] Epistent Investigators. Randomised placebo-controlled and balloon-angioplasty-controlled trial to assess safety of coronary stenting with use of platelet glycoprotein-IIb/IIIa blockade. Lancet 1998;352(9122):87–92.
[243] Lindemann S, et al. Activated platelets mediate inflammatory signaling by regulated interleukin 1beta synthesis. J Cell Biol 2001;154(3):485–90.
[244] Clemetson KJ, et al. Functional expression of CCR1, CCR3, CCR4, and CXCR4 chemokine receptors on human platelets. Blood 2000;96(13):4046–54.
[245] Neumann FJ, et al. Effect of glycoprotein IIb/IIIa receptor blockade on platelet-leukocyte interaction and surface expression of the leukocyte integrin Mac-1 in acute myocardial infarction. J Am Coll Cardiol 1999;34(5):1420–6.
[246] Lincoff AM, et al. Abciximab suppresses the rise in levels of circulating inflammatory markers after percutaneous coronary revascularization. Circulation 2001;104(2):163–7.
[247] Hu H, Zhang W, Li N. Glycoprotein IIb/IIIa inhibition attenuates platelet-activating factor-induced platelet activation by reducing protein kinase C activity. J Thromb Haemost 2003;1(8):1805–12.
[248] Liverani E, et al. LPS-induced systemic inflammation is more severe in P2Y12 null mice. J Leukoc Biol 2014;95(2):313–23.
[249] Jia LX, et al. Inhibition of platelet activation by clopidogrel prevents hypertension-induced cardiac inflammation and fibrosis. Cardiovasc Drugs Ther 2013;27(6):521–30.
[250] Xiao Z, Theroux P. Clopidogrel inhibits platelet-leukocyte interactions and thrombin receptor agonist peptide-induced platelet activation in patients with an acute coronary syndrome. J Am Coll Cardiol 2004;43(11):1982–8.

[251] Van Tassell BW, et al. Interleukin-1 trap attenuates cardiac remodeling after experimental acute myocardial infarction in mice. J Cardiovasc Pharmacol 2010;55(2):117–22.
[252] Toldo S, et al. Interleukin-1beta blockade improves cardiac remodelling after myocardial infarction without interrupting the inflammasome in the mouse. Exp Physiol 2013;98(3):734–45.
[253] Abbate A, et al. Interleukin-1 blockade with anakinra to prevent adverse cardiac remodeling after acute myocardial infarction (Virginia Commonwealth University Anakinra Remodeling Trial [VCU-ART] Pilot study). Am J Cardiol 2010;105(10): 1371-1377.e1.
[254] Zhao X, et al. Endothelial cells overexpressing IL-8 receptor reduce cardiac remodeling and dysfunction following myocardial infarction. Am J Physiol Heart Circ Physiol 2013;305(4):H590–8.
[255] Velasquez IM, et al. Association of interleukin 8 with myocardial infarction: results from the Stockholm Heart Epidemiology Program. Int J Cardiol 2014;172(1):173–8.
[256] Krishnamurthy P, et al. IL-10 inhibits inflammation and attenuates left ventricular remodeling after myocardial infarction via activation of STAT3 and suppression of HuR. Circ Res 2009;104(2):e9–e18.
[257] Stumpf C, et al. Interleukin-10 improves left ventricular function in rats with heart failure subsequent to myocardial infarction. Eur J Heart Fail 2008;10(8):733–9.
[258] Seki K, et al. Interleukin-33 prevents apoptosis and improves survival after experimental myocardial infarction through ST2 signaling. Circ Heart Fail 2009;2(6):684–91.
[259] Yin H, et al. IL-33 attenuates cardiac remodeling following myocardial infarction via inhibition of the p38 MAPK and NF-kappaB pathways. Mol Med Rep 2014;9(5):1834–8.
[260] Gurevitch J, et al. Anti-tumor necrosis factor-alpha improves myocardial recovery after ischemia and reperfusion. J Am Coll Cardiol 1997;30(6):1554–61.
[261] Sun M, et al. Excessive tumor necrosis factor activation after infarction contributes to susceptibility of myocardial rupture and left ventricular dysfunction. Circulation 2004;110(20):3221–8.
[262] Ramani R, et al. Inhibition of tumor necrosis factor receptor-1-mediated pathways has beneficial effects in a murine model of postischemic remodeling. Am J Physiol Heart Circ Physiol 2004;287(3):H1369–77.
[263] Zhang Y, et al. Tumor necrosis factor-alpha and lymphotoxin-alpha mediate myocardial ischemic injury via TNF receptor 1, but are cardioprotective when activating TNF receptor 2. PLoS One 2013;8(5):e60227.
[264] Sugano M, et al. In vivo transfer of soluble TNF-alpha receptor 1 gene improves cardiac function and reduces infarct size after myocardial infarction in rats. FASEB J 2004;18(7):911–3.
[265] Padfield GJ, et al. Cardiovascular effects of tumour necrosis factor alpha antagonism in patients with acute myocardial infarction: a first in human study. Heart 2013;99(18):1330–5.
[266] Gurantz D, et al. Etanercept or intravenous immunoglobulin attenuates expression of genes involved in post-myocardial infarction remodeling. Cardiovasc Res 2005;67(1):106–15.
[267] http://clinicaltrials.gov/show/NCT01372930.
[268] Fraccarollo D, Galuppo P, Bauersachs J. Novel therapeutic approaches to post-infarction remodelling. Cardiovasc Res 2012;94(2):293–303.
[269] Zidar N, et al. MicroRNAs, innate immunity and ventricular rupture in human myocardial infarction. Dis Markers 2011;31(5):259–65.
[270] Li J, et al. Altered microRNA expression profile with miR-146a upregulation in CD4+ T cells from patients with rheumatoid arthritis. Arthritis Res Ther 2010;12(3):R81.
[271] Taganov KD, et al. NF-kappaB-dependent induction of microRNA miR-146, an inhibitor targeted to signaling proteins of innate immune responses. Proc Natl Acad Sci U S A 2006;103(33):12481–6.
[272] Tili E, et al. Modulation of miR-155 and miR-125b levels following lipopolysaccharide/TNF-alpha stimulation and their possible roles in regulating the response to endotoxin shock. J Immunol 2007;179(8):5082–9.
[273] Yao R, et al. The altered expression of inflammation-related microRNAs with microRNA-155 expression correlates with Th17 differentiation in patients with acute coronary syndrome. Cell Mol Immunol 2011;8(6):486–95.
[274] Heymans S, et al. Macrophage microRNA-155 promotes cardiac hypertrophy and failure. Circulation 2013;128(13):1420–32.
[275] Kaisho T, Akira S. Toll-like receptor function and signaling. J Allergy Clin Immunol 2006;117(5):979–87, quiz 988.
[276] Feng Y, Chao W. Toll-like receptors and myocardial inflammation. Int J Inflam 2011;2011:170352.
[277] Sakata Y, et al. Toll-like receptor 2 modulates left ventricular function following ischemia-reperfusion injury. Am J Physiol Heart Circ Physiol 2007;292(1):H503–9.
[278] Shishido T, et al. Toll-like receptor-2 modulates ventricular remodeling after myocardial infarction. Circulation 2003;108(23):2905–10.
[279] Lu C, et al. Toll-like receptor 3 plays a role in myocardial infarction and ischemia/reperfusion injury. Biochim Biophys Acta 2014;1842(1):22–31.
[280] Timmers L, et al. Toll-like receptor 4 mediates maladaptive left ventricular remodeling and impairs cardiac function after myocardial infarction. Circ Res 2008;102(2):257–64.
[281] Chinenov Y, Gupte R, Rogatsky I. Nuclear receptors in inflammation control: repression by GR and beyond. Mol Cell Endocrinol 2013;380(1–2):55–64.
[282] Brown M, et al. Cardiac-specific blockade of NF-kappaB in cardiac pathophysiology: differences between acute and chronic stimuli in vivo. Am J Physiol Heart Circ Physiol 2005;289(1):H466–76.
[283] Lawrence T, et al. Possible new role for NF-kappaB in the resolution of inflammation. Nat Med 2001;7(12):1291–7.
[284] Jolly SR, et al. Canine myocardial reperfusion injury. Its reduction by the combined administration of superoxide dismutase and catalase. Circ Res 1984;54(3):277–85.
[285] Gallagher KP, et al. Failure of superoxide dismutase and catalase to alter size of infarction in conscious dogs after 3 hours of occlusion followed by reperfusion. Circulation 1986;73(5):1065–76.
[286] Flaherty JT, et al. Recombinant human superoxide dismutase (h-SOD) fails to improve recovery of ventricular function in patients undergoing coronary angioplasty for acute myocardial infarction. Circulation 1994;89(5):1982–91.
[287] Sia YT, et al. Improved post-myocardial infarction survival with probucol in rats: effects on left ventricular function, morphology, cardiac oxidative stress and cytokine expression. J Am Coll Cardiol 2002;39(1):148–56.

[288] Miida T, et al. Probucol markedly reduces HDL phospholipids and elevated prebeta1-HDL without delayed conversion into alpha-migrating HDL: putative role of angiopoietin-like protein 3 in probucol-induced HDL remodeling. Atherosclerosis 2008;200(2):329–35.

[289] Rodrigo R, et al. Molecular basis of cardioprotective effect of antioxidant vitamins in myocardial infarction. Biomed Res Int 2013;2013:437613.

[290] Ridker PM. Testing the inflammatory hypothesis of atherothrombosis: scientific rationale for the cardiovascular inflammation reduction trial (CIRT). J Thromb Haemost 2009;7(Suppl 1):332–9.

[291] Gerards AH, et al. Inhibition of cytokine production by methotrexate. Studies in healthy volunteers and patients with rheumatoid arthritis. Rheumatology (Oxford) 2003;42(10):1189–96.

[292] Chan ES, Cronstein BN. Molecular action of methotrexate in inflammatory diseases. Arthritis Res 2002;4(4):266–73.

[293] Micha R, et al. Systematic review and meta-analysis of methotrexate use and risk of cardiovascular disease. Am J Cardiol 2011;108(9):1362–70.

[294] Asanuma H, et al. Methotrexate and MX-68, a new derivative of methotrexate, limit infarct size via adenosine-dependent mechanisms in canine hearts. J Cardiovasc Pharmacol 2004;43(4):574–9.

[295] Everett BM, et al. Rationale and design of the cardiovascular inflammation reduction trial: a test of the inflammatory hypothesis of atherothrombosis. Am Heart J 2013;166(2): 199-207.e15.

[296] Moreira DM, et al. Rationale and design of the TETHYS trial: the effects of methotrexate therapy on myocardial infarction with ST-segment elevation. Cardiology 2013;126(3):167–70.

[297] Maiese K, Li F, Chong ZZ. New avenues of exploration for erythropoietin. JAMA 2005;293(1):90–5.

[298] van der Meer P, et al. Erythropoietin induces neovascularization and improves cardiac function in rats with heart failure after myocardial infarction. J Am Coll Cardiol 2005;46(1):125–33.

[299] Parsa CJ, et al. Cardioprotective effects of erythropoietin in the reperfused ischemic heart: a potential role for cardiac fibroblasts. J Biol Chem 2004;279(20):20655–62.

[300] Liu X, et al. Mechanism of the cardioprotection of rhEPO pretreatment on suppressing the inflammatory response in ischemia-reperfusion. Life Sci 2006;78(19):2255–64.

[301] Li Y, et al. Reduction of inflammatory cytokine expression and oxidative damage by erythropoietin in chronic heart failure. Cardiovasc Res 2006;71(4):684–94.

[302] Daskalopoulos EP, et al. Targeting the Wnt/frizzled signaling pathway after myocardial infarction: a new tool in the therapeutic toolbox? Trends Cardiovasc Med 2013;23:121–7.

[303] Masckauchan TN, et al. Wnt/beta-catenin signaling induces proliferation, survival and interleukin-8 in human endothelial cells. Angiogenesis 2005;8(1):43–51.

[304] Pereira C, et al. Wnt5A/CaMKII signaling contributes to the inflammatory response of macrophages and is a target for the antiinflammatory action of activated protein C and interleukin-10. Arterioscler Thromb Vasc Biol 2008;28(3):504–10.

[305] Blumenthal A, et al. The Wingless homolog WNT5A and its receptor Frizzled-5 regulate inflammatory responses of human mononuclear cells induced by microbial stimulation. Blood 2006;108(3):965–73.

[306] Schaale K, et al. Wnt signaling in macrophages: augmenting and inhibiting mycobacteria-induced inflammatory responses. Eur J Cell Biol 2011;90(6–7):553–9.

[307] Ueland T, et al. Dickkopf-1 enhances inflammatory interaction between platelets and endothelial cells and shows increased expression in atherosclerosis. Arterioscler Thromb Vasc Biol 2009;29(8):1228–34.

[308] Barandon L, et al. sFRP-1 improves postinfarction scar formation through a modulation of inflammatory response. Arterioscler Thromb Vasc Biol 2011;31(11):e80–7.

[309] Klune JR, et al. HMGB1: endogenous danger signaling. Mol Med 2008;14(7–8):476–84.

[310] Lotze MT, Tracey KJ. High-mobility group box 1 protein (HMGB1): nuclear weapon in the immune arsenal. Nat Rev Immunol 2005;5(4):331–42.

[311] Popovic PJ, et al. High mobility group B1 protein suppresses the human plasmacytoid dendritic cell response to TLR9 agonists. J Immunol 2006;177(12):8701–7.

[312] Erlandsson Harris H, Andersson U. Mini-review: the nuclear protein HMGB1 as a proinflammatory mediator. Eur J Immunol 2004;34(6):1503–12.

[313] Yao HC, et al. Correlation between serum high-mobility group box-1 levels and high-sensitivity C-reactive protein and troponin I in patients with coronary artery disease. Exp Ther Med 2013;6(1):121–4.

[314] Xu H, et al. Endogenous HMGB1 contributes to ischemia-reperfusion-induced myocardial apoptosis by potentiating the effect of TNF-α/JNK. Am J Physiol Heart Circ Physiol 2011;300(3):H913–21.

[315] Takahashi K, et al. Modulated inflammation by injection of high-mobility group box 1 recovers post-infarction chronically failing heart. Circulation 2008;118(14 Suppl):S106–14.

[316] Kohno T, et al. Role of high-mobility group box 1 protein in post-infarction healing process and left ventricular remodelling. Cardiovasc Res 2009;81(3):565–73.

[317] Delerive P, Fruchart JC, Staels B. Peroxisome proliferator-activated receptors in inflammation control. J Endocrinol 2001;169(3):453–9.

[318] Ohshima K, Mogi M, Horiuchi M. Role of peroxisome proliferator-activated receptor-gamma in vascular inflammation. Int J Vasc Med 2012;2012:508416.

[319] Wayman NS, et al. Ligands of the peroxisome proliferator-activated receptors (PPAR-gamma and PPAR-alpha) reduce myocardial infarct size. FASEB J 2002;16(9):1027–40.

[320] Yue TL, et al. Activation of peroxisome proliferator-activated receptor-alpha protects the heart from ischemia/reperfusion injury. Circulation 2003;108(19):2393–9.

[321] Yue TL, et al. In vivo activation of peroxisome proliferator-activated receptor-delta protects the heart from ischemia/reperfusion injury in Zucker fatty rats. J Pharmacol Exp Ther 2008;325(2):466–74.

CHAPTER

4

Cross Talk Between Inflammation and Extracellular Matrix Following Myocardial Infarction

Yonggang Ma[1,2], *Rugmani Padmanabhan Iyer*[1,2], *Lisandra E. de Castro Brás*[1,2], *Hiroe Toba*[1,2,3], *Andriy Yabluchanskiy*[1,2], *Kristine Y. Deleon-Pennell*[1,2], *Michael E. Hall*[1,2,4], *Richard A. Lange*[1,5], *Merry L. Lindsey*[1,2,6]

[1]San Antonio Cardiovascular Proteomics Center, Jackson, MS, USA
[2]Mississippi Center for Heart Research, Department of Physiology and Biophysics, University of Mississippi Medical Center, Jackson, MS, USA
[3]Department of Clinical Pharmacology, Division of Pathological Sciences, Kyoto Pharmaceutical University, Kyoto, Japan
[4]Cardiology Division, University of Mississippi Medical Center, Jackson, MS, USA
[5]Paul L. Foster School of Medicine, Texas Tech University Health Sciences Center El Paso, El Paso, TX, USA
[6]Research Services, G.V. (Sonny) Montgomery Veterans Affairs Medical Center, Jackson, MS, USA

4.1 INTRODUCTION

Cardiovascular disease is a leading cause of mortality, with myocardial infarction (MI) significantly contributing to the majority of morbidity and mortality [1]. Immediate survival in post-MI patients has greatly improved over the past three decades, mainly due to coronary reperfusion therapy and other therapeutic strategies. However, the incidence of chronic heart failure post-MI has substantially increased as a direct consequence.

Post-MI, the left ventricle (LV) undergoes a series of remodeling events, with inflammation initiating healing and scar formation [2]. While activation of the inflammatory response is important for wound healing, it is also harmful in generating additional injury that further promotes LV dilation. Neutrophils, macrophages, and fibroblasts are key cells involved in extracellular matrix (ECM) remodeling.

In the absence of reperfusion, neutrophils are the initial leukocytes that respond to the ischemic stimulus and provide chemotactic signals for subsequent macrophage recruitment [3]. Migrated macrophages engulf dead tissue and apoptotic neutrophils and generate matrix metalloproteinases (MMPs) and inflammatory mediators, such as cytokines, chemokines, and growth factors [4]. MMPs are responsible for ECM breakdown and coordinate many aspects of the inflammatory response through direct and indirect effects. Chemokines signal additional leukocyte migration to the site of injury by altering cell adhesiveness and regulating directional movement [5]. In response to inflammatory mediators, the fibroblast population within the injured myocardium differentiates into myofibroblasts and synthesizes a multitude of ECM proteins, including collagens. ECM breakdown by MMPs and ECM synthesis by fibroblasts set up a yin-yang relationship for scar formation.

Inflammation in Heart Failure
http://dx.doi.org/10.1016/B978-0-12-800039-7.00004-9

© 2015 Elsevier Inc. All rights reserved.

An appropriate extent of inflammation and ECM accumulation is important for post-MI infarct healing. An imbalance between ECM deposition and degradation can result in adverse LV remodeling. In this chapter, we summarize the current literature on the roles of leukocytes and fibroblasts to oversee the inflammatory and ECM responses in the infarcted LV and discuss research issues to be addressed in future studies.

4.2 ROLES OF INFLAMMATION IN THE MI SETTING

After MI, neutrophils and macrophages are the predominant leukocytes recruited to the site of ischemia. They play key roles in regulating the inflammatory, angiogenic, and fibrotic responses by secreting a multitude of inflammatory mediators and proteinases.

4.2.1 Neutrophil Degranulation

We have previously reviewed in detail the roles of neutrophils in LV remodeling post-MI [3]. Briefly, neutrophils are recruited into the infarcted tissue within minutes (for ischemia/reperfused myocardium) to hours (for myocardium supplied by a persistently occluded coronary artery) after MI, and infiltration peaks at days 1-3 post-MI [3]. Neutrophil depletion results in reduced infarct size and a lesser extent of ischemic myocardial injury, indicating that neutrophil infiltration can extend injury past the initial ischemia [6,7]. When degranulated, the neutrophil releases multiple proteases (e.g., serine elastase and MMPs) necessary for degrading the ECM surrounding the necrotic cardiomyocytes [3]. Neutrophil elastase hydrolyzes several ECM proteins (e.g., elastin, fibronectin, and collagen types III, IV, and VIII) and plasma proteins (e.g., complement and clotting factors), providing a mechanism of communication between leukocytes and ECM components [3,8]. Neutrophils are a rich source of MMP-9 early post-MI, which degrades ECM to provide a favorable environment for additional leukocyte (e.g., neutrophil and macrophage) infiltration [9]. Both neutrophil-derived elastase and MMP-8 have been shown to cleave elastin and fibronectin, respectively, to generate peptide fragments that are chemotactic for monocytes [10,11].

4.2.2 Macrophage Activation

Adherence of monocytes to infarcted tissue initiates their conversion to macrophages by inducing expression of factors such as macrophage colony stimulating factor and tumor necrosis factor (TNF)-α [4]. Following the influx of neutrophils, macrophages dominate the cellular infiltrate for the first 2 weeks after MI, thus exerting key roles in cardiac repair [12]. The exact roles of macrophages in the healing myocardium have not been fully elucidated, in part, because the heterogeneity of the macrophage in this setting has only recently been appreciated. In permanent coronary occlusion animal models of MI, macrophage numbers have been negatively and positively associated with LV remodeling, depending on the timing of the evaluation [13–16]. Specifically, inhibiting macrophage recruitment by anti-monocyte chemoattractant protein-1 gene therapy attenuates LV remodeling and dysfunction post-MI [13]. To the contrary, activated macrophages also improve LV healing and function post-MI [15]. Recent identification of the existence of macrophage population subsets, mainly pro- (M1 macrophages) and anti-inflammatory (M2 macrophages) phenotypes, has opened this research arena to new concepts.

In general, M1 macrophages promote inflammation and ECM destruction, while M2 macrophages facilitate angiogenesis, cell proliferation, and ECM reconstruction [4]. The M1 phenotype is also known as the classically activated macrophage, and these cells are characterized by upregulated secretion of proinflammatory mediators such as interleukin (IL)-1β, IL-6, and TNF-α. M1 macrophages are the source of several MMPs, including MMP-1, -3, -7, -10, -12, -14, and -25 [17]. In contrast, activation of the M2 phenotype is characterized by increased secretion of anti-inflammatory mediators such as IL-10, arginase, and transforming growth factor (TGF)-β1. In addition, M2 macrophages produce lower levels of MMP-2, -8, and -19 but increased levels of MMP-11, -12, and -25, compared to M1 macrophages [17]. Therefore, the timing of when the inflammatory response is studied post-MI is crucial, since macrophages will have different functions and express different molecules at distinct times along the wound-healing continuum.

4.3 CYTOKINE AND CHEMOKINE ROLES IN LV REMODELING

4.3.1 Cytokines Regulate Fibroblast Phenotype and Function

Acute MI initiates a cytokine and chemokine cascade in both the infarcted and noninfarcted myocardium. Early post-MI, cytokines such as IL-1β, IL-6, and TNF-α, which are present at very low levels in the normal heart, are substantially upregulated [18,19]. This robust upregulation may be short term if the infarcted area is small or reperfused [18]. However, in the case of a large infarcted region, or with permanent coronary occlusion, this upregulation may be sustained, corresponding to the chronic remodeling phase. IL-1β and TNF-α affect LV remodeling by activating fibroblasts and inducing their migration through MAP kinase pathways [20]. Both cytokines also activate endothelial cells, which facilitate the recruitment and migration of leukocytes into the injured myocardium, forming a positive feedback mechanism [21]. In a permanent coronary occlusion animal model of MI, IL-1β and IL-6 robustly increased in the infarcted area and in the remote region starting at 3h post-MI [18]. These changes preceded the increase in MMP-9 levels in the infarct area and collagen expression in the noninfarcted myocardium, suggesting that IL-1β and IL-6 act as upstream molecules to promote degradation of the necrotic tissue, ECM remodeling, and early induction of fibrosis after MI.

TGF-β1 is a 25-kDa cytokine expressed by neutrophils, macrophages, and fibroblasts and participates in several processes during LV remodeling post-MI. Latent TGF-β1 is activated by a number of molecules including integrins, MMP-2, MMP-9, plasmin, reactive oxygen species, and thrombospondin (TSP)-1 [22]. The common denominator in the variety of components that can activate latent TGF-β1 is the fact that they all indicate changes in ECM homeostasis.

Post-MI, TGF-β1 is upregulated early. TGF-β1 is a dual regulator of the inflammatory response. In the early stage of MI, TGF-β1 elicits a direct chemotactic response to neutrophils and monocytes [23,24]. Conversely, TGF-β1 deactivates inflammatory macrophages at the late stage of MI, facilitating repair response [22]. TGF-β1 influences LV fibrosis by stimulating fibroblast differentiation to myofibroblasts, and enhancing ECM protein synthesis [25]. TGF-β1 also inhibits MMP expression and induces the expression of tissue inhibitor of metalloproteinases [26]. Therefore, TGF-β1 has been considered as a master switch controlling inflammation and ECM deposition and as a target for therapies to prevent adverse LV remodeling [22].

4.3.2 Chemokines Regulate LV Remodeling

The chemokines CCL2/monocyte chemoattractant protein-1, CXCL8/IL-8, and CXCL10/interferon-γ-inducible protein-10 are consistently upregulated in experimental models of MI [27]. CCL2 is produced by a variety of cell types; however, macrophages and endothelial cells are the major contributors [28]. CCL2, a potent monocyte chemoattractant, recruits monocytes to the site of injury and forms a positive feedback loop for additional cell infiltration, which is a critical event for acute inflammation initiation. Additionally, CCL2 has been associated with modulation of cardiac fibroblast phenotype and activation by inducing collagen, TGF-β1, and MMPs [29].

In both experimental animals and in human MI studies, serum CCL2, CCL3, and CCL5 levels increase and are higher in patients who progress to heart failure [30]. CC chemokine receptor 5 (CCR5) is the natural receptor for CCL2, CCL3, and CCL4, making it important for macrophage recruitment into the infarcted myocardium. In a murine MI model, CCR5 deletion impaired LV remodeling by inhibiting macrophage activation [31]. Macrophages isolated from infarct LV in CCR5 null mice showed a >50% reduction in gene expression levels of the proinflammatory cytokines IL-1β, IL-6, and TNF-α, highlighting the pathogenic effects of CCR5 in LV remodeling.

The balance in post-MI inflammatory and fibrotic responses is crucial for appropriate scar formation. On one hand, cytokines and chemokines attract leukocytes that remove the necrotic tissue, induce angiogenesis, and trigger new ECM synthesis by myofibroblasts. Conversely, these cells can secrete MMPs and reactive oxygen species that compromise tissue integrity. The dynamic balance between the two constitutes the remodeling process.

4.4 MMP ROLES IN THE INFARCTED MYOCARDIUM

4.4.1 MMPs

Inflammatory cells secrete a group of MMPs responsible for degradation of the ECM surrounding necrotic cardiomyocytes. For the most part, MMPs consist of four domains: a pro domain, a catalytic domain, a hinge region, and a hemopexin domain [32]. MMPs are secreted in the zymogen form and are activated by a cysteine switch mechanism involving zinc dissociation to expose the MMP active site [33]. The exception to this mechanism is MMP-23, which lacks the cysteine switch motif. MMPs have been classified into five groups based roughly on localization of where the MMP was first identified and initial *in vitro* substrate specificity profiles. The groups include collagenases, gelatinases, stromelysins, matrilysins, and membrane-type (MT)-MMPs. Some MMPs have not been assigned to the above groups and are cataloged as other MMPs. However, with the growing knowledge about MMPs, this cataloging is no longer useful. For example, MMP-14 is a collagenase but also is a MT-MMP. In the following section, we will focus only on those MMPs that have been evaluated in the post-MI setting (Table 4.1).

4.4.2 MMP-1

MMP-1, also known as collagenase-1, was the first MMP identified by Gross and Lapiere in 1962 [34]. Humans express MMP-1 while rodents have two MMP-1 isoforms—namely, MMP-1a and -1b. MMP-1 cleaves both ECM and non-ECM substrates such as collagen, gelatin, laminin, complement C1q, IL-1β, and TNF-α, suggesting a crucial role in inflammatory and fibrotic responses [32]. MMP-1 can also activate MMP-2 and -9, initiating an activation cascade [32]. MMP-1 activity increases 4.5-fold at day 2 following MI and peaks at day 7 [35]. Plasma MMP-1 concentration positively correlates with LV dilation and dysfunction, suggesting that it may be due, at least in part, to increased collagen degradation by MMP-1 [36].

4.4.3 MMP-2

MMP-2, also known as gelatinase A, is secreted by cardiomyocytes, fibroblasts, and myofibroblasts. MMP-2 has a wide range of substrates, which include collagen, elastin, endothelin, fibroblast growth factor, MMP-9, MMP-13, plasminogen, and TGF-β, indicating comprehensive roles of MMP-2 [32]. MMP-2 activity increases at day 4 post-MI and reaches a maximum by day 7 [37]. In animal models of acute MI, MMP-2 deficiency attenuates LV rupture and improves survival, which is strongly related to reduced macrophage recruitment and ECM degradation [38]. MMP-2 deficiency results in decreased ECM degradation, thus limiting macrophage migration into the necrotic tissue and LV rupture.

4.4.4 MMP-3

MMP-3 was first named stromelysin-1 and is secreted by cardiomyocytes, fibroblasts, and macrophages [39]. MMP-3 actively interacts with other MMPs such as MMP-1, -3, -7, -8, -9, and -13 [40]. Post-MI, MMP-3 expression is markedly increased. In patients with acute MI, plasma MMP-3 concentration correlates with patient age and gender (e.g., higher levels in males), creatinine concentration, and hypertension, but negatively associates with LV function [41]. As such, MMP-3 may serve as a novel predictor for adverse LV remodeling after MI.

TABLE 4.1 MMPs in the MI Setting

Name	Post-MI Expression	Post-MI Roles
MMP		
1	↑	Predicts LV dysfunction and mortality in MI patients
2	↑	↓ Survival, ↑ LV rupture, ↑ macrophage infiltration, ↑ ECM degradation
3	↑	Correlates with LV dysfunction and mortality in MI patients, activates MMP-1, -3, -7, -8, -9, -13
7	↑	↓ Survival, ↓ conduction velocity by connexin-43 cleavage, activates MMP-1, -2, -9
8	↑	Degrades collagen, ↑ neutrophil infiltration, ↑ LV rupture
9	↑	↑ LV dilation, ↓/↑ LV function, ↑/↓ inflammation, ↑ collagen deposition
13	↑	Activates MMP-9
14	↑	↑ Fibrosis, ↓ survival, ↓ LV function, activates MMP-2 and -13
28	↓ in cardiomyocyte ↑ in macrophage	↓ Rupture, ↓ mortality, ↑ LV function, ↑ M2 macrophage differentiation, ↑ collagen deposition and cross-linking

MMPs, matrix metalloproteinases; MI, myocardial infarction; LV, left ventricle; ECM, extracellular matrix

4.4.5 MMP-7

MMP-7, the smallest MMP member, is expressed in cardiomyocytes and macrophages. The ECM substrates for MMP-7 include collagen type IV, fibronectin, laminin, and non-ECM substrates, such as MMP-1, -2, and -9 and TNF-α [42]. Cardiac MMP-7 expression increases in the infarct and border region during the first week post-MI and returns to basal levels by 8 weeks [37,42]. Our laboratory demonstrated improved survival rate and conduction velocity in MMP-7 null mice subjected to MI through interaction with connexin-43, fibronectin, and tenascin-C [42].

4.4.6 MMP-8

MMP-8, also known as neutrophil collagenase, is expressed not only by neutrophils, as it was suggested before, but also by macrophages. Substrates for MMP-8 include both ECM (e.g., aggrecan and collagen) and non-ECM (e.g., angiotensin and plasminogen) proteins [32]. At 1 day post-MI, infiltrated neutrophils release abundant MMP-8, which, in turn, increases additional neutrophil infiltration by cleaving collagen, suggesting a positive feedback loop [43]. MMP-8 expression remains elevated in the infarcted regions through 8 weeks post-MI, indicating that other cell types, including fibroblasts and endothelial cells, may be rich sources of MMP-8 [37].

4.4.7 MMP-9

MMP-9, also known as gelatinase B, is secreted by a wide range of cells, such as the cardiomyocyte, fibroblast, neutrophil, macrophage, vascular smooth muscle cell, and endothelial cell [39,44]. Early increase in MMP-9 after MI is derived from infiltrated neutrophils, while late increase is mainly due to recruited macrophages [3]. Known substrates for MMP-9 activity include aggrecan, collagen, gelatin, laminin, and a variety of non-ECM substrates such as angiotensin II, casein, plasminogen, and TGF-β1 [32,39]. Plasma MMP-9 concentration correlates with the development of LV dysfunction and survival post-MI, and is thus identified as a novel predictor of mortality in MI patients [45]. MMP-9 null mice have smaller LV dimensions, reduced collagen deposition, decreased macrophage numbers, and increased expression of MMP-2, MMP-13, and tissue inhibitor of metalloproteinases-1 following MI in comparison to wild-type mice [46]. Interestingly, MMP-9 overexpression in macrophages also attenuates the inflammatory response and improves LV function in animal models of acute MI [47]. This indicates complicated roles of MMP-9 in LV remodeling, depending on the cellular source and temporal and spatial expression.

4.4.8 MMP-13

MMP-13, or collagenase 3, degrades casein, collagen, fibrinogen, and gelatin. MMP-13 is expressed in cardiac fibroblasts and macrophages [32,39]. MMP-13 tissue levels were elevated by 1 week post-MI and remained elevated up to 8 weeks in the infarct region in rats [39]. MMP-13 is an activator of MMP-9. As a compensatory mechanism, MMP-9 deficiency induces synthesis of MMP-13 in mice, which, in turn, tries to activate MMP-9. This may explain why MMP-9 null mice show higher expression of MMP-13 post-MI [46].

4.4.9 MMP-14

Unlike classic MMPs, MMP-14 belongs to a MT-MMP, also named MT1-MMP. In addition to cleaving the ECM (e.g., collagen, fibronectin, and gelatin), MMP-14 can activate MMP-2 and -13 [32]. MMP-14 level increases post-MI in cardiac fibroblasts and cardiomyocytes and positively correlates with cardiac fibrosis, LV dysfunction, and poor survival [39].

4.4.10 MMP-28

MMP-28, or epilysin, is released from cardiomyocytes under normal conditions. MMP-28 cleaves casein, neural cell adhesion molecule, and Nogo-A (a myelin component) [48]. Post-MI, MMP-28 tissue levels decrease, due to the loss of cardiomyocytes; however, macrophage-derived MMP-28 is upregulated [49]. In mice studies of acute MI, MMP-28 deletion is associated with a higher incidence of LV dysfunction, cardiac rupture and mortality, as well as impaired ECM deposition and M2 macrophage activation [49]. In supporting of the above findings, peritoneal macrophages isolated from MMP-28 null mice have impaired M2 macrophage polarization in response to IL-4 stimulation [49]. This challenges our perception that all MMPs adversely affect LV remodeling and that inhibiting their action is beneficial.

In summary, individual MMPs play different and even opposite roles in post-MI LV repair. Further, depending on timing and when different substrates are present, the same MMP potentially also has opposite roles. This may partly explain the failure of broad spectrum MMP inhibitors to improve outcomes in large-scale clinical trials. These studies suggest that selective MMP inhibitors are needed to specifically suppress individual MMP without impacting other members and that we need more information on optimal timing strategies.

4.5 ECM ROLES IN THE MI SETTING

The ECM is composed mainly of structural proteins and nonstructural matricellular proteins, as well as proteinase enzymes and their inhibitors. The cardiac ECM provides structural support by serving as a scaffold for cells and transduces mechanical, chemical, and biological signaling to regulate LV homeostasis and the response to MI.

4.5.1 Structural ECM

Collagens, fibronectin, and laminins are structural ECM proteins that maintain architecture and function in the normal heart. They also play a critical role in regulating the inflammatory response and LV repair after MI (Table 4.2) [50].

4.5.2 Collagens

Collagen is the major cardiac ECM protein, with collagen type I being the most abundant fibrillar collagen in the normal heart [51]. Macrophage and cardiac fibroblast production of collagen type I, III, IV, V, and VI increases following MI, and collagen type I and III are major components of the scar [52]. Several studies demonstrate that collagen type III is deposited in the earlier phases of LV remodeling, while collagen type I deposition predominates in the

TABLE 4.2 Roles of ECM Proteins Post-MI

Name	Post-MI Roles
Structural	
Collagen type I, III	Scar components
Collagen type IV	Basement membrane component, ↑ angiogenesis
Collagen type V	Regulate collagen type I fibril assembly
Collagen type VI	↓ LV function, ↑ cardiomyocyte apoptosis, ↑ collagen deposition
Fibronectin-EDA	↑ LV dilation, ↓ LV function, ↑ collagen synthesis, ↑ inflammation, ↑ MMP activity, ↑ myofibroblast accumulation
Laminin	Basement membrane component, correlates negatively with LV function
Matricellular	
CCN-1	↑ Apoptosis, ↓ inflammation, ↓ fibrosis
CCN-2/CTGF	↓ Ischemia/reperfusion infarct size
CCN-4	↑ Hypertrophy, ↓ cardiomyocyte apoptosis, ↑ fibroblast proliferation
Osteopontin	↓ Remodeling, ↓ LV dilation, ↑ collagen, ↑ angiogenesis
Periostin	↑ LV function, ↓ LV rupture, ↓ fibrosis, ↑ cardiac regeneration
SPARC	↑ LV function, ↓ LV rupture, ↓ mortality, ↑ macrophage infiltration, ↑ scar quality
Tenascin-C	↑ LV remodeling, ↓ LV function, ↑ fibrosis, ↑ fibroblast migration and differentiation
TSP-1	↓ LV remodeling, ↑ LV function, ↓ inflammation, ↓ angiogenesis

Note that each of these proteins also has additional possible roles when degraded by MMPs to form ECM-derived fragments. ECM, extracellular matrix; MI, myocardial infarction; MMPs, matrix metalloproteinases; LV, left ventricle; EDA, extra domain A; CTGF, connective tissue growth factor; SPARC, secreted protein acidic and rich in cysteine; TSP-1, thrombospondin-1.

intermediate and late remodeling phases [52,53]. The process of collagen turnover after MI may last months or even years before good scar quality is achieved [54].

Cardiomyocytes are surrounded by a basement membrane that is mainly composed of collagen type IV. Collagen type IV has a critical role in the regulation of angiogenesis. Collagen type IV induces the formation of neovessels, stabilizes neovascular outgrowth, and prevents vascular regression [55]. Post-MI, collagen type IV expression is elevated in the border zone at day 3, peaking at days 7-11 [56]. In the infarcted LV, collagen type IV expression only starts to increase at day 10. These data suggest that the absence of collagen type IV in the earlier stages of LV remodeling facilitate leukocyte and fibroblast migration, while at a later stage collagen type IV is necessary for scar vascularization.

Collagen type V is a low abundance but widely distributed protein. Collagen type V is necessary for collagen type I fibril assembly and can regulate fibril size and organization, thus possibly playing a critical role during scar formation [57]. Collagen type V knockout mice are embryonic lethal due to cardiovascular failure [58]. Although the specific function of collagen type V in the heart is not yet defined, it is secreted in the myocardium by cardiac fibroblasts and vascular smooth-muscle cells suggestive of an important role during fibrosis and angiogenesis.

Collagen type VI is a nonfibrillar collagen and forms a microfilament network that organizes the fibrillar collagen type I and III, anchoring them to the basement membrane [59]. *In vitro*, collagen type VI can induce myofibroblast differentiation, indicating possible roles in post-MI LV repair [60]. Interestingly, collagen type VI deletion has been shown to attenuate LV remodeling and function post-MI by inhibiting apoptosis and collagen deposition [61].

4.5.3 Fibronectin

Fibronectin regulates cell morphology and movement by adhering to other ECM proteins. In response to MI, fibronectin is generated by endothelial cells, fibroblasts, and macrophages. Fibronectin contains an alternatively spliced exon encoding type III repeat extra domain A (EDA). EDA acts as an endogenous ligand for toll-like receptor (TLR)-2 and -4, resulting in induction of proinflammatory gene expression and monocyte activation [62]. EDA injection into murine joints induces an inflammatory response via the activation of nuclear factor-κB [63]. In addition to its proinflammatory properties, EDA can bind to several integrins and regulate cell adhesion and proliferation [64,65]. In EDA null mice, less LV dilation and enhanced LV systolic performance are observed after MI compared with wild type mice. EDA null mice also exhibit reduced inflammation, MMP- 2 and -9 activities, and myofibroblast accumulation [66]. Proteomics studies from our laboratory demonstrated that fibronectin is an *in vivo* substrate for both MMP-7 and -9, which may partially explain the beneficial effects observed in mice deficient in MMP-7 or -9 [67,68].

4.5.4 Laminins

Laminins, the first ECM glycoproteins detectable in the embryo, are found in basement membrane. Laminins consist of three peptide chains: α, β, and γ. Laminin protein is detected in the infarct area at day 3 post-MI, peaks in concentration at days 7-11, and then returns to baseline levels [69]. The wide existence of laminins throughout the infarct area suggests that they may directly regulate LV repair post-MI [69]. In patients with acute MI, serum laminin level is higher than in patients with stable coronary artery disease and those without coronary artery disease [70]. This report suggests the possibility that serum laminin could be a potential prognostic marker for MI patients.

4.5.5 Matricellular Proteins

Although matricellular proteins do not serve a direct structural role in the ECM, they are capable of interacting with cell surface receptors, proteinases, growth factors, and other ECM proteins. Matricellular proteins are expressed at very low concentrations in the LV at baseline but increase substantially following MI, implying their involvement in LV remodeling and repair [50]. Matricellular proteins include CCN family, osteopontin (OPN), periostin, secreted protein acidic and rich in cysteine (SPARC), tenascins, and TSPs [50]. Their roles in LV repair post-MI are summarized in Table 4.2.

4.5.6 CCNs

The CCN family consists of cysteine-rich protein 61 (CCN1), connective tissue growth factor (CTGF, CCN2), nephroblastoma overexpressed protein (CCN3), and Wnt-inducible secreted proteins CCN4, CCN5, and CCN6. CCN proteins modulate cell adhesion, migration, and survival. CCN1, CCN2, and CCN3 promote migration of inflammatory cells, endothelial cells, and fibroblasts, while CCN4 and CCN5 inhibit their migration [71,72]. CCN1 induces proinflammatory polarization of macrophages, and promotes the expression of genes responsible for adhesion, angiogenesis, and ECM turnover [73]. CCN1 is highly induced in end-stage ischemic cardiomyopathy and regulates apoptosis, angiogenesis, inflammation, and the fibrotic response [74,75]. In an ischemia reperfusion model, CCN2 overexpression in cardiomyocytes reduces infarct size by regulating the Akt/p70S6 kinase/GSK-3β kinase pathway [76]. *In vitro*, CCN4 stimulates cardiomyocyte hypertrophy, inhibits cardiomyocyte apoptosis, and facilitates cardiac fibroblast proliferation [77]. After MI, CCN4 induction in the infarct border and remote regions peaks at 24 h [77].The actions of other CCN members in the MI setting have not been investigated.

4.5.7 Osteopontin

OPN, originally identified as a bone matrix protein, is a phosphorylated acidic glycoprotein expressed in many immune cells (e.g., macrophages and dendritic cells). OPN is significantly increased post-MI [78]. OPN signaling is achieved through integrin- or CD44 mediated pathways and modulates cell adhesion, survival, and gene expression [50,79]. OPN promotes migration of monocytes, T cells, endothelial cells, and smooth muscle cells [79]. In the post-MI setting, OPN co-localizes with macrophages infiltrating the infarct area, indicating OPN originates from or cross talks with macrophages. OPN can activate macrophages via induction of IL-12 synthesis and suppression of IL-10 expression [80]. In studies conducted in animals, OPN deficiency exacerbates MI-induced LV dilation and remodeling, without influencing infarct size or survival [81]; the underlying mechanisms are associated with reduction in TGF-β1 signaling, collagen levels, and angiogenesis.

4.5.8 Periostin

Periostin binds to the ECM to mediate smooth muscle cell migration and invasion of mesenchymal cushion cells via αvβ3 and αvβ5 signaling [82]. Tissue periostin level is markedly increased post-MI and mainly localizes to fibroblasts in the infarct area [83]. Fibrogenic activity of periostin is attributed to enhancement of the TGF-β/Smad signaling pathway [84]. Periostin null mice show an increased incidence of cardiac rupture in the first 10 days after MI, compared to wild-type mice [85]. Interestingly, myocardial administration of periostin in post-MI mice resulted in improved cardiac function, cardiomyocyte proliferation, and reduced cardiac fibrosis, suggesting that periostin may enhance the regenerative capacity of the myocardium without inducing fibrosis [86]. Dissecting the specific pathway responsible for the proregenerative activities of periostin may identify selective intervention targets.

4.5.9 SPARC

SPARC regulates ECM deposition and proteinase activity. LV SPARC expression is markedly increased after MI, and is mainly localized to infiltrating macrophages, endothelial cells, and myofibroblasts [87]. Post-MI, SPARC null mice exhibit fewer macrophages in the infarct region, suggesting that SPARC may regulate macrophage viability and the chronic immune response [87]. SPARC increases the expression of TGF-β in cardiac fibroblasts and regulate macrophage clearance through interaction with scavenger receptor stabilin-1 [50,79,88]. In the post-MI setting, deletion of the SPARC gene results in the formation of disorganized, immature scar tissue and rupture phenotype, and infusion of TGF-β rescues SPARC null hearts from rupture [87]. Mice studies from our laboratory indicate that SPARC deletion preserves LV function at day 3 post-MI but may be detrimental for the long-term response due to impaired fibroblast activation [89].

4.5.10 Tenascin-C

Tenascin-C is an oligometric glycoprotein exclusively expressed in the chordae tendineae and base of valve leaflets in the normal heart. In animal models of MI, tenascin-C expression can be detected in the infarct border zone, and is thought to loosen the strong adhesion of surviving cardiomyocytes to connective tissue [90]. Accordingly, tenascin-C aggravates LV remodeling and dysfunction after MI in mice; its deletion attenuates adverse LV fibrosis and dysfunction, without affecting infarct sizes or survival rates [91]. *In vitro*, tenascin-C fosters fibroblast migration and differentiation, and collagen gel contraction [92]. In patients with MI, serum concentration of tenascin-C positively correlates with the incidence of adverse cardiac remodeling and worse clinical outcomes [93].

4.5.11 Thrombospondin-1

TSP-1 is a multimodular, calcium-binding glycoprotein and an endogenous inhibitor of angiogenesis. It inhibits endothelial function, stimulates apoptosis, and suppresses nitric oxide and vascular endothelial growth factor signaling [94]. TSP-1 activates TGF-β1 by preventing its latency-associated peptide from deactivating the mature domain of TGF-β1 [95]. TSP-1 mRNA and protein are induced markedly after MI [96]. Absence of TSP-1 exacerbates MI-induced LV adverse remodeling and dysfunction via enhancing inflammation duration, extent, and infarct expansion [96]. Hence, TSP-1 protects adverse cardiac repair by limiting the inflammatory response.

4.6 MATRICRYPTINS: ECM FRAGMENTS WITH BIOLOGICAL ACTIVITY

Biologically active fragments of the ECM, termed matricryptins, are produced by various mechanisms, such as enzymatic degradation and denaturation. Matricryptins exert important actions post-MI, including regulating the inflammatory reaction and angiogenesis [97]. Peptide fragments from collagen type IV exhibit chemotactic properties to neutrophils, while a fragment from the α3 chain of collagen type IV inhibits neutrophil activation [98,99]. MMP-9 is reported to generate tumstatin, which inhibits angiogenesis, by cleaving the α3 chain of collagen type IV [100]. Furthermore, MMPs generate both pro- and anti-angiogenic fragments from collagen type IV [101].

Endostatin, the carboxy-terminal fragment of collagen type XVIII, exhibits a potent anti-angiogenic activity by inhibiting the proliferation and migration of endothelial cells [102]. In a rat MI model, tissue endostatin levels are elevated; neutralization of endostatin results in marked increase in the expression and activity of MMP-2 and -9, adverse LV remodeling, and higher mortality [103].

Fibronectin fragments, derived from MMP cleavage, are also reported to increase MMP activity, suggesting a positive feedback loop [104]. Macrophages stimulated with fibronectin fragments *in vitro* produce soluble factors that protect hypoxic cardiomyocytes from apoptosis [105]. Neutrophil elastase digests laminin to generate fragments chemotactic to neutrophils [106]. Laminin α5 fragment induces cytokine production from macrophages, resulting in a chemotactic response [107]. In addition, MMP-9 expression in macrophages is enhanced by laminin α1 peptide [108]. Both laminin and its fragments may play an important role in regulating inflammation in the post-MI setting.

In summary, ECM proteins not only constitute the major components of reparative scar, but also directly regulate LV repair by regulating inflammation, angiogenesis, and fibrosis.

4.7 FUTURE DIRECTIONS

The inflammatory response initiates the repair process and tightly regulates the healing response after MI. Appropriate inflammation facilitates favorable infarct healing, while insufficient or excessive inflammation may impair stable scar formation and damage viable myocardium [49]. Balanced ECM turnover, through regulation of synthesis by myofibroblasts and degradation by MMPs is critical for appropriate cardiac repair post-MI. Matricellular proteins regulate ECM deposition, inflammation, angiogenesis, and fibrosis. Hence, inflammatory leukocytes and mediators and ECM molecules together coordinate the repair response post-MI.

Despite the great progress in understanding the mechanisms of LV remodeling post-MI, additional studies are still necessary to further clarify issues (Table 4.3). First, the key molecules and signaling pathways that most influence extension and resolution of inflammation need to be identified. For example, enumerating TGF-β1 downstream signals may help identify therapeutic targets. Second, key molecules and signals responsible for communication between neutrophils, macrophages, and fibroblasts need to be elucidated. Previous studies showed that neutrophils and macrophages are sequentially recruited to the ischemia site, and fibroblasts are activated by leukocyte secreted mediators; however, the precise molecular mechanisms responsible for these processes are still poorly understood. Third, additional information regarding matricryptins and their roles in infarct healing are needed. Simply measuring levels of MMPs and ECM proteins following MI is not insufficient; an in-depth knowledge of the temporal sequences of these changes and the resultant downstream effects and mechanisms is required.

TABLE 4.3 Research Directions for Future Studies

1. Identify the primary pathways that regulate post-myocardial infarction inflammation
2. Identify the key signals between leukocytes and fibroblasts that regulate intercellular communication
3. Elucidate how matricryptin signaling influences the post-myocardial infarction response

4.8 CONCLUSIONS

In conclusion, MI remains a major cause of congestive heart failure. LV repair following MI is highly dependent on the inflammatory response and scar formation. Elucidating the events regulating the transition of MI to heart failure may help to identify novel interventions that promote a favorable wound-healing response.

Acknowledgments

We acknowledge support from the American Heart Association for 14POST18770012 to RPI, 14SDG18860050 to LECB, and 13POST14350034 to KYD-P; from the Rapoport Foundation for Cardiovascular Research to RAL; from National Institutes of Health (NIH)/NIH Heart, Lung and Blood Institute HHSN 268201000036C (N01-HV-00244) for the San Antonio Cardiovascular Proteomics Center, HL075360 and HL051971; from NIH for GM104357; and from the Biomedical Laboratory Research and Development Service of the Veterans Affairs Office of Research and Development Award 5I01BX000505 to MLL.

References

[1] Anderson L. Candidate-based proteomics in the search for biomarkers of cardiovascular disease. J Physiol 2005;563:23–60.
[2] Matsui Y, Morimoto J, Uede T. Role of matricellular proteins in cardiac tissue remodeling after myocardial infarction. World J Biol Chem 2010;1:69–80.
[3] Ma Y, Yabluchanskiy A, Lindsey ML. Neutrophil roles in left ventricular remodeling following myocardial infarction. Fibrogenesis Tissue Repair 2013;6:11.
[4] Lambert JM, Lopez EF, Lindsey ML. Macrophage roles following myocardial infarction. Int J Cardiol 2008;130:147–58.
[5] Vaday GG, Franitza S, Schor H, Hecht I, Brill A, Cahalon L, et al. Combinatorial signals by inflammatory cytokines and chemokines mediate leukocyte interactions with extracellular matrix. J Leukoc Biol 2001;69:885–92.
[6] Romson JL, Hook BG, Kunkel SL, Abrams GD, Schork MA, Lucchesi BR. Reduction of the extent of ischemic myocardial injury by neutrophil depletion in the dog. Circulation 1983;67:1016–23.
[7] Jolly SR, Kane WJ, Hook BG, Abrams GD, Kunkel SL, Lucchesi BR. Reduction of myocardial infarct size by neutrophil depletion: effect of duration of occlusion. Am Heart J 1986;112:682–90.
[8] Janoff A. Elastase in tissue injury. Annu Rev Med 1985;36:207–16.
[9] Lindsey M, Wedin K, Brown MD, Keller C, Evans AJ, Smolen J, et al. Matrix-dependent mechanism of neutrophil-mediated release and activation of matrix metalloproteinase 9 in myocardial ischemia/reperfusion. Circulation 2001;103:2181–7.
[10] Nowak D, Glowczynska I, Piasecka G. Chemotactic activity of elastin-derived peptides for human polymorphonuclear leukocytes and their effect on hydrogen peroxide and myeloperoxidase release. Arch Immunol Ther Exp (Warsz) 1989;37:741–8.
[11] Norris DA, Clark RA, Swigart LM, Huff JC, Weston WL, Howell SE. Fibronectin fragment(s) are chemotactic for human peripheral blood monocytes. J Immunol 1982;129:1612–8.
[12] Nahrendorf M, Swirski FK, Aikawa E, Stangenberg L, Wurdinger T, Figueiredo JL, et al. The healing myocardium sequentially mobilizes two monocyte subsets with divergent and complementary functions. J Exp Med 2007;204:3037–47.
[13] Hayashidani S, Tsutsui H, Shiomi T, Ikeuchi M, Matsusaka H, Suematsu N, et al. Anti-monocyte chemoattractant protein-1 gene therapy attenuates left ventricular remodeling and failure after experimental myocardial infarction. Circulation 2003;108:2134–40.
[14] Maekawa Y, Anzai T, Yoshikawa T, Sugano Y, Mahara K, Kohno T, et al. Effect of granulocyte-macrophage colony-stimulating factor inducer on left ventricular remodeling after acute myocardial infarction. J Am Coll Cardiol 2004;44:1510–20.
[15] Leor J, Rozen L, Zuloff-Shani A, Feinberg MS, Amsalem Y, Barbash IM, et al. Ex vivo activated human macrophages improve healing, remodeling, and function of the infarcted heart. Circulation 2006;114:I94–I100.
[16] van Amerongen MJ, Harmsen MC, van Rooijen N, Petersen AH, van Luyn MJ. Macrophage depletion impairs wound healing and increases left ventricular remodeling after myocardial injury in mice. Am J Pathol 2007;170:818–29.
[17] Huang WC, Sala-Newby GB, Susana A, Johnson JL, Newby AC. Classical macrophage activation up-regulates several matrix metalloproteinases through mitogen activated protein kinases and nuclear factor-kappaB. PLoS One 2012;7:e42507.
[18] Deten A, Volz HC, Briest W, Zimmer HG. Cardiac cytokine expression is upregulated in the acute phase after myocardial infarction. Experimental studies in rats. Cardiovasc Res 2002;55:329–40.
[19] Irwin MW, Mak S, Mann DL, Qu R, Penninger JM, Yan A, et al. Tissue expression and immunolocalization of tumor necrosis factor-alpha in postinfarction dysfunctional myocardium. Circulation 1999;99:1492–8.
[20] Mitchell MD, Laird RE, Brown RD, Long CS. IL-1beta stimulates rat cardiac fibroblast migration via MAP kinase pathways. Am J Physiol Heart Circ Physiol 2007;292:H1139–47.
[21] Montgomery KF, Osborn L, Hession C, Tizard R, Goff D, Vassallo C, et al. Activation of endothelial-leukocyte adhesion molecule 1 (ELAM-1) gene transcription. Proc Natl Acad Sci U S A 1991;88:6523–7.
[22] Dobaczewski M, Chen W, Frangogiannis NG. Transforming growth factor (TGF)-beta signaling in cardiac remodeling. J Mol Cell Cardiol 2011;51:600–6.
[23] Fava RA, Olsen NJ, Postlethwaite AE, Broadley KN, Davidson JM, Nanney LB, et al. Transforming growth factor beta 1 (TGF-beta 1) induced neutrophil recruitment to synovial tissues: implications for TGF-beta-driven synovial inflammation and hyperplasia. J Exp Med 1991;173:1121–32.
[24] Wahl SM, Hunt DA, Wakefield LM, McCartney-Francis N, Wahl LM, Roberts AB, et al. Transforming growth factor type beta induces monocyte chemotaxis and growth factor production. Proc Natl Acad Sci U S A 1987;84:5788–92.

[25] Lijnen P, Petrov V, Rumilla K, Fagard R. Transforming growth factor-beta 1 promotes contraction of collagen gel by cardiac fibroblasts through their differentiation into myofibroblasts. Methods Find Exp Clin Pharmacol 2003;25:79–86.
[26] Schiller M, Javelaud D, Mauviel A. TGF-beta-induced SMAD signaling and gene regulation: consequences for extracellular matrix remodeling and wound healing. J Dermatol Sci 2004;35:83–92.
[27] Liehn EA, Postea O, Curaj A, Marx N. Repair after myocardial infarction, between fantasy and reality: the role of chemokines. J Am Coll Cardiol 2011;58:2357–62.
[28] Kanda H, Tateya S, Tamori Y, Kotani K, Hiasa K, Kitazawa R, et al. MCP-1 contributes to macrophage infiltration into adipose tissue, insulin resistance, and hepatic steatosis in obesity. J Clin Invest 2006;116:1494–505.
[29] Gharaee-Kermani M, Denholm EM, Phan SH. Costimulation of fibroblast collagen and transforming growth factor beta1 gene expression by monocyte chemoattractant protein-1 via specific receptors. J Biol Chem 1996;271:17779–84.
[30] Parissis JT, Adamopoulos S, Venetsanou KF, Mentzikof DG, Karas SM, Kremastinos DT. Serum profiles of C-C chemokines in acute myocardial infarction: possible implication in postinfarction left ventricular remodeling. J Interferon Cytokine Res 2002;22:223–9.
[31] Zamilpa R, Kanakia R, Cigarroa JT, Dai Q, Escobar GP, Martinez H, et al. CC chemokine receptor 5 deletion impairs macrophage activation and induces adverse remodeling following myocardial infarction. Am J Physiol Heart Circ Physiol 2011;300:H1418–26.
[32] Visse R, Nagase H. Matrix metalloproteinases and tissue inhibitors of metalloproteinases: structure, function, and biochemistry. Circ Res 2003;92:827–39.
[33] Van Wart HE, Birkedal-Hansen H. The cysteine switch: a principle of regulation of metalloproteinase activity with potential applicability to the entire matrix metalloproteinase gene family. Proc Natl Acad Sci U S A 1990;87:5578–82.
[34] Gross J, Lapiere CM. Collagenolytic activity in amphibian tissues: a tissue culture assay. Proc Natl Acad Sci U S A 1962;48:1014–22.
[35] Cleutjens JP, Kandala JC, Guarda E, Guntaka RV, Weber KT. Regulation of collagen degradation in the rat myocardium after infarction. J Mol Cell Cardiol 1995;27:1281–92.
[36] Papadopoulos DP, Moyssakis I, Makris TK, Poulakou M, Stavroulakis G, Perrea D, et al. Clinical significance of matrix metalloproteinases activity in acute myocardial infarction. Eur Cytokine Netw 2005;16:152–60.
[37] Vanhoutte D, Schellings M, Pinto Y, Heymans S. Relevance of matrix metalloproteinases and their inhibitors after myocardial infarction: a temporal and spatial window. Cardiovasc Res 2006;69:604–13.
[38] Matsumura S, Iwanaga S, Mochizuki S, Okamoto H, Ogawa S, Okada Y. Targeted deletion or pharmacological inhibition of MMP-2 prevents cardiac rupture after myocardial infarction in mice. J Clin Invest 2005;115:599–609.
[39] Lindsey ML, Zamilpa R. Temporal and spatial expression of matrix metalloproteinases and tissue inhibitors of metalloproteinases following myocardial infarction. Cardiovasc Ther 2012;30:31–41.
[40] Van Hove I, Lemmens K, Van de Velde S, Verslegers M, Moons L. Matrix metalloproteinase-3 in the central nervous system: a look on the bright side. J Neurochem 2012;123:203–16.
[41] Kelly D, Khan S, Cockerill G, Ng LL, Thompson M, Samani NJ, et al. Circulating stromelysin-1 (MMP-3): a novel predictor of LV dysfunction, remodelling and all-cause mortality after acute myocardial infarction. Eur J Heart Fail 2008;10:133–9.
[42] Lindsey ML, Escobar GP, Mukherjee R, Goshorn DK, Sheats NJ, Bruce JA, et al. Matrix metalloproteinase-7 affects connexin-43 levels, electrical conduction, and survival after myocardial infarction. Circulation 2006;113:2919–28.
[43] Wang W, McKinnie SM, Patel VB, Haddad G, Wang Z, Zhabyeyev P, et al. Loss of Apelin exacerbates myocardial infarction adverse remodeling and ischemia-reperfusion injury: therapeutic potential of synthetic Apelin analogues. J Am Heart Assoc 2013;2:e000249.
[44] Yabluchanskiy A, Ma Y, Iyer RP, Hall ME, Lindsey ML. Matrix metalloproteinase-9: many shades of function in cardiovascular disease. Physiology (Bethesda) 2013;28:391–403.
[45] Blankenberg S, Rupprecht HJ, Poirier O, Bickel C, Smieja M, Hafner G, et al. Plasma concentrations and genetic variation of matrix metalloproteinase 9 and prognosis of patients with cardiovascular disease. Circulation 2003;107:1579–85.
[46] Ducharme A, Frantz S, Aikawa M, Rabkin E, Lindsey M, Rohde LE, et al. Targeted deletion of matrix metalloproteinase-9 attenuates left ventricular enlargement and collagen accumulation after experimental myocardial infarction. J Clin Invest 2000;106:55–62.
[47] Zamilpa R, Ibarra J, de Castro Bras LE, Ramirez TA, Nguyen N, Halade GV, et al. Transgenic overexpression of matrix metalloproteinase-9 in macrophages attenuates the inflammatory response and improves left ventricular function post-myocardial infarction. J Mol Cell Cardiol 2012;53:599–608.
[48] Ma Y, Chiao YA, Zhang J, Manicone AM, Jin YF, Lindsey ML. Matrix metalloproteinase-28 deletion amplifies inflammatory and extracellular matrix responses to cardiac aging. Microsc Microanal 2012;18:81–90.
[49] Ma Y, Halade GV, Zhang J, Ramirez TA, Levin D, Voorhees A, et al. Matrix metalloproteinase-28 deletion exacerbates cardiac dysfunction and rupture after myocardial infarction in mice by inhibiting M2 macrophage activation. Circ Res 2013;112:675–88.
[50] Ma Y, Halade GV, Lindsey ML. Extracellular matrix and fibroblast communication following myocardial infarction. J Cardiovasc Transl Res 2012;5:848–57.
[51] Carver W, Terracio L, Borg TK. Expression and accumulation of interstitial collagen in the neonatal rat heart. Anat Rec 1993;236:511–20.
[52] Shamhart PE, Meszaros JG. Non-fibrillar collagens: key mediators of post-infarction cardiac remodeling? J Mol Cell Cardiol 2010;48:530–7.
[53] Wei S, Chow LT, Shum IO, Qin L, Sanderson JE. Left and right ventricular collagen type I/III ratios and remodeling post-myocardial infarction. J Card Fail 1999;5:117–26.
[54] Willems IE, Havenith MG, De Mey JG, Daemen MJ. The alpha-smooth muscle actin-positive cells in healing human myocardial scars. Am J Pathol 1994;145:868–75.
[55] Bonanno E, Iurlaro M, Madri JA, Nicosia RF. Type IV collagen modulates angiogenesis and neovessel survival in the rat aorta model. In Vitro Cell Dev Biol Anim 2000;36:336–40.
[56] Yamanishi A, Kusachi S, Nakahama M, Ninomiya Y, Watanabe T, Kumashiro H, et al. Sequential changes in the localization of the type IV collagen alpha chain in the infarct zone: immunohistochemical study of experimental myocardial infarction in the rat. Pathol Res Pract 1998;194:413–22.
[57] Wenstrup RJ, Florer JB, Brunskill EW, Bell SM, Chervoneva I, Birk DE. Type V collagen controls the initiation of collagen fibril assembly. J Biol Chem 2004;279:53331–7.

[58] Wenstrup RJ, Florer JB, Davidson JM, Phillips CL, Pfeiffer BJ, Menezes DW, et al. Murine model of the Ehlers-Danlos syndrome. col5a1 haploinsufficiency disrupts collagen fibril assembly at multiple stages. J Biol Chem 2006;281:12888–95.
[59] Keene DR, Engvall E, Glanville RW. Ultrastructure of type VI collagen in human skin and cartilage suggests an anchoring function for this filamentous network. J Cell Biol 1988;107:1995–2006.
[60] Naugle JE, Olson ER, Zhang X, Mase SE, Pilati CF, Maron MB, et al. Type VI collagen induces cardiac myofibroblast differentiation: implications for postinfarction remodeling. Am J Physiol Heart Circ Physiol 2006;290:H323–30.
[61] Luther DJ, Thodeti CK, Shamhart PE, Adapala RK, Hodnichak C, Weihrauch D, et al. Absence of type VI collagen paradoxically improves cardiac function, structure, and remodeling after myocardial infarction. Circ Res 2012;110:851–6.
[62] Schoneveld AH, Hoefer I, Sluijter JP, Laman JD, de Kleijn DP, Pasterkamp G. Atherosclerotic lesion development and toll like receptor 2 and 4 responsiveness. Atherosclerosis 2008;197:95–104.
[63] Gondokaryono SP, Ushio H, Niyonsaba F, Hara M, Takenaka H, Jayawardana ST, et al. The extra domain A of fibronectin stimulates murine mast cells via toll-like receptor 4. J Leukoc Biol 2007;82:657–65.
[64] Liao YF, Gotwals PJ, Koteliansky VE, Sheppard D, Van De Water L. The EIIIA segment of fibronectin is a ligand for integrins alpha 9beta 1 and alpha 4beta 1 providing a novel mechanism for regulating cell adhesion by alternative splicing. J Biol Chem 2002;277:14467–74.
[65] Manabe R, Oh-e N, Sekiguchi K. Alternatively spliced EDA segment regulates fibronectin-dependent cell cycle progression and mitogenic signal transduction. J Biol Chem 1999;274:5919–24.
[66] Arslan F, Smeets MB, Riem Vis PW, Karper JC, Quax PH, Bongartz LG, et al. Lack of fibronectin-EDA promotes survival and prevents adverse remodeling and heart function deterioration after myocardial infarction. Circ Res 2011;108:582–92.
[67] Chiao YA, Zamilpa R, Lopez EF, Dai Q, Escobar GP, Hakala K, et al. In vivo matrix metalloproteinase-7 substrates identified in the left ventricle post-myocardial infarction using proteomics. J Proteome Res 2010;9:2649–57.
[68] Zamilpa R, Lopez EF, Chiao YA, Dai Q, Escobar GP, Hakala K, et al. Proteomic analysis identifies in vivo candidate matrix metalloproteinase-9 substrates in the left ventricle post-myocardial infarction. Proteomics 2010;10:2214–23.
[69] Morishita N, Kusachi S, Yamasaki S, Kondo J, Tsuji T. Sequential changes in laminin and type IV collagen in the infarct zone—immunohistochemical study in rat myocardial infarction. Jpn Circ J 1996;60:108–14.
[70] Dinh W, Bansemir L, Futh R, Nickl W, Stasch JP, Coll-Barroso M, et al. Increased levels of laminin and collagen type VI may reflect early remodelling in patients with acute myocardial infarction. Acta Cardiol 2009;64:329–34.
[71] Grzeszkiewicz TM, Kirschling DJ, Chen N, Lau LF. CYR61 stimulates human skin fibroblast migration through Integrin alpha vbeta 5 and enhances mitogenesis through integrin alpha vbeta 3, independent of its carboxyl-terminal domain. J Biol Chem 2001;276:21943–50.
[72] Lake AC, Bialik A, Walsh K, Castellot Jr JJ. CCN5 is a growth arrest-specific gene that regulates smooth muscle cell proliferation and motility. Am J Pathol 2003;162:219–31.
[73] Bai T, Chen CC, Lau LF. Matricellular protein CCN1 activates a proinflammatory genetic program in murine macrophages. J Immunol 2010;184:3223–32.
[74] Hilfiker-Kleiner D, Kaminski K, Kaminska A, Fuchs M, Klein G, Podewski E, et al. Regulation of proangiogenic factor CCN1 in cardiac muscle: impact of ischemia, pressure overload, and neurohumoral activation. Circulation 2004;109:2227–33.
[75] Jun JI, Lau LF. The matricellular protein CCN1 induces fibroblast senescence and restricts fibrosis in cutaneous wound healing. Nat Cell Biol 2010;12:676–85.
[76] Ahmed MS, Gravning J, Martinov VN, von Lueder TG, Edvardsen T, Czibik G, et al. Mechanisms of novel cardioprotective functions of CCN2/CTGF in myocardial ischemia-reperfusion injury. Am J Physiol Heart Circ Physiol 2011;300:H1291–302.
[77] Colston JT, de la Rosa SD, Koehler M, Gonzales K, Mestril R, Freeman GL, et al. Wnt-induced secreted protein-1 is a prohypertrophic and profibrotic growth factor. Am J Physiol Heart Circ Physiol 2007;293:H1839–46.
[78] Lindsey ML, Mann DL, Entman ML, Spinale FG. Extracellular matrix remodeling following myocardial injury. Ann Med 2003;35:316–26.
[79] Frangogiannis NG. Matricellular proteins in cardiac adaptation and disease. Physiol Rev 2012;92:635–88.
[80] Ashkar S, Weber GF, Panoutsakopoulou V, Sanchirico ME, Jansson M, Zawaideh S, et al. Eta-1 (osteopontin): an early component of type-1 (cell-mediated) immunity. Science 2000;287:860–4.
[81] Trueblood NA, Xie Z, Communal C, Sam F, Ngoy S, Liaw L, et al. Exaggerated left ventricular dilation and reduced collagen deposition after myocardial infarction in mice lacking osteopontin. Circ Res 2001;88:1080–7.
[82] Butcher JT, Norris RA, Hoffman S, Mjaatvedt CH, Markwald RR. Periostin promotes atrioventricular mesenchyme matrix invasion and remodeling mediated by integrin signaling through Rho/PI 3-kinase. Dev Biol 2007;302:256–66.
[83] Shimazaki M, Nakamura K, Kii I, Kashima T, Amizuka N, Li M, et al. Periostin is essential for cardiac healing after acute myocardial infarction. J Exp Med 2008;205:295–303.
[84] Sidhu SS, Yuan S, Innes AL, Kerr S, Woodruff PG, Hou L, et al. Roles of epithelial cell-derived periostin in TGF-beta activation, collagen production, and collagen gel elasticity in asthma. Proc Natl Acad Sci U S A 2010;107:14170–5.
[85] Oka T, Xu J, Kaiser RA, Melendez J, Hambleton M, Sargent MA, et al. Genetic manipulation of periostin expression reveals a role in cardiac hypertrophy and ventricular remodeling. Circ Res 2007;101:313–21.
[86] Kuhn B, del Monte F, Hajjar RJ, Chang YS, Lebeche D, Arab S, et al. Periostin induces proliferation of differentiated cardiomyocytes and promotes cardiac repair. Nat Med 2007;13:962–9.
[87] Schellings MW, Vanhoutte D, Swinnen M, Cleutjens JP, Debets J, van Leeuwen RE, et al. Absence of SPARC results in increased cardiac rupture and dysfunction after acute myocardial infarction. J Exp Med 2009;206:113–23.
[88] Workman G, Sage EH. Identification of a sequence in the matricellular protein SPARC that interacts with the scavenger receptor stabilin-1. J Cell Biochem 2011;112:1003–8.
[89] McCurdy SM, Dai Q, Zhang J, Zamilpa R, Ramirez TA, Dayah T, et al. SPARC mediates early extracellular matrix remodeling following myocardial infarction. Am J Physiol Heart Circ Physiol 2011;301:H497–505.
[90] Imanaka-Yoshida K, Hiroe M, Nishikawa T, Ishiyama S, Shimojo T, Ohta Y, et al. Tenascin-C modulates adhesion of cardiomyocytes to extracellular matrix during tissue remodeling after myocardial infarction. Lab Invest 2001;81:1015–24.
[91] Nishioka T, Onishi K, Shimojo N, Nagano Y, Matsusaka H, Ikeuchi M, et al. Tenascin-C may aggravate left ventricular remodeling and function after myocardial infarction in mice. Am J Physiol Heart Circ Physiol 2010;298:H1072–8.

[92] Tamaoki M, Imanaka-Yoshida K, Yokoyama K, Nishioka T, Inada H, Hiroe M, et al. Tenascin-C regulates recruitment of myofibroblasts during tissue repair after myocardial injury. Am J Pathol 2005;167:71–80.
[93] Sato A, Aonuma K, Imanaka-Yoshida K, Yoshida T, Isobe M, Kawase D, et al. Serum tenascin-C might be a novel predictor of left ventricular remodeling and prognosis after acute myocardial infarction. J Am Coll Cardiol 2006;47:2319–25.
[94] Ma Y, Yabluchanskiy A, Lindsey ML. Thrombospondin-1: the good, the bad, and the complicated. Circ Res 2013;113:1272–4.
[95] Schultz-Cherry S, Ribeiro S, Gentry L, Murphy-Ullrich JE. Thrombospondin binds and activates the small and large forms of latent transforming growth factor-beta in a chemically defined system. J Biol Chem 1994;269:26775–82.
[96] Frangogiannis NG, Ren G, Dewald O, Zymek P, Haudek S, Koerting A, et al. Critical role of endogenous thrombospondin-1 in preventing expansion of healing myocardial infarcts. Circulation 2005;111:2935–42.
[97] Ricard-Blum S, Ballut L. Matricryptins derived from collagens and proteoglycans. Front Biosci (Landmark Ed) 2011;16:674–97.
[98] Senior RM, Hinek A, Griffin GL, Pipoly DJ, Crouch EC, Mecham RP. Neutrophils show chemotaxis to type IV collagen and its 7S domain and contain a 67 kD type IV collagen binding protein with lectin properties. Am J Respir Cell Mol Biol 1989;1:479–87.
[99] Monboisse JC, Garnotel R, Bellon G, Ohno N, Perreau C, Borel JP, et al. The alpha 3 chain of type IV collagen prevents activation of human polymorphonuclear leukocytes. J Biol Chem 1994;269:25475–82.
[100] Hamano Y, Zeisberg M, Sugimoto H, Lively JC, Maeshima Y, Yang C, et al. Physiological levels of tumstatin, a fragment of collagen IV alpha3 chain, are generated by MMP-9 proteolysis and suppress angiogenesis via alphaV beta3 integrin. Cancer Cell 2003;3:589–601.
[101] Chang C, Werb Z. The many faces of metalloproteases: cell growth, invasion, angiogenesis and metastasis. Trends Cell Biol 2001;11:S37–43.
[102] Marneros AG, Olsen BR. Physiological role of collagen XVIII and endostatin. Faseb J 2005;19:716–28.
[103] Isobe K, Kuba K, Maejima Y, Suzuki J, Kubota S, Isobe M. Inhibition of endostatin/collagen XVIII deteriorates left ventricular remodeling and heart failure in rat myocardial infarction model. Circ J 2010;74:109–19.
[104] Schedin P, Strange R, Mitrenga T, Wolfe P, Kaeck M. Fibronectin fragments induce MMP activity in mouse mammary epithelial cells: evidence for a role in mammary tissue remodeling. J Cell Sci 2000;113(Pt 5):795–806.
[105] Trial J, Rossen RD, Rubio J, Knowlton AA. Inflammation and ischemia: macrophages activated by fibronectin fragments enhance the survival of injured cardiac myocytes. Exp Biol Med (Maywood) 2004;229:538–45.
[106] Mydel P, Shipley JM, Adair-Kirk TL, Kelley DG, Broekelmann TJ, Mecham RP, et al. Neutrophil elastase cleaves laminin-332 (laminin-5) generating peptides that are chemotactic for neutrophils. J Biol Chem 2008;283:9513–22.
[107] Adair-Kirk TL, Atkinson JJ, Kelley DG, Arch RH, Miner JH, Senior RM. A chemotactic peptide from laminin alpha 5 functions as a regulator of inflammatory immune responses via TNF alpha-mediated signaling. J Immunol 2005;174:1621–9.
[108] Faisal Khan KM, Laurie GW, McCaffrey TA, Falcone DJ. Exposure of cryptic domains in the alpha 1-chain of laminin-1 by elastase stimulates macrophages urokinase and matrix metalloproteinase-9 expression. J Biol Chem 2002;277:13778–86.

CHAPTER

5

Cross Talk Between Brain and Inflammation

Regien G. Schoemaker[1,2], *Uli L.M. Eisel*[2,3]

[1]Department of Cardiology, University Medical Centre Groningen, Groningen, The Netherlands
[2]Department of Molecular Neurobiology, University of Groningen, Groningen, The Netherlands
[3]Department of Psychiatry, University Medical Centre Groningen, Groningen, The Netherlands

5.1 CARDIOVASCULAR DISEASE AND BRAIN DISORDERS

5.1.1 Introduction

Cardiovascular diseases are found to frequently coincide with disorders of the brain, such as depression, anxiety, and cognitive decline. Cardiovascular disease and major depression are two of the most prevalent illnesses in Western populations, affecting a large part of the population and leading to a high economic burden. The lifetime prevalence of major depression is 8.3-16.2% [1–3], while cardiovascular disease still is the leading cause of death worldwide. In recent years, there has been more interest in the link between cardiovascular disease and brain disorders. It is known that patients with cardiovascular disease, including heart failure and acute myocardial infarction, have an increased risk of developing depression and cognitive decline. On the other hand, patients suffering from depression or chronic stress are more likely to develop cardiovascular disease, including myocardial infarction. Cardiovascular disease with co-morbid depression is associated with lower adherence to medication and lifestyle advises, increased morbidity, and substantially worse prognosis. Although the underlying mechanism of the interaction between these distinct pathologies is still largely unknown, inflammation is thought to play a major role. Both cardiovascular disease and depression share an increased expression of cytokines, with substantial overlap [4]. Optimal cardiovascular therapy improves cardiovascular prognosis without major effects on depression. Alternatively, antidepressant therapy in cardiovascular disease is associated with modest improvement in depressive symptoms, however, without improvement in cardiac outcome. Neuroimmune interactions may provide a novel way to view these severely ill patients.

5.1.2 Brain Disorders Leading to Cardiovascular Disease

Major depression is a common disorder with a lifetime prevalence of 13% [4]. In the past few years, the comorbidity of depression and congestive heart failure has been thoroughly investigated. Epidemiological data clearly suggests that depression is an independent risk factor for acute myocardial infarction and heart diseases in general. Patients with depression have a greater mortality risk due to cardiovascular related conditions up to 10 years after the diagnosis [5], which accounts for a mild as well as for major depression [6].

In addition, prospective studies with depressed individuals showed that a history of a major depressive episode was associated with a higher risk of acute myocardial infarction, even after correction for major coronary risk factors [7,8]. Also, in patients who already have developed congestive heart disease, the impact of depression is of great importance. A prospective population-based cohort study, investigated age- and sex-adjusted hazard ratios for death from all causes. Results from this study showed that patients with only depressive symptoms had a higher odds ratio (OR: 2.10) compared with patients with only congestive heart disease (OR: 1.67). However, patients with both

http://dx.doi.org/10.1016/B978-0-12-800039-7.00005-0
© 2015 Elsevier Inc. All rights reserved.

congestive heart disease and depressive symptoms displayed an additive hazard ratio for death (OR: 4.99) [9]. Junger and coworkers [10] found similar results; they concluded that "...depression score predicts mortality independent of physical parameters in congestive heart failure patients not treated for depression. Its prognostic power increases over time and should, thus, be accounted for in risk stratification and therapy."

Apart from depression, chronic stress is a well-recognized risk factor for cardiovascular disease, by deregulation of complex physiological systems promoting hypertension, dyslipidemia, diabetes, and atherosclerosis. Moreover, the brain-heart interaction may even have direct effects on the myocardium, the cardiac myocyte. The so-called Takotsubo cardiomyopathy, consisting of transient left-ventricular dysfunction triggered by acute emotional or physical stress, presents like an acute myocardial infarction, including mild inflammatory cell infiltration and considerable increase in extracellular matrix proteins [7].

5.1.3 Cardiovascular Disease Leading to Brain Disorders

It is commonly known that patients suffering from cardiovascular disease have a greater chance of developing various complications. According to Triposkiadis *et al.* [11], inflammation could be regarded as a common denominator of heart failure and associated peripheral noncardiac comorbidities. In line with this, brain-associated comorbidities, such as depression, anxiety, and also fatigue and inactivity, are associated with inflammation [12].

Depression is a complication, which is of particular interest because of its recognized high impact on the quality of life. Up to 65% of patients recovering from acute myocardial infarction show symptoms associated with depression [13], and from those patients, 15-30% develop a major depression [14,15]. These data are consistent with a meta-study, where one in three patients hospitalized for acute myocardial infarction had "at least mild-to-moderate symptoms of depression" [16]. Both clinical depression and elevated levels of subclinical depressive symptoms are common in the weeks following acute coronary syndrome (including myocardial infarction or unstable angina) [16] and predict recurrent cardiac events and cardiovascular mortality [17].

Although interest in cognitive impairment in cardiovascular disease is growing, this could also be regarded as a consequence of depression, that is, as depression independently predicted cognitive outcome [18]. Moreover, when depression is subdivided into cognitive and somatic subscales, the somatic rather than the cognitive scores are predictive for prognosis and association with inflammation [19].

While optimal treatment of cardiovascular disease usually has no major effects on depression, treatment of depression in these patients, though associated with modest improvement in depressive symptoms, does not improve cardiovascular prognosis [20]. Interestingly, in women when the depression is treated with anxiolytics and/or antidepressants, the mortality due to cardiovascular events is even significantly higher compared to untreated women [21]. The mechanism behind this has yet to be discovered but might indicate depression/anxiety as a functional response.

5.1.4 Inflammation as the Link in Neurocardiac Interaction

Abnormal levels of inflammatory proteins, and in particular increased proinflammatory cytokines and products of inflammatory processes [22,23], have been found in heart failure as well as in depression [24], with clear overlap [4]. The proinflammatory cytokines TNF-α, IL1-β, IL-2, and IL-6 are reported to be elevated in heart failure as well as in depression. The coexistence of depression and heart failure has been found to be related to higher levels of circulating proinflammatory proteins [25,26]. However, due to the wide variation in expression of symptoms of depression and the variable etiology of heart failure, detailed information on more specific inflammatory markers associated with symptoms of depression in heart failure is still lacking. In this regard, the newly described circulating indicator for heart failure [27] and depression [28], as well as the combination [29]; neutrophil gelatinase associated lipocalin (Lipocalin-2/NGAL), might provide an interesting candidate. Depression is associated with excessive secretion of TNF-α, IL1-β, and interferon. Although these proinflammatory cytokines are associated with rather nonspecific immune responses, however, when TNF-α is administered to humans, it results in depressive symptoms, such as fatigue, malaise, lethargy, and anorexia [30].

Myocardial infarction, as the most common cause of heart failure, can be regarded as a healing wound, as reviewed by Frantz *et al.* [31]. Proper healing of this wound is essential for survival. Activation of the innate immune system by cardiac injury evokes release of several inflammatory mediators and inflammatory cells to the site of injury. Certain products of tissue injury, including reactive oxygen species (ROS) and intracellular proteins released from necrotic cells, initiate an inflammatory response, leading to pattern recognition receptors such as toll-like receptors (TLRs) and the transcription factor nuclear factor kappa B (NF-κB). Studies in KO mice support the role for TLRs and NF-κB in activation of inflammatory cells after cardiac injury as important initiators of inflammation and healing. Subsequently, NF-κB leads to production of proinflammatory cytokines, including TNF-α, IL1-β, IL-2, and IL6, which in turn stimulates NF-κB (positive feedback) and activates danger-associated-molecular patterns (DAMPS). Systemic TNF-α, as a necessary and sufficient mediator of local and systemic inflammation, depresses cardiac output, inducing microvascular thrombosis and mediate systemic capillary leakage [32,33]. Similarly, in the brain TNF-α induces leakage of the blood-brain barrier [34]. TNF-α amplifies and prolongs the inflammatory response by activating other cells to release cytokines, such as IL-1, and other mediators, such as eicosanoids, nitric oxide, and ROS, which promote further inflammation and tissue injury [35]. TNF-α is essential for the complete expression of inflammation, and self-limited inflammation is normally characterized by decreasing TNF-α activity. In a successful inflammatory response, the duration and magnitude of TNF-α release is limited, its beneficial and protective effects predominate, and it is not released systemically [32]. The prolonged and exaggerated TNF-α levels in plasma from rats with myocardial infarction as well as in heart failure patients, hence, is suggestive for a disturbed inflammatory process in heart failure. Repair of the infarcted myocardium must be well balanced and timely suppressed, for instance, by the release of the anti-inflammatory cytokines, such as IL-10. It is clear that defects in this initially beneficial response may contribute to prolonged and extended inflammation, which can contribute to the progression of heart failure [36].

There are several hypotheses with respect to the source of proinflammatory cytokines in cardiovascular disease [4]: activation of the immune system in response to tissue injury; the failing heart itself becomes the source of TNF-α production and elevated TNF-α levels represent spillover of cytokines that were produced locally within the myocardium, leading to secondary activation of the immune system; the decreased cardiac output leads to the elaboration of TNF-α by underperfusion of systemic tissues. An extension of this hypothesis is that gut wall oedema allows translocation of endotoxin, which activates cytokine production. Circulating cytokines then may enter the brain, facilitating neuronal dysfunction.

On the other hand, proinflammatory cytokines infused into various brain areas result in significant hemodynamic and neurohormonal responses that are typical for cardiovascular diseases [37]. Central infusions of TNF-α or IL1-β were found to increase blood pressure, sympathetic activity, and synthesis of renin, aldosterone, atrial natriuretic peptide, and vasopressin. Inhibition of TNF-α synthesis by pentoxiphyllin or inhibition of TNF-α by etanercept in rats with post-myocardial infarction-induced heart failure resulted in reduced stimulation of neurons in the paraventricular nucleus (PVN) of the hypothalamus, decreased renal sympathetic nerve activation, and lower plasma catecholamines [38,39]. Chronic central blockade of TNF-α in these rats had similar effects. Accordingly, etanercept reversed depressive behavior in these rats [40,41]. Alternatively, increase in the brain concentration of the anti-inflammatory cytokines, IL-1ra or IL-10, exerted the opposite effects; cerebroventricular transfer of IL-10 gene reduces hemodynamic and neurohumoral indices of heart failure in infarcted rats, while central infusion of IL-1ra decreased the hypertensive response to acute stressors in healthy rats.

What is interesting in this regard, is cytokines in the brain exert their action by the influence on the synthesis of other mediators, including eicosanoids, nitric oxide, Angiotensin II, and their receptors [37,42], the latter being factors of the Renin Angiotensin System, which is now well recognized to play a major role in the pathophysiology of heart failure and hypertension.

5.2 CROSS TALK BETWEEN BRAIN AND CARDIOVASCULAR SYSTEM

5.2.1 Cardiovascular Regulation and the Brain

Recently, the interaction between heart and brain has received more attention. However, even the ancient Greeks appreciated this interaction as they regarded the heart as the location of memory, while the brain was indicated to cool the blood. As reviewed by de la Torre [43], several researchers in the late 1970s became aware of an intriguing link between a sick heart and the start of cognitive deterioration, "cardiogenic dementia" [44], which was later put aside as a myth [45]. This was further established in studies on vascular Alzheimer's disease. Hypoperfusion of the brain attributed to cardiovascular disease was associated with cognitive decline and dementia [43], and may be mediated by alterations in axons and myelin, which are suggested to contribute to the multiple autonomic and neurophysiological symptoms in heart failure [46]. Today, under normal physiological conditions, regulation of the cardiovascular system by the central nervous system is well recognized, implicating extensive cross talk between heart and blood vessels and the brain. This cross talk is indicated to involve several pathways, including neuronal paths as well as circulating factors, such as saturable transport across the blood-brain barrier; brain circumventricular organs; cytokines binding to brain endothelial cells, evoking release of paracrine factors; and cytokines activating peripheral sensory nerves [47].

One of the first experimental studies showing a direct effect of cardiovascular disease on the brain is reported by the group of Patel [48], showing increased hexokinase activity in the hypothalamus of rats with chronic myocardial infarction. This area (PVN) is specifically associated with the connection of sympathetic outflow to the body with the vasopressin-mediated reflection to higher brain areas. Later on, the PVN is broadly acknowledged for its altered expression of mediators linking cardiovascular dysfunction to the brain [38,49–52]. This includes increased expression of TNF-α [38], IL1-β [53], and cyclooxygenase II [53], but also factors of the renin angiotensin system and sympathetic nervous system (noradrenaline) [54]. However, how this increased expression relates to changes in brain areas is associated with depression and cognition is far from clear. In most of the above mentioned studies, expression measured in the cortex was not altered, indicating area-specific responses. The main areas in the brain that are involved in cardiovascular regulation in health and disease are thoroughly reviewed by Szczepanska *et al.* [37,42] indicating a central role for the nucleus tractus solitaries in the interaction between hypothalamic regulation of the peripheral cardiovascular system and the brain areas involved in cognition and emotion. In patients with heart failure, specific areas, such as mammillary bodies and fornix fibers [55] and putamen [56], are reduced in volume, and areas associated with autonomic, emotional, and cognitive regulation are found injured [57]. The main mechanisms involved in the cross talk between heart and brain are summarized in Figure 5.1.

5.2.2 Inflammation

5.2.2.1 *Circulating Cytokines*

Recently, more attention has been paid to the role of inflammation in the cross talk between heart and brain. Is it possible that the peripheral inflammatory response evoked by cardiovascular events, such as myocardial infarction, can induce neuroinflammation?

Animal models of myocardial infarction-induced heart failure have reported increased levels of, among others, circulating TNF-α [59,60]. However, although the inflammatory response in the heart is meant to remove dead cells and initiate repair and scar formation, which is merely finished 3-4 weeks after infarction in rats, TNF-α levels kept increasing for at least 4 weeks, and may contribute to neuroinflammation. Experiments investigating

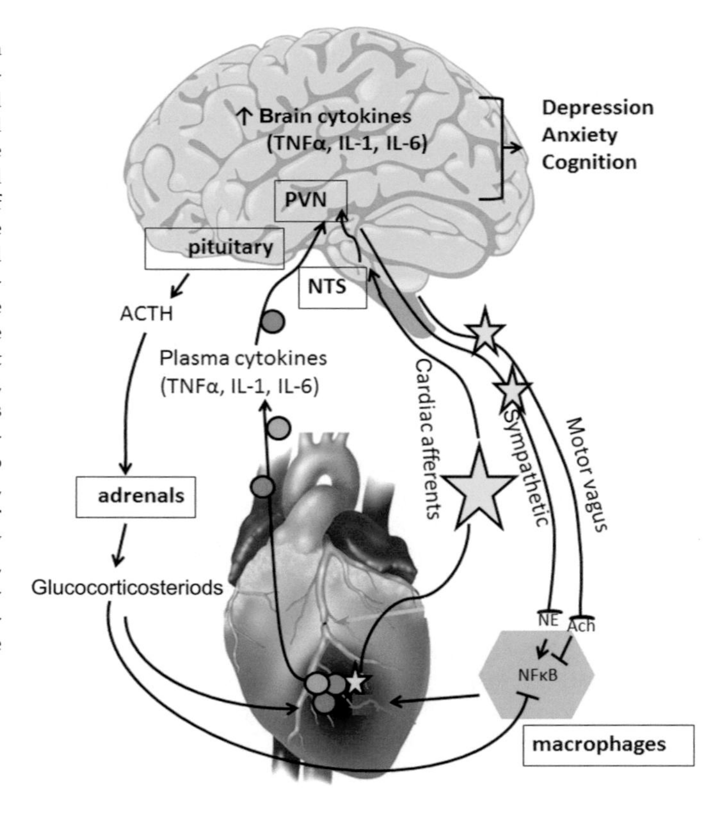

FIGURE 5.1 Mechanisms involved in the cross talk between heart and brain in normal regulation as well as in case of a myocardial infarction. Occlusion of the coronary artery (blue arrowhead) will deprive the area perfused by that artery from blood flow. Damaged cardiac tissue evokes wound-healing processes to repair or replace the damaged tissue. These processes include production of cytokines and chemokines, stimulation of cardiac sensory nerves, and infiltration of inflammatory cells (neutrophils, macrophages), that further contribute to the production/release of cytokines. Cytokines released or produced in circulation may enter the brain through leaky regions and/or afferent nervous, stimulating the local inflammatory cells in the brain, the microglia and astrocytes, to produce cytokines. Subsequently, these cytokines lead to stimulation of sympathetic as well as vagal efferent stimulation, with a balanced effect on macrophage activity. Moreover, sympathetic stimulation activates the HPA axis. Glucocorticosteriods act on the infarct as well as on the macrophages. All activated mechanisms initially are aimed to repair the damaged cardiac tissue and to constrain the inflammatory response to this site of action. However, in case of heart failure the inflammatory response seems to persist or even progresses into the circulation and into the brain, where it may cause neuronal changes that could be associated with brain disorders, such as depression, anxiety, and cognitive decline. PVN, paraventricular nucleus hypothalamus; NTS, nucleus solitary tract; ACTH, adrenocorticotrophic hormone; NE, norepinephrine; Ach, acetylcholine (free to Raison *et al.* [58]).

the effects of myocardial infarction on cerebral activity showed a selective regional endothelial leakage mainly in the prefrontal cortex and most severe in the anterior cingulate cortex, an effect that can be mimicked by intravenous infusion of the proinflammatory cytokine TNFα [61]. The effects of TNF-α on blood-brain barrier function in cardiovascular disease was recently thoroughly reviewed by Liu *et al.* [34]. The findings suggest that after infarction, increased circulating proinflammatory cytokines may enter the brain; for instance, TNF-α is observed to be increased in the heart as well as in the hypothalamus in the first hours after myocardial infarction [60]. However, circulating cytokines crossing the blood-brain barrier may not be the only way to explain increased brain levels of cytokines, because PVN mRNA levels for TNF-α had also increased until at least 4 weeks after myocardial infarction, indicating local production of TNF-α [60]. This effect is not limited to TNF-α, as IL1-β, COX-2 mRNA, as well as protein levels are elevated in the hypothalamus [53]. The proinflammatory activity in the PVN is further supported by the observed microglia activation in this area [49,62]. Microglia activation in the PVN is reported to depend on the distance from the 3rd ventricle [49], suggesting "diffusion processes" from the cerebrospinal fluid into brain tissue to play a role as well.

Rats with myocardial infarction show depressive behavior as well [40,63–66]. Indeed, reduction of circulating proinflammatory cytokines seems to improve this depressive behavior [40,65]. These results match the clinical data regarding antidepressant effects of TNF-α blockade [4] as well as the induction of depressive behavior by intravenous infusion of TNF-α [30].

The biological mechanisms involved in the relationship between depression and heart disease are thoroughly reviewed by Grippo and Johnston [12]. Proinflammatory cytokines have been found to have profound effects on the peripheral and brain serotonergic systems. Serotonin plays a major role in depression. Administration of IL-1β, INF-γ, or TNF-α increases extracellular serotonin concentration in several brain areas, including the hypothalamus, hippocampus, and cortex [67]. IL-1β modulates the activity of the serotonin transporter, which plays a central role in serotonergic neurotransmission by reuptake of serotonin. Proinflammatory cytokines, such as IL-1 and INF-γ, may induce activity of indoleamine-2,3 dioxygenase, which converts tryptophan, a precursor of serotonin, to kynurenic acid and quinolinic acid. Induction of indoleamine-2,3 dioxygenase due to inflammation may be detrimental because of depletion of plasma tryptophan and reduced synthesis of serotonin in the brain. Although the exact cellular targets of proinflammatory cytokines in the brain are still elusive, it is evident from animal studies that receptors for IL-1, IL-2, IL-6, and TNF-α have been localized in the brain, with the highest density in the hippocampus and the hypothalamus [4].

Intriguing in this regard is the newly discovered plasma marker in patients with depression and/or Alzheimer's Disease—namely, NGAL. This proinflammatory cytokine has a history as a marker for acute tubular damage, renal disease, and heart failure [27]. NGAL is produced after stimulation with TNF-α in neurons, astrocytes, and microglia, among others [68], but also in cardiomyocytes after myocardial infarction in rats [69]. This latter phenomenon is observed both at mRNA as well as protein level and lasts at least until 9 weeks after infarction, hence, substantially longer than the initial healing phase of the infarct (scar formation). NGAL is noticed as a regulator of inflammatory processes [70–72]. Recently, we showed that in heart failure patients, depression is significantly correlated with plasma NGAL levels, irrespective of measures for cardiac dysfunction and/or renal dysfunction [29].

5.2.2.2 Autonomic Nervous System

Indeed, retrograde transneuronal viral labeling with pseudo-rabies virus injection in different parts of the heart showed bilateral supraspinal infections in brain areas, including the nucleus of the soletary tract, area postrema, raphe nuclei, and also the hypothalamus, amygdala, and cortical areas, including the prefrontal cortex [61]. Right atrial stretch activates neurons in several key autonomic regions, PVN, NTS, and caudal ventrolateral medulla (CVLM), which project directly to the rostral ventrolateral medulla, a critical region in the generation of sympathetic vasomotor tone [73]. As the heart and blood vessels are innervated by sympathetic as well as parasympathetic nerves, the autonomic nervous system carefully balances heart rate, blood pressure, and tissue perfusion to its physiological requirements—regulation by neuronal function, supported by circulating hormones. The rostral ventrolateral medulla and the nucleus of the solitary tract, located in the brain stem, form the main centers for this cardiovascular regulation. The hypothalamic-pituitary-adrenal (HPA) axis supports the cardiovascular regulation; hypothalamic neurons synthesize and release corticotrophin-releasing hormone to the hypothalamo-hypophyseal portal system and induce the release of adrenocorticotropic hormone, which evokes glucocorticoids secretion from the adrenals. Similar mechanisms may play a role in the cardiovascular regulation in cardiovascular disease and its link with brain disorders. Stress is most extensively studied in this regard as it is associated with stimulation of the HPA axis. Using magnetic resonance T2 relaxometry across the entire brain revealed brain injury in autonomic, emotional, and cognitive regulatory areas in patients with heart failure [57].

5.2.2.2.1 SYMPATHETIC ACTIVATION

It is generally acknowledged that cardiovascular disease is accompanied by increased sympathetic drive, or a shift in autonomic balance toward more sympathetic and/or less parasympathetic tone. Although this alteration is initially suggested to have a functional role in compensating the loss of function in the cardiovascular system, when activation is persistent, it may contribute to further progression of heart failure and associated comorbidities.

Indeed in rats with heart failure, sympathetic activation is consistently shown and is related to alteration at the level of the hypothalamus [74,75]. Persistent alterations in heart rate variability, baroreflex sensitivity, and anxiety-like behaviors in rats with heart failure support long-term sympathetic activation [41]. The interrelation between the sympathetic nervous system and inflammation is rather complex as it may integrate signals from different pathways. Sympathetic outflow to the cardiovascular system is mainly regulated in the PVN. TNF-α in the PVN contributes to sympathoexcitation in heart failure by modulating AT1 receptors and neurotransmitters [38,39]. The increase in PVN noradrenaline levels as well as the increased renal sympathetic nerve activity after myocardial infarction can be completely prevented by the cytokine synthase inhibitor pentoxyfillin or etanercept. Felder [76] revealed that brain mineralocorticoid receptors might influence sympathetic discharge by regulating the release of proinflammatory cytokines into the circulation. Alternatively, inflammation is associated with increased sympathetic drive in cardiovascular disease [53]. Blood borne proinflammatory cytokines act upon receptors in the microvasculature of the brain to induce COX2 activity and the production of prostaglandin E2, which penetrates the blood-brain barrier to activate the sympathetic nervous system. Brain mineralocorticoid receptors may influence sympathetic drive by upregulating the activity of the brain renin angiotensin system [76]. Treating heart failure in rats for 6 weeks with the mineralocorticoid receptor blocker Eplenerone, as proven to be effective therapy in heart failure treatment in patients, reduces plasma levels of TNF-α, IL1-β, and IL-6, and is associated with fewer PVN neurons stained positive for TNF-α, IL1-β, or CRH. However, this could not be associated with improved cardiac output [54].

5.2.2.2.2 PARASYMPATHETIC INHIBITION

Neural reflex circuits sense peripheral inflammation and provide regulatory feedback through specific nervous signals and humoral factors—namely, the inflammatory reflex [77]. Sensory vagal fibers innervate many organs, including the cardiovascular system. They are activated by cytokines induced by tissue damage or PAMPs in the periphery and transmit signals to the NTS in the brain stem. Polysynaptic relays connect to the vagal motor neurons in the dorsal vagal motor nucleus and nucleus ambiguous and sympathoexcitatory neurons in the rostral ventrolateral medulla. Efferent vagal nerve signals travel to the celiac plexus and directly to target organs and suppress innate immune responses. Activation of afferent vagal signals also activates the HPA axis, which promotes glucocorticoid release from the adrenal glands. Hence, the central nervous system can rapidly inhibit the release of macrophage TNF-α and attenuate systemic inflammatory responses acting through the vagus (parasympathetic) nerve [47]. This physiological mechanism, termed the "cholinergic anti-inflammatory pathway" [78] has major implications in immunology and therapeutics [79]. The main vagal neurotransmitter, acetylcholine, inhibits lipopolysaccharide-induced TNF-α, IL1-β, and IL6 release, but not the anti-inflammatory cytokine IL-10. Peripheral vagal stimulation inhibits liver TNF-α production, attenuates peak serum TNF-α, and prevents development of shock during lethal endotoxemia in rats [78]. As indicated by Fernandez and Acuna-Castillo [47], the spleen plays a crucial role in vagal nerve control of inflammation. The spleen is the major source of serum TNF-α during endotoxemia [80]. In splenectomized rats, serum TNF-α is reduced by 70% and vagal nerve stimulation does not further suppress TNF-α. The spleen is innervated by the splenic nerve, which is composed mainly of catecholaminergic fibers, which terminate in close proximity to immune cells [81]. Hence, vagus nerve stimulation-induced inhibition of splenic TNF-α production is mediated by norepinephrine released from splenic nerve endings. This stresses the importance of adrenergic transmitters in the regulation of the immune response. Still, all immune cells possess the essential components of cholinergic pathways independent of cholinergic innervation; acetylcholine acts as immunomodulator via both muscarinic and nicotinic receptors [82,83]. Nicotine inhibits cardiac apoptosis induced by LPS in rats [84]. Much less is known about the effects of the immune system on the brain. In response to plasma levels of TNF-α, vagal immunosensory activity increases or decreases vagal motor activity [85,86]. Inflammation and endothelial dysfunction play an important role in hypertension-induced tissue damage. A role for α7nAChR in this process was evidenced from increased end-organ damage, and higher proinflammatory cytokine levels in α7nAChR deficient mice subjected to hypertension [87]. Chronic administration of a α7nAChR antagonist reduced levels of TNF-α, IL1-β, and IL6 and decreased end-organ damage in spontaneously hypertensive rats [88].

5.2.2.2.3 SYNERGISM

High sympathetic activity and, consequently, increases in catecholamines stimulate the beta-receptor-dependent release of IL 10, a potent anti-inflammatory cytokine from monocytes [89]. Thus, the anti-inflammatory effects of the sympathetic and parasympathetic nervous system seem to be synergistic in this setting [32]. In addition to the well-known reciprocal actions, synergistic actions of sympathetic and parasympathetic activation were described in 1982 [90]. Cardiac output increases more at co-stimulation of sympathetic and parasympathetic nerves than by stimulation of either nerve alone; an effect that is originated in the hypothalamus. Fight or flight activation of sympathetic responses also stimulates increased vagus output. The combined action of these neural systems is significantly anti-inflammatory and is positioned anatomically to constrain local inflammation by preventing spillover of potentially lethal toxins into the circulation through both local (neural) and systemic (humoral) anti-inflammatory mechanisms [32]. Neuronal regulation of discrete, distributed, localized inflammatory sites provides a mechanism for integrating responses in real time. It is intriguing to consider that, in addition to the development of immunological memory, the involvement of the cholinergic anti-inflammatory pathway might also modulate processes that promote neural memory of the peri-inflammatory events, such as "sissing of the snake" [32]. Accordingly, Clark *et al.* [91] indicted that electrical stimulation of the vagus nerve in humans significantly enhanced word recognition memory, indicating that memory function and vagus nerve activity are closely linked.

5.2.2.3 Cytokine Functions in the Brain

Cytokines, such as TNF-α, Il-1β, and Il-6, have been shown to be elevated in patients with heart failure in plasma, circulating leukocytes, and myocardial tissue. Although there are strong indications for an involvement of inflammatory processes, it is far from clear whether immune modulatory treatment of heart failure would be useful. Given the overall importance of TNF in heart failure, clinical trials have concentrated mainly on anti-TNF treatment strategies. So far, anti-TNF treatments using Etanercept (recombinant sTNFR2) have yielded mixed results. Infliximab, a monoclonal anti-TNF antibody, even showed adverse effects, which might be due to its binding and blocking soluble as well as membrane-bound TNF [92]. It is now generally believed that other molecules of the TNF and TNF receptor superfamily (OPG, RANK, RANK ligand) and other cytokines, such as IL-1β, IL-15, and IL-18, may be involved in heart failure, and future studies may resolve the complex interaction of immune mediated signaling in heart failure. Triggered by observations in animal and human sickness behavior, it became obvious that sickness behavior and depression share remarkable similarities [93]. For example, intracisternal injection of LPS resulted in animal models in neuroinflammation and induced distinct behavioral changes resulting in anhedonia, social withdrawal, and other aspects similar to depression [94,95]. In several studies, increased cytokine levels have been observed to be related with depression. Cytokines can enter the brain directly across the blood-brain barrier by a saturable transport mechanism; for example, via the interaction of cytokine with circumventricular organs such as the orgnum vasculosum of the lamina terminalis (OVLT) and area postrema, which lack a blood-brain barrier, or by activation of afferent neurons of the vagal nerve [96]. But nearly all types of brain cells like microglia, astrocytes, and neurons can also produce cytokines. In fact, the innate immune response is absolutely important for the maintenance of the brain homeostasis. Therefore, the neuroinflammatory response has to be considered as protective in its very essence. Upon stress, microglia gets activated, their morphology is changed, and they start to secrete proinflammatory cytokines. Proinflammatory cytokines such as interleukin (IL-1), TNFα, and IFNγ coordinate the local and systemic inflammatory response to pathogens. As an example: upon local challenges such as ischemia (in stroke) or amyloid precipitations (in Alzheimer's disease), TNFα and its receptors become strongly expressed locally. TNFα does not necessarily, however, damage the brain tissue via activating TNF receptor 1 (TNFR1). In contrast, by stimulation of TNF receptor 2 (TNFR2) through the membrane form of TNFα, it antagonizes TNFR1 death signals by inducing a neuroprotective signaling cascade that requires the activation of protein kinase B/Akt and nuclear factor kappa B (NF-κB) [97]. It seems that under normal conditions, the TNFR2 protective effect mainly triggered through the membrane standing form of TNF is dominant over the proapoptotic signaling of TNFR1, which is mainly due to the soluble form of TNF [97,98]. A gene array study analyzing neuronal cells treated with TNF revealed that among those factors upregulated selectively via TNFR1 is Lipocalin-2/NGAL. A protein released mainly from astrocytes. Also microglia and neurons have been shown to antagonize the protective TNFR2 function and to sensitize neurons against amyloid beta [68]. It was shown that this protein is upregulated in the hippocampus of Alzheimer patients. Interestingly, it was also found to be increased in patients with heart failure and depression [29].

Whereas the role of proinflammatory cytokines in tissue damage can be thoroughly studied in rather easy experimentally accessible models, the investigation of inflammatory signals on behavioral and cognitive functions is more difficult. Although we know about the effects of inflammation on mood and social behavior (simply because everyone has experienced a common cold during his or her life), it took a surprisingly long time until the link between sickness behavior and depression became obvious. Although the molecular mechanisms for how cytokines contribute to sickness behavior and depression are far from being understood, one clearly identified mechanism may influence serotonin metabolism. It was shown that LPS and cytokines, such as interferons or TNF, influence the production of indoleamine,2,3-dioxygenase (IDO). IDO is involved in the catabolism of tryptophan and may therefore lower the availability of serotonin. In addition, it was shown that some of the resulting catabolites also influence neuronal function. In an experimental setting, it was shown that blockade of IDO prevents depressive behavior in mice besides leaving the neuroinflammatory response to LPS unaffected [94]. However, lower serotonin levels or tryptophan catabolites alone probably do not explain the development of depressive symptoms associated with neuroinflammation. Other inflammatory mediators and mechanisms may be involved in what we consider heart failure associated depression. Hwang and coworkers [99] convincingly showed enhanced cell proliferation and neuroblast differentiation in the hippocampus after myocardial infarction in rats, which they associate with neuronal damage in the limbic system, such as the amygdala. Iosif *et al.* [100] indicated a strong role for TNFα in this hippocampal neurogenesis, with a suppressive effect of the TNFR1 receptor and a neutral or stimulating effect of the TNFR2 receptor. As depression may be associated with a proinflammatory cytokine profile, this would indicate a negative regulation of progenitor proliferation in hippocampal neurogenesis. Although underlying mechanisms of neurocardiac interaction in heart failure are not widely investigated in patients, Woo and coworkers [57], showed a thorough overview of brain areas that are injured in heart failure, providing an interesting base for further animal research. It was also shown that cytokines could directly influence neuronal functions such as long-term potentiation [101,102]. This may explain some behavioral and perhaps also cognitive changes observed related to sickness, as a result from direct interaction with cytokines and their receptors in neuronal and glial cells. In fact, neuroimmunological influences are discussed not only in response to neurological diseases such as Alzheimer's disease or Parkinson's disease but also may be involved in the initiation of the disease by upregulation of factors such as Lipocalin-2/NGAL. NGAL was shown to be unregulated by TNF via TNFR1 and to sensitize neurons against Aβ [68]. It was also shown that NGAL levels increase with age and might link aging with neurodegenerative diseases [28].

5.3 CONCLUSIONS

As most cardiovascular diseases and their comorbidities can be associated with peripheral inflammation, this inflammatory response may well be reflected in the brain as neuroinflammation. This process can be attributed to direct diffusion of inflammatory mediators from circulation into the brain, passing the blood-brain barrier, or at leaky regions, induced by circulating inflammatory mediators. Moreover, peripheral inflammation stimulates afferent neurons from the autonomic nervous system, thereby activating inflammatory regulation in the brain (hypothalamus). Under normal circumstances, the inflammatory responses are activated to save/repair tissue and should subside when the process is finished. For some reason, as in cardiovascular disease, the proinflammatory state persists or even progresses. In brain tissue, persistent inflammation is associated with neuronal damage, which can lead to neurodegeneration in vulnerable areas in the brain. Neurodenegeration can be associated with loss of function, which becomes apparent as psychiatric disorders such as major depression or cognitive decline, negatively affecting prognosis. Because therapy in heart failure patients usually is not aimed at concomitantly treating cardiovascular disease as well as associated brain disorders, (neuro)inflammation may provide a novel target for therapy in this regard.

References

[1] Bourdon KH, Rae DS, Locke BZ, Narrow WE, Regier DA. Estimating the prevalence of mental disorders in U.S. adults from the epidemiologic catchment area survey. Public Health Rep 1992;107(6):663–8.
[2] Kessler RC, McGonagle KA, Zhao S, Nelson CB, Hughes M, Eshleman S, Wittchen HU, Kendler KS. Lifetime and 12-month prevalence of DSM-III-R psychiatric disorders in the United States. Results from the national comorbidity survey. Arch Gen Psychiatry 1994;51(1):8–19.
[3] Kessler RC, Berglund P, Demler O, Jin R, Koretz D, Merikangas KR, Rush AJ, Walters EE, Wang PS, National Comorbidity Survey Replication. The epidemiology of major depressive disorder: results from the national comorbidity survey replication (NCS-R). JAMA 2003;289(23):3095–105.
[4] Pasic J, Levy WC, Sullivan MD. Cytokines in depression and heart failure. Psychosom Med 2003;65(2):181–93.
[5] Barefoot JC, Schroll M. Symptoms of depression, acute myocardial infarction, and total mortality in a community sample. Circulation 1996;93(11):1976–80.
[6] Penninx BW, Beekman AT, Honig A, Deeg DJ, Schoevers RA, van Eijk JT, van Tilburg W. Depression and cardiac mortality: results from a community-based longitudinal study. Arch Gen Psychiatry 2001;58(3):221–7.
[7] Pereira VH, Cerqueira JJ, Palha JA, Sousa N. Stressed brain, diseased heart: a review on the pathophysiologic mechanisms of neurocardiology. Int J Cardiol 2013;166(1):30–7.

[8] Lesperance F, Frasure-Smith N, Theroux P, Irwin M. The association between major depression and levels of soluble intercellular adhesion molecule 1, interleukin-6, and C-reactive protein in patients with recent acute coronary syndromes. Am J Psychiatry 2004;161(2):271–7.
[9] Nabi H, Shipley MJ, Vahtera J, Hall M, Korkeila J, Marmot MG, Kivimaki M, Singh-Manoux A. Effects of depressive symptoms and coronary heart disease and their interactive associations on mortality in middle-aged adults: the Whitehall II cohort study. Heart 2010;96(20):1645–50.
[10] Junger J, Schellberg D, Muller-Tasch T, Raupp G, Zugck C, Haunstetter A, Zipfel S, Herzog W, Haass M. Depression increasingly predicts mortality in the course of congestive heart failure. Eur J Heart Fail 2005;7(2):261–7.
[11] Triposkiadis FK, Skoularigis J. Prevalence and importance of comorbidities in patients with heart failure. Curr Heart Fail Rep 2012;9(4):354–62.
[12] Grippo AJ, Johnson AK. Biological mechanisms in the relationship between depression and heart disease. Neurosci Biobehav Rev 2002;26(8):941–62.
[13] Guck TP, Kavan MG, Elsasser GN, Barone EJ. Assessment and treatment of depression following myocardial infarction. Am Fam Physician 2001;64(4):641–8.
[14] Frasure-Smith N, Lesperance F. Depression and other psychological risks following myocardial infarction. Arch Gen Psychiatry 2003;60(6):627–36.
[15] Lane D, Carroll D, Ring C, Beevers DG, Lip GY. Mortality and quality of life 12 months after myocardial infarction: effects of depression and anxiety. Psychosom Med 2001;63(2):221–30.
[16] Thombs BD, Bass EB, Ford DE, Stewart KJ, Tsilidis KK, Patel U, Fauerbach JA, Bush DE, Ziegelstein RC. Prevalence of depression in survivors of acute myocardial infarction. J Gen Intern Med 2006;21(1):30–8.
[17] Meijer A, Conradi HJ, Bos EH, Thombs BD, van Melle JP, de Jonge P. Prognostic association of depression following myocardial infarction with mortality and cardiovascular events: a meta-analysis of 25 years of research. Gen Hosp Psychiatry 2011;33(3):203–16.
[18] Garcia S, Spitznagel MB, Cohen R, Raz N, Sweet L, Colbert L, Josephson R, Hughes J, Rosneck J, Gunstad J. Depression is associated with cognitive dysfunction in older adults with heart failure. Cardiovasc Psychiatry Neurol 2011;2011:368324.
[19] Duivis HE, Vogelzangs N, Kupper N, de Jonge P, Penninx BW. Differential association of somatic and cognitive symptoms of depression and anxiety with inflammation: findings from the Netherlands study of depression and anxiety (NESDA). Psychoneuroendocrinology 2013;38(9):1573–85.
[20] Thombs BD, de Jonge JP, Coyne JC, Whooley MA, Frasure-Smith N, Mitchell AJ, Zuidersma M, Eze-Nliam C, Lima BB, Smith CG, et al. Depression screening and patient outcomes in cardiovascular care: a systematic review. JAMA 2008;300(18):2161–71.
[21] Krantz DS, Whittaker KS, Francis JL, Rutledge T, Johnson BD, Barrow G, McClure C, Sheps DS, York K, Cornell C, et al. Psychotropic medication use and risk of adverse cardiovascular events in women with suspected coronary artery disease: outcomes from the women's ischemia syndrome evaluation (WISE) study. Heart 2009;95(23):1901–6.
[22] Anker SD, von Haehling S. Inflammatory mediators in chronic heart failure: an overview. Heart 2004;90(4):464–70.
[23] Mommersteeg PM, Kupper N, Schoormans D, Emons W, Pedersen SS. Health-related quality of life is related to cytokine levels at 12 months in patients with chronic heart failure. Brain Behav Immun 2010;24(4):615–22.
[24] Krishnadas R, Cavanagh J. Depression: an inflammatory illness? J Neurol Neurosurg Psychiatry 2012;83(5):495–502.
[25] Johansson P, Lesman-Leegte I, Svensson E, Voors A, van Veldhuisen DJ, Jaarsma T. Depressive symptoms and inflammation in patients hospitalized for heart failure. Am Heart J 2011;161(6):1053–9.
[26] Kupper N, Widdershoven JW, Pedersen SS. Cognitive/affective and somatic/affective symptom dimensions of depression are associated with current and future inflammation in heart failure patients. J Affect Disord 2012;136(3):567–76.
[27] van Deursen VM, Damman K, Voors AA, van der Wal MH, Jaarsma T, van Veldhuisen DJ, Hillege HL. Prognostic value of plasma neutrophil gelatinase-associated lipocalin for mortality in patients with heart failure. Circ Heart Fail 2014;7(1):35–42.
[28] Naude PJ, Eisel UL, Comijs HC, Groenewold NA, De Deyn PP, Bosker FJ, Luiten PG, den Boer JA, Oude Voshaar RC. Neutrophil gelatinase-associated lipocalin: a novel inflammatory marker associated with late-life depression. J Psychosom Res 2013;75(5):444–50.
[29] Naude PJ, Mommersteeg PM, Zijlstra WP, Gouweleeuw L, Kupper N, Eisel UL, Kop WJ, Schoemaker RG. Neutrophil gelatinase-associated lipocalin and depression in patients with chronic heart failure. Brain Behav Immun 2014;38:59–65.
[30] Spriggs DR, Sherman ML, Michie H, Arthur KA, Imamura K, Wilmore D, Frei 3rd E, Kufe DW. Recombinant human tumor necrosis factor administered as a 24-hour intravenous infusion. A phase I and pharmacologic study. J Natl Cancer Inst 1988;80(13):1039–44.
[31] Frantz S, Bauersachs J, Ertl G. Post-infarct remodelling: contribution of wound healing and inflammation. Cardiovasc Res 2009;81(3):474–81.
[32] Tracey KJ. The inflammatory reflex. Nature 2002;420(6917):853–9.
[33] Tracey KJ. Physiology and immunology of the cholinergic antiinflammatory pathway. J Clin Invest 2007;117(2):289–96.
[34] Liu H, Luiten PG, Eisel UL, Dejongste MJ, Schoemaker RG. Depression after myocardial infarction: TNF-alpha-induced alterations of the blood-brain barrier and its putative therapeutic implications. Neurosci Biobehav Rev 2013;37(4):561–72.
[35] Wang H, Bloom O, Zhang M, Vishnubhakat JM, Ombrellino M, Che J, Frazier A, Yang H, Ivanova S, Borovikova L, et al. HMG-1 as a late mediator of endotoxin lethality in mice. Science 1999;285(5425):248–51.
[36] Ahn J, Kim J. Mechanisms and consequences of inflammatory signaling in the myocardium. Curr Hypertens Rep 2012;14(6):510–6.
[37] Szczepanska-Sadowska E, Cudnoch-Jedrzejewska A, Ufnal M, Zera T. Brain and cardiovascular diseases: common neurogenic background of cardiovascular, metabolic and inflammatory diseases. J Physiol Pharmacol 2010;61(5):509–21.
[38] Kang YM, Wang Y, Yang LM, Elks C, Cardinale J, Yu XJ, Zhao XF, Zhang J, Zhang LH, Yang ZM, et al. TNF-alpha in hypothalamic paraventricular nucleus contributes to sympathoexcitation in heart failure by modulating AT1 receptor and neurotransmitters. Tohoku J Exp Med 2010;222(4):251–63.
[39] Kang YM, Gao F, Li HH, Cardinale JP, Elks C, Zang WJ, Yu XJ, Xu YY, Qi J, Yang Q, et al. NF-kappaB in the paraventricular nucleus modulates neurotransmitters and contributes to sympathoexcitation in heart failure. Basic Res Cardiol 2011;106(6):1087–97.
[40] Grippo AJ, Francis J, Weiss RM, Felder RB, Johnson AK. Cytokine mediation of experimental heart failure-induced anhedonia. Am J Physiol Regul Integr Comp Physiol 2003;284(3):R666–73.

[41] Henze M, Hart D, Samarel A, Barakat J, Eckert L, Scrogin K. Persistent alterations in heart rate variability, baroreflex sensitivity, and anxiety-like behaviors during development of heart failure in the rat. Am J Physiol Heart Circ Physiol 2008;295(1):H29–38.
[42] Zera T, Ufnal M, Szczepanska-Sadowska E. Central TNF-alpha elevates blood pressure and sensitizes to central pressor action of angiotensin II in the infarcted rats. J Physiol Pharmacol 2008;59(Suppl 8):117–21.
[43] de la Torre JC. Cardiovascular risk factors promote brain hypoperfusion leading to cognitive decline and dementia. Cardiovasc Psychiatry Neurol 2012;2012:367516.
[44] Cardiogenic dementia. Lancet 1977;1(8001):27–8.
[45] Emerson TR, Milne JR, Gardner AJ. Cardiogenic dementia-a myth? Lancet 1981;2(8249):743–4.
[46] Kumar R, Woo MA, Macey PM, Fonarow GC, Hamilton MA, Harper RM. Brain axonal and myelin evaluation in heart failure. J Neurol Sci 2011;307(1–2):106–13.
[47] Fernandez R, Acuna-Castillo C. Neural reflex control of inflammation during sepsis syndromes. In: Azevedo L, editor. Sepsis—an ungoing and significant challenge. INTECH, Chile; 2012; 6:133–56.
[48] Patel KP, Zhang PL, Krukoff TL. Alterations in brain hexokinase activity associated with heart failure in rats. Am J Physiol 1993;265(4):R923–8.
[49] Rana I, Stebbing M, Kompa A, Kelly DJ, Krum H, Badoer E. Microglia activation in the hypothalamic PVN following myocardial infarction. Brain Res 2010;1326:96–104.
[50] Kang YM, Ma Y, Elks C, Zheng JP, Yang ZM, Francis J. Cross-talk between cytokines and renin-angiotensin in hypothalamic paraventricular nucleus in heart failure: role of nuclear factor-kappaB. Cardiovasc Res 2008;79(4):671–8.
[51] Zhang K, Li YF, Patel KP. Reduced endogenous GABA-mediated inhibition in the PVN on renal nerve discharge in rats with heart failure. Am J Physiol Regul Integr Comp Physiol 2002;282(4):R1006–15.
[52] Kang YM, Zhang ZH, Xue B, Weiss RM, Felder RB. Inhibition of brain proinflammatory cytokine synthesis reduces hypothalamic excitation in rats with ischemia-induced heart failure. Am J Physiol Heart Circ Physiol 2008;295(1):H227–36.
[53] Yu Y, Zhang ZH, Wei SG, Serrats J, Weiss RM, Felder RB. Brain perivascular macrophages and the sympathetic response to inflammation in rats after myocardial infarction. Hypertension 2010;55(3):652–9.
[54] Kang YM, Zhang ZH, Johnson RF, Yu Y, Beltz T, Johnson AK, Weiss RM, Felder RB. Novel effect of mineralocorticoid receptor antagonism to reduce proinflammatory cytokines and hypothalamic activation in rats with ischemia-induced heart failure. Circ Res 2006;99(7):758–66.
[55] Kumar R, Woo MA, Birrer BV, Macey PM, Fonarow GC, Hamilton MA, Harper RM. Mammillary bodies and fornix fibers are injured in heart failure. Neurobiol Dis 2009;33(2):236–42.
[56] Kumar R, Nguyen HD, Ogren JA, Macey PM, Thompson PM, Fonarow GC, Hamilton MA, Harper RM, Woo MA. Global and regional putamen volume loss in patients with heart failure. Eur J Heart Fail 2011;13(6):651–5.
[57] Woo MA, Kumar R, Macey PM, Fonarow GC, Harper RM. Brain injury in autonomic, emotional, and cognitive regulatory areas in patients with heart failure. J Card Fail 2009;15(3):214–23.
[58] Raison CL, Capuron L, Miller AH. Cytokines sing the blues: Inflammation and the pathogenesis of depression. Trends Immunol 2006;27(1):24–31.
[59] Felder RB, Francis J, Zhang ZH, Wei SG, Weiss RM, Johnson AK. Heart failure and the brain: new perspectives. Am J Physiol Regul Integr Comp Physiol 2003;284(2):R259–76.
[60] Francis J, Chu Y, Johnson AK, Weiss RM, Felder RB. Acute myocardial infarction induces hypothalamic cytokine synthesis. Am J Physiol Heart Circ Physiol 2004;286(6):H2264–71.
[61] Ter Horst GJ. Neuroanatomy of cardiac activity-regulating circuitry: a transneuronal retrograde viral labelling study in the rat. Eur J Neurosci 1996;8(10):2029–41.
[62] Dworak M, Stebbing M, Kompa AR, Rana I, Krum H, Badoer E. Sustained activation of microglia in the hypothalamic PVN following myocardial infarction. Auton Neurosci 2012;169(2):70–6.
[63] Schoemaker RG, Smits JF. Behavioral changes following chronic myocardial infarction in rats. Physiol Behav 1994;56(3):585–9.
[64] Schoemaker RG, Kalkman EA, Smits JF. 'Quality of life' after therapy in rats with myocardial infarction: dissociation between hemodynamic and behavioral improvement. Eur J Pharmacol 1996;298(1):17–25.
[65] Bah TM, Kaloustian S, Rousseau G, Godbout R. Pretreatment with pentoxifylline has antidepressant-like effects in a rat model of acute myocardial infarction. Behav Pharmacol 2011;22(8):779–84.
[66] Wann BP, Bah TM, Kaloustian S, Boucher M, Dufort AM, Le Marec N, Godbout R, Rousseau G. Behavioural signs of depression and apoptosis in the limbic system following myocardial infarction: effects of sertraline. J Psychopharmacol 2009;23(4):451–9.
[67] Clement HW, Buschmann J, Rex S, Grote C, Opper C, Gemsa D, Wesemann W. Effects of interferon-gamma, interleukin-1 beta, and tumor necrosis factor-alpha on the serotonin metabolism in the nucleus raphe dorsalis of the rat. J Neural Transm 1997;104(10):981–91.
[68] Naude PJ, Nyakas C, Eiden LE, Ait-Ali D, van der Heide R, Engelborghs S, Luiten PG, De Deyn PP, den Boer JA, Eisel UL. Lipocalin 2: novel component of proinflammatory signaling in alzheimer's disease. FASEB J 2012;26(7):2811–23.
[69] Yndestad A, Landro L, Ueland T, Dahl CP, Flo TH, Vinge LE, Espevik T, Froland SS, Husberg C, Christensen G, et al. Increased systemic and myocardial expression of neutrophil gelatinase-associated lipocalin in clinical and experimental heart failure. Eur Heart J 2009;30(10):1229–36.
[70] Lee S, Lee WH, Lee MS, Mori K, Suk K. Regulation by lipocalin-2 of neuronal cell death, migration, and morphology. J Neurosci Res 2012;90(3):540–50.
[71] Lee S, Lee J, Kim S, Park JY, Lee WH, Mori K, Kim SH, Kim IK, Suk K. A dual role of lipocalin 2 in the apoptosis and deramification of activated microglia. J Immunol 2007;179(5):3231–41.
[72] Shashidharamurthy R, Machiah D, Aitken JD, Putty K, Srinivasan G, Chassaing B, Parkos CA, Selvaraj P, Vijay-Kumar M. Differential role of lipocalin 2 during immune complex-mediated acute and chronic inflammation in mice. Arthritis Rheum 2013;65(4):1064–73.
[73] Kantzides A, Owens NC, De Matteo R, Badoer E. Right atrial stretch activates neurons in autonomic brain regions that project to the rostral ventrolateral medulla in the rat. Neuroscience 2005;133(3):775–86.
[74] Patel KP, Zhang K, Carmines PK. Norepinephrine turnover in peripheral tissues of rats with heart failure. Am J Physiol Regul Integr Comp Physiol 2000;278(3):R556–62.

[75] Badoer E. Role of the hypothalamic PVN in the regulation of renal sympathetic nerve activity and blood flow during hyperthermia and in heart failure. Am J Physiol Renal Physiol 2010;298(4):F839–46.
[76] Felder RB. Mineralocorticoid receptors, inflammation and sympathetic drive in a rat model of systolic heart failure. Exp Physiol 2010;95(1):19–25.
[77] Olofsson PS, Rosas-Ballina M, Levine YA, Tracey KJ. Rethinking inflammation: neural circuits in the regulation of immunity. Immunol Rev 2012;248(1):188–204.
[78] Borovikova LV, Ivanova S, Zhang M, Yang H, Botchkina GI, Watkins LR, Wang H, Abumrad N, Eaton JW, Tracey KJ. Vagus nerve stimulation attenuates the systemic inflammatory response to endotoxin. Nature 2000;405(6785):458–62.
[79] Rosas-Ballina M, Tracey KJ. The neurology of the immune system: neural reflexes regulate immunity. Neuron 2009;64(1):28–32.
[80] Mignini F, Streccioni V, Amenta F. Autonomic innervation of immune organs and neuroimmune modulation. Auton Autacoid Pharmacol 2003;23(1):1–25.
[81] Felten DL, Ackerman KD, Wiegand SJ, Felten SY. Noradrenergic sympathetic innervation of the spleen: I. Nerve fibers associate with lymphocytes and macrophages in specific compartments of the splenic white pulp. J Neurosci Res 1987;18(1):28–36 118–21.
[82] Kawashima K, Fujii T. The lymphocytic cholinergic system and its biological function. Life Sci 2003;72(18–19):2101–9.
[83] Kawashima K, Fujii T. Extraneuronal cholinergic system in lymphocytes. Pharmacol Ther 2000;86(1):29–48.
[84] Suzuki J, Bayna E, Dalle Molle E, Lew WY. Nicotine inhibits cardiac apoptosis induced by lipopolysaccharide in rats. J Am Coll Cardiol 2003;41(3):482–8.
[85] Emch GS, Hermann GE, Rogers RC. Tumor necrosis factor-alpha inhibits physiologically identified dorsal motor nucleus neurons in vivo. Brain Res 2002;951(2):311–5.
[86] Emch GS, Hermann GE, Rogers RC. TNF-alpha activates solitary nucleus neurons responsive to gastric distension. Am J Physiol Gastrointest Liver Physiol 2000;279(3):G582–6.
[87] Bautista LE. Inflammation, endothelial dysfunction, and the risk of high blood pressure: epidemiologic and biological evidence. J Hum Hypertens 2003;17(4):223–30.
[88] Li DJ, Evans RG, Yang ZW, Song SW, Wang P, Ma XJ, Liu C, Xi T, Su DF, Shen FM. Dysfunction of the cholinergic anti-inflammatory pathway mediates organ damage in hypertension. Hypertension 2011;57(2):298–307.
[89] van der Poll T, Coyle SM, Barbosa K, Braxton CC, Lowry SF. Epinephrine inhibits tumor necrosis factor-alpha and potentiates interleukin 10 production during human endotoxemia. J Clin Invest 1996;97(3):713–9.
[90] Koizumi K, Terui N, Kollai M, Brooks CM. Functional significance of coactivation of vagal and sympathetic cardiac nerves. Proc Natl Acad Sci U S A 1982;79(6):2116–20.
[91] Clark KB, Naritoku DK, Smith DC, Browning RA, Jensen RA. Enhanced recognition memory following vagus nerve stimulation in human subjects. Nat Neurosci 1999;2(1):94–8.
[92] Gullestad L, Ueland T, Vinge LE, Finsen A, Yndestad A, Aukrust P. Inflammatory cytokines in heart failure: mediators and markers. Cardiology 2012;122:23–35.
[93] Dantzer R. Cytokine, sickness behavior, and depression. Immunol Allergy Clin North Am 2009;29(2):247–64.
[94] Dobos N, de Vries EF, Kema IP, Patas K, Prins M, Nijholt IM, Dierckx RA, Korf J, den Boer JA, Luiten PG, et al. The role of indoleamine 2,3-dioxygenase in a mouse model of neuroinflammation-induced depression. J Alzheimers Dis 2012;28(4):905–15.
[95] van Heesch F, Prins J, Korte-Bouws GA, Westphal KG, Lemstra S, Olivier B, Kraneveld AD, Korte SM. Systemic tumor necrosis factor-alpha decreases brain stimulation reward and increases metabolites of serotonin and dopamine in the nucleus accumbens of mice. Behav Brain Res 2013;253:191–5.
[96] Hosoi T, Okuma Y, Nomura Y. The mechanisms of immune-to-brain communication in inflammation as a drug target. Curr Drug Targets Inflamm Allergy 2002;1(3):257–62.
[97] Marchetti L, Klein M, Schlett K, Pfizenmaier K, Eisel UL. Tumor necrosis factor (TNF)-mediated neuroprotection against glutamate-induced excitotoxicity is enhanced by N-methyl-D-aspartate receptor activation. essential role of a TNF receptor 2-mediated phosphatidylinositol 3-kinase-dependent NF-kappa B pathway. J Biol Chem 2004;279(31):32869–81.
[98] Fontaine V, Mohand-Said S, Hanoteau N, Fuchs C, Pfizenmaier K, Eisel U. Neurodegenerative and neuroprotective effects of tumor necrosis factor (TNF) in retinal ischemia: Opposite roles of TNF receptor 1 and TNF receptor 2. J Neurosci 2002;22(7):RC216.
[99] Hwang IK, Yoo KY, Han TH, Lee CH, Choi JH, Yi SS, Lee SY, Ryu PD, Yoon YS, Won MH. Enhanced cell proliferation and neuroblast differentiation in the rat hippocampal dentate gyrus following myocardial infarction. Neurosci Lett 2009;450(3):275–80.
[100] Iosif RE, Ekdahl CT, Ahlenius H, Pronk CJ, Bonde S, Kokaia Z, Jacobsen SE, Lindvall O. Tumor necrosis factor receptor 1 is a negative regulator of progenitor proliferation in adult hippocampal neurogenesis. J Neurosci 2006;26(38):9703–12.
[101] Curran BP, O'Connor JJ. The inhibition of long-term potentiation in the rat dentate gyrus by pro-inflammatory cytokines is attenuated in the presence of nicotine. Neurosci Lett 2003;344(2):103–6.
[102] Balschun D, Randolf A, Pitossi F, Schneider H, Del Rey A, Besedovsky HO. Hippocampal interleukin-1 beta gene expression during long-term potentiation decays with age. Ann N Y Acad Sci 2003;992:1–8.

CHAPTER

6

Translation of Animal Models into Clinical Practice: Application to Heart Failure

Robrecht Thoonen,[a] *Sara Vandenwijngaert,*[a] *Jonathan Beaudoin,*[a] *Emmanuel Buys, Marielle Scherrer-Crosbie*

Massachusetts General Hospital, Boston, Massachusetts, USA

6.1 INTRODUCTION

Animal models remain an invaluable tool in the study of cardiovascular diseases. Over the last few decades, the advent of genetic manipulation has allowed extensive characterization of the pathophysiological impact of gain or loss of function of virtually every known gene or coding sequence. Mice in particular have proven to be amenable to this type of approach and more than 100 million mice are used each year for research. There are many other advantages to mouse research, including a great similarity of mouse and human genomes (more than 99% of the of mouse genes have a homologue in the human genome [1]) reasonable closeness of biochemical pathways, physiological, pharmacological, and organ functions, short gestation period (19 days), large litters (5-8 pups) and low cost (see Table 6.1). Advanced techniques such as high-resolution echocardiography, magnetic resonance imaging, positron emission tomography, invasive hemodynamic measurements, and electrocardiography have been developed to allow for the accurate evaluation of heart function, infarct size, and quantitative characterization of post-infarct remodeling and electrophysiological aspects in these small species [2].

In addition to mice, other mammals, including rats, rabbits, dogs, pigs, goats, sheep, cows, and primates, have been extensively studied in models of cardiovascular disease. Large-animal models often allow for a more straightforward translation to humans due to the greater similarity between LV geometry, loading conditions, heart rate, oxygen consumption, and associated valvular disease. Other advantages of studying larger animals include the possibility to serially harvest samples and perform procedures that are not always possible in smaller organisms. Finally, studies in large animals are required before new drug treatments, surgical therapies, and devices can be tested in clinical trials.

There are limitations to all animal models. First, most experimental studies are undertaken in young, healthy animals, which may not reflect the age and comorbidities of most human patients. Furthermore, the models themselves imperfectly reflect the development of human pathologies.

In mice studies, many examples exist of genetic diseases in humans whose equivalent in genetically modified mice results in very different phenotypes or no phenotype at all. This discrepancy may reflect differences between murine and human genome or physiology. For instance, although calcium removal from the cytosol is dominated by the activity of the SR Ca^{2+}-ATPase in humans and rodents, the Na^{+}/Ca^{2+}-exchanger activity is less relevant in rodents than in humans [3]. The cardiomyocyte action potentials of rat and mouse have a very short duration and lack a plateau phase and the force-frequency relation is inverse [4]. The resting heart rate is five times higher in mice than in humans. Oxygen consumption, adrenergic receptor ratios, and the response to loss of regulatory

[a] Authors contributed equally to the work.

http://dx.doi.org/10.1016/B978-0-12-800039-7.00006-2
© 2015 Elsevier Inc. All rights reserved.

TABLE 6.1 Advantages and Limitations of Animal Models

	Advantages	Limitations
Choice of animals		Often young and healthy animals
Models		May not represent clinical evolution
Mice	Genetic modifications: investigating one gene	Genetic modifications: not always relevant to humans (multigenic)
	Genome similar	Some differences in genome
	Inexpensive	
	Fast breeders	
		Compensatory mechanisms
		Genetic background may induce different phenotypes
Rats/rabbits		Possible genetic modifications but less versatile
Large animals	Closer in anatomy and physiology to humans	Differences exist (ex: coronary collaterals in dogs)
		Less mechanistic insights and genetic modifications
		Genetic variations among animals
		Expensive

proteins are different when mice are compared to humans [5]. In addition, the expression of contractile proteins—specifically the predominant myosin isoforms—is different between the two species leading to a different excitation-contraction coupling process [5]. As a result, it is difficult to extrapolate murine systems when making interpretations of human heart failure pathophysiology, particularly after induction of cardiovascular stress. Therefore, large-animal models of heart failure, which more closely resemble human physiology, function, and anatomy, are essential to bridge the discoveries from murine models into clinical therapies and interventions for heart failure.

Alternatively, compensatory mechanisms may develop in mice that lack a specific gene. The generation of inducible genetic modifications [6], in which the modification can be induced after embryonic life, may reduce the latter issue. In other occasions, a phenotype only becomes apparent when a mouse is exposed to particular stressors. The phenotypical abnormalities attributed to the null mutation in several studies can sometimes result from interaction of effects caused by the mutation with the effects of strain specific genes or alleles, demonstrating the critical importance of genetic background.

Regarding large-animal studies, cost as well as the regulatory burden associated with researching larger animal models can be prohibitive. Furthermore, genetic modifications techniques are not available in large animals, thus pathway analysis is often less extensive in larger animals than in mice, resulting in a poorer understanding of disease mechanisms in larger species relative to mice. Finally, unlike studies in mice on fixed genetic backgrounds, studies in larger animals may yield noisy results due to genetic variation.

6.2 ANIMAL MODELS OF ACQUIRED CARDIOMYOPATHY

6.2.1 Models of Ischemia-Reperfusion and Infarction

Coronary heart disease, leading to the development of myocardial infarction (MI) or ischemia, is the principal cause of congestive heart failure (CHF) in the majority of patients [7]. With the wide application of reperfusion therapies in the treatment of patients with acute MI, cardiac dysfunction resulting from ischemia/reperfusion (I/R) injury is an increasingly important pathology. A number of animal models have been developed to study the processes of MI and I/R injury. These models employ either a catheterization to perform coronary embolization or a balloon occlusion, or, more frequently, a surgical intervention to place a ligature around a coronary artery after a thoracotomy. The ligation is either performed in a permanent way to induce a MI or in a temporary way to induce I/R injury. The MI model is used to investigate changes in the myocardium that occur over an extended period of time, whereas the I/R model is mostly used to examine short-term consequences of ischemic injury to the heart [8].

6.2.1.1 Small Rodent Models

Myocardial infarction and I/R in mice have been extensively studied and nonreperfused MI is a recognized model used to study the development of heart failure. Limitations include the technical challenge, the fact that only the left anterior descending artery (LAD) is large enough to be ligated, and the large variations in infarct size (the standard deviation is usually approximately 10% [9]).

The mouse model of LAD ligation to induce MI was first described three decades ago by Zolotareva *et al.* [10] and more recently I/R in mice was described by Michael with others [11]. In the latter publication, the highly variable anatomy of the mouse LAD and the variable MI size were identified. Specific approaches have been developed to keep the variability of the infarct size to a minimum [12,13]. Mortality associated with MI induction in mice depends on the MI size and can vary between 20% and 50% [9,11]. Most cases of death occur within the first days after MI. Recently, a new method was developed without mechanical ventilation in which survival rates were improved by 20% in comparison to the classic method [14].

The induction of left ventricular MI in rats was first established by Pfeffer *et al.* [15]. Rats with MIs greater than 46% of the LV may develop overt CHF after 3-6 weeks, however, this HF is not uniform. Interestingly, LV remodeling and failure depends on the strain. In a comparative study by Liu *et al.*, the mortality of Sprague-Dawley rats was 36% whereas in Lewis inbred rats it was only 16% [16]. In addition, Liu *et al.* reported that the position and the branching of the LAD in Sprague-Dawley rats is more heterogeneous compared to Lewis inbred rats. Concomitantly, infarct sizes were more uniform in Lewis inbred rats after LAD ligation [16].

Similar to the MI models, the induction of I/R involves a ligation of the LAD, but to allow for an easier and safe relief of the occlusion, a small plastic tube is placed between the ligated vessel and the knot [11]. Additionally, a "closed chest" ligation model of myocardial I/R injury has been developed to minimize interfering inflammatory effects induced by surgery, however, this model has not been extensively used [17]. In comparison to the MI model, the I/R injury model is associated with smaller MI sizes, a higher infiltration of inflammatory cells, enhanced neovascularization in the infarct area, and attenuated fibrotic remodeling [18].

6.2.1.2 Large-Animal Models

Canine models of ischemia and myocardial infarctions have provided critical data about the ischemic cascade ultimately leading to necrosis following coronary occlusion [19], subsequent LV remodeling, and heart failure following MI and potential impact of early reperfusion [20]. Permanent or temporary coronary occlusion by either open-chest coronary ligation or closed-chest percutaneous procedure are commonly used; these models are representative of the MI seen clinically (followed or not by reperfusion). However, canine models of myocardial ischemia and infarction have the drawback that extensive collateral circulation can develop after ligation, which in turn induces variable reperfusion and salvage of the ischemic tissue, making inconsistent degrees of myocardial injury and post-MI remodeling [20]. To circumvent this problem, in canines the repetitive left-sided coronary artery microembolization model can be performed [21]. The canine microembolization model recapitulates the clinical phenotype of ischemic cardiomyopathy, however, it does so through multiple sites of infarction and remodeling as opposed to single, discrete lesion of a large, focal MI. This multiplicity and heterogeneity of the myocardial response to the microembolizations can make the interpretation of the biological responses difficult. The CHF induced by this model is irreversible, but malignant dysrhythmias contribute to high mortality rates. A major advantage of this model is that varying degrees of dysfunction can be regulated by the number of embolic events.

The canine models have helped to describe the inflammatory response following ischemic events and associated reperfusion injury. Proinflammatory cytokines such as IL-1 and TNF-alpha have been linked to contractile dysfunction in ischemic myocardium [22,23], with evidence of neutrophil-mediated injury [24]. Several inflammation-based therapies such as corticosteroids, alteration of complement system, or interaction with neutrophils were found to reduce infarct size in dogs and nonhuman primates, but subsequent clinical trials have triggered disappointing results [25–27]. The presence of collateral circulation in canine models limits the creation of consistent infarct size and location, and therefore increases the variability in the resulting phenotype, an important factor to account for since these large-animal studies have typically limited sample size. In part, and for that reason, similar models have been created in pigs and sheep [28]. Both pigs and sheep display consistent coronary anatomy, lack of preformed collateral vessels, and provide the ability to create infarctions of predictable size and location [28,29]. The coronary artery anatomy and gross anatomical structure of porcine heart, in particular, is very similar to that of humans, and hence has been the subject of translational studies [28]. A difficulty in pig MI models is the predisposition of pigs to refractory arrhythmogenesis, necessitating aggressive airway protection and ventilatory management, electrolyte supplementation, and antiarrhythmetic administration.

Ovine models are often used to study the pathogenesis and impact of functional regurgitation in the setting of ischemic and nonischemic heart failure, in part, because of anatomic similarities with human mitral apparatus and papillary muscles configuration. Such studies help to understand the relation between infarct location and secondary functional mitral regurgitation [29], as well as the impact of secondary mitral regurgitation on subsequent LV remodeling [30].

6.2.2 Animal Models of Pressure and Volume Overload

6.2.2.1 The Mouse Model of Transverse Aortic Constriction

Rockman *et al.* were the first investigators to describe the transverse aortic constriction (TAC) model [31], in which a mechanical obstacle distal to the LV is produced by constriction of the transverse aorta. An advantage of this model is the ability to quantify the pressure gradient across the aortic constriction. Typically, the diameter of the aorta is reduced by approximately 50%, and a 50-60 mmHg systolic pressure gradient is created between LV and aortic pressure. The gradient varies depending on the hemodynamic conditions, and the severity of the ligation itself can be variable, necessitating superior technical skills. TAC causes a sudden onset of hypertension leading to an approximately 50% increase in LV mass within 2 weeks. The development of LV dysfunction, heart failure, and its timing are highly dependent on the experimental conditions, and differing results have been reported for the same genetic modification depending on the TAC surgery [32,33]. One limitation of the model is that the acute onset of severe hypertension that characterizes this model lacks direct clinical relevance. Models of suprarenal aortic constriction and of angiotensin chronic infusion have also been developed but are rarely used for the investigation of CHF.

6.2.2.2 The Rat Model of Aortic Constriction

Suprarenal constriction of the aorta is often used in rats and produces LV hypertrophy and subsequent CHF. After 20 weeks of aortic banding, two distinctive groups can be identified, including rats with and without a significant reduction in LV systolic pressure [34]. Ascending aortic stenosis (which is not frequently used in mice) is a severe model of pressure overload and rapidly induces overt heart failure [35] when acutely performed in adult animals. Another technique that may better reflect clinical conditions is to place a suture around the ascending aorta in rat weanlings, which progressively constricts the aorta as the animal grows [36].

6.2.2.3 Spontaneous Hypertensive Rats

The spontaneous hypertensive rat (SHR) has been used as a well-established model of genetic hypertension and age-dependent LV dysfunction. More than 50% of the animals show clinical signs of HF at the age of 18-24 months. During the development of LF dysfunction, increased fibrosis and an altered LV geometry are observed [37,38]. The model of SHR rats is a good model to reproduce hypertension-induced HF in humans and to study the transition from hypertrophy to HF. An additional advantage of this model is that it does not require surgical procedures or pharmacological intervention.

6.2.2.4 Large-Animal Models

A comparatively smaller number of studies have used large-animal models to create pressure overload; however, models of aortic valvular stenosis, renovascular hypertension, or aortic banding have been successfully used before [39–41], with a resulting phenotype characterized by concentric hypertrophy, myocardial fibrosis, and diastolic dysfunction.

A larger number of studies have used large-animal models of volume overload mimicking the effect of acute or chronic valvular diseases. Mitral subvalvular chordaes disruption has been employed to create severe regurgitation [42], with subsequent LV eccentric hypertrophy and loss of systolic function. Other approaches have utilized the implantation of an LV-to-LA shunt with the advantage of predictable and standardized degrees of volume overload [43]; systemic arterio-venous shunts have been also described [44]. These models can be used alone or in combination with infarction to reproduce different physiologies seen clinically. Canine and ovine models of mitral regurgitation are regularly employed to investigate the effect of medical or surgical therapies on LV remodeling and failure [43,45]. Of note, a similar model of volume overload by aorta-caval fistula has been described in rats [46]; it is difficult, although feasible, to reliably reproduce it in mice [47].

Tachycardia-induced CHF through prolonged rapid pacing is the most commonly utilized model of nonischemic dilated cardiomyopathy (DCM). Rapid pacing (>200 bpm over 3-4 weeks) consistently result in global LV dilatation, systolic dysfunction, and humoral activation seen in heart failure [48]. Monocyte infiltration suggestive of active inflammation was shown following rapid pacing in dogs [49]. One advantage of this approach is the possibility of changing the pacing rate and duration to obtain the desired degree of LV dysfunction. This model has been utilized to investigate mechanisms and therapies for nonischemic heart failure irrespective of the underlying etiology. While tachycardia-induced systolic dysfunction is by itself a clinically relevant disease for which rapid pacing models are perfectly suited, it should be remembered that the causes of dilated cardiomyopathies are various and heterogeneous, and therefore caution is needed when translating the findings of tachycardia-induced HF models to other causes of heart failure.

6.2.3 Animal Models of Toxic Cardiomyopathy

The clinical relevance of chemotherapy-induced cardiomyopathy has prompted the development of animal models. Much effort has been made to reproduce features of cardiotoxicity induced by the antineoplastic anthracycline drugs in laboratory animals, encompassing rabbits, rats, mice, pigs, nonhuman primates, and dogs.

The first animal studies to investigate chemotherapy-induced cardiomyopathy were performed in rabbits. Generally, cardiotoxicity was induced by repeatedly administering anthracyclines over a period of 10-18 weeks with cumulative doses of 20-30 mg/kg [50]. Initial rodent studies showed signs of CHF in Fischer rats several weeks after chronic doxorubicin exposure [51]. In search of an improved rat model, it was discovered that spontaneously hypertensive rats were more sensitive to doxorubicin-associated cardiotoxicity than normotensive rats [52], a finding similar to humans [53]. Because of this increased sensitivity and the high reproducibility of anthracycline-induced cardiac lesions, spontaneously hypertensive rats have been considered more suitable than other rat strains as small-animal models of anthracycline-induced cardiotoxicity. In addition to rats, mice have also been used to study toxic cardiomyopathy. A drawback of this model is the technical difficulty associated with multiple intraperitoneal or intravenous injections of corrosive drugs. The magnitude of left-ventricular dysfunction in these models is generally mild; however, sensitive, noninvasive myocardial deformation parameters have been validated in this species, allowing detection of subtle anthracycline-induced cardiac dysfunction [54].

In addition to rodents, large-animal models have been used to study anthracycline-induced cardiomyopathy because of a greater similarity in cardiac anatomy and physiology with humans. These large-animal models include pigs (full-size and miniature), in which characteristic cardiac lesions were observed following chronic exposure to lower doxorubicin doses [55]. In addition, cardiotoxicity has been induced in nonhuman primates by anthracycline administration [56,57]. The resulting cardiac alterations resembled those found in patients, although cardiotoxicity was detected at lower cumulative doses. Finally, dogs represent a valuable animal model of anthracycline-induced cardiotoxicity because of the uniformity of the cardiotoxic response in different animals and the lack of significant toxicity in noncardiac tissues [56]. The main obstacle in using large animals is the expense involved in the experiments.

Both (sub)acute and chronic cardiotoxicity have been investigated in laboratory animals. Acute cardiotoxicity is generally evaluated within hours or days after a single administration of a relatively high anthracycline dose. Chronic cardiotoxicity is induced by repeatedly administering lower drug doses and is evaluated weeks to months after cessation of anthracycline treatment. Although the chronic model is clinically more relevant, it is associated with higher costs and extended study time frames.

6.2.4 Models of Sepsis-Associated LV Dysfunction

Although translation of findings in animal models of systemic inflammation to the clinic has been marred by failure, they have considerably advanced our understanding of the pathological mechanisms contributing to the morbidity and mortality associated with sepsis [58]. Here, we will focus on how animal models have helped us understand the cardiovascular sequelae of sepsis and the role of nitric oxide (NO) signaling therein.

It is now recognized that both refractory hypotension and myocardial depression contribute significantly to the morbidity and mortality of sepsis and septic shock. The life-threatening cardiovascular dysfunction in septic shock originates, at least in part, from NO overproduction [59], historically believed to derive from transcriptionally regulated inducible NO synthase (NOS2). Administration of a highly selective NOS2 inhibitor prevented systemic hypotension and LV systolic and diastolic dysfunction associated with endotoxin challenge in mice [60]. Cardiac dysfunction and systemic hypotension were less pronounced in NOS2-deficient mice challenged with high doses of endotoxin than in WT mice, suggesting a critical role of NOS2 in the pathogenesis of LV dysfunction associated with endotoxemia [61]. In contrast, administration of L-NAME, an inhibitor of all three NOS isoforms, exacerbated mortality and cardiac dysfunction in mice challenged with endotoxin [60]. These observations suggest that although inhibition of NOS2 may be beneficial in septic shock, concurrent inhibition of NOS1 and NOS3 may be deleterious, implying that NO can act as a double-edged sword during septic shock. Protective effects of NOS1 and/or NOS3 against the LV dysfunction associated with endotoxemia in mice may help to explain the generally disappointing results reported from clinical trials of patients with septic shock treated with nonselective NOS inhibitors [62]. It is of note that the serious adverse events in a multicenter trial with L-NMMA appeared to be primarily of cardiac origin [62].

There are limitations to the septic shock animal models. In patients, the course of infection is slower in onset and less fulminant than after administration of endotoxin (the most utilized model of sepsis). Furthermore, different animal models are often difficult to compare, not only due to obvious species-related differences but also because of variances in methodology, including anesthesia protocol, dosage of the compound injected, timing, duration and route of administration, and regime of treatment. In human studies, supportive therapies in the form of fluids or antibiotics and the presence or absence of comorbidities that are not typical in animal models may influence the outcome. Lastly, the unfortunate reality that promising therapeutic strategies tested in animal models get lost in translation is also due to flawed patient selection. For example, lessons learned from models of sepsis that involve administration of bacteria to young adult mice may not necessarily be relevant to patients of different age and ethnicity with a very diverse array of initial events (including burns, trauma, surgical procedures).

6.3 ANIMAL MODELS OF GENETIC CARDIOMYOPATHIES

Familial cardiomyopathy results from inherited defects in genes encoding a broad range of proteins, including sarcomeric, cytoskeletal, desmosomal, and nuclear membrane proteins, as well as proteins involved in Ca^{2+}-metabolism and energy production and regulation [63]. Animal models of familial cardiomyopathy have been instrumental in confirming causality of gene mutations and dissecting out key molecular pathways involved in the development of cardiomyopathy and its sequelae, including heart failure and sudden death.

Animal models of familial hypertrophic cardiomyopathy (HCM) mainly harbor mutations in the cardiac β-myosin heavy chain (MYH7), troponin T (TNNT2), and myosin-binding protein C (MYBPC3) genes, since mutations in these three genes are encountered in more than 70% of HCM patients [64]. The first mutation associated with HCM in humans was a missense mutation, R403Q, in MYH7, causing severe, early-onset disease in patients [65]. In the representative mouse model, this mutation was introduced in the α-myosin heavy chain gene (MYH6) instead of the MYH7 gene, because of the disparate myosin heavy chain isoform content in humans and mice. These knock-in mice provided definitive evidence that a mutation in a sarcomeric gene leads to HCM [66]. Since the sarcomeric protein composition in rabbits more closely resembles that of humans, transgenic rabbits expressing the MYH7 R403Q mutant were also generated [66,67].

In contrast to individuals with MYH7 mutations, ventricular hypertrophy is generally mild or absent in patients with TNNT2 mutations. Nevertheless, these mutations herald a high incidence of premature sudden cardiac death [68]. A TNNT2 mutation, resulting in a truncated cardiac troponin T protein in patients, was mimicked in both a mouse and a rat model [69,70]. In addition, mice were genetically modified to express human TNNT2 containing the missense mutation I79N or R92Q, revealing that an increased Ca^{2+}-sensitivity of cardiac muscle contraction is involved in the pathogenesis of HCM associated with these TNNT2 mutations [71].

In addition to MYH7 and TNNT2 mutations, mutations in MYBPC3 also represent a frequent genetic cause of human HCM, and are associated with a late onset of disease, lower incidence of hypertrophy, and a good prognosis [72]. A comparable disease presentation was observed in mice expressing a truncated myosin-binding protein C, lacking the titin- and myosin-binding domains [73]. Moreover, mice with MYBPC3 ablation showed that, although myosin-binding C protein is not essential for formation and maintenance of the sarcomere ultrastructure, its absence results in HCM [74]. Mutations in genes encoding cardiac troponin I, α-tropomyosin, and ventricular regulatory myosin light chain have also been identified in HCM patients and were successfully reproduced in animal models [75–77].

Besides this artificially induced HCM in laboratory animals, naturally occurring HCM (and the resulting CHF and cardiac death) has been reported in a colony of Maine coon cats, although its genetic basis remains to be determined [78].

With regard to familial DCM, mutations have been identified in 30-35% of the patients, with the majority occurring in the titin (TTN), lamin A/C (LMNA), β-myosin heavy chain (MYH7), and cardiac troponin T (TNNT2) genes.

In order to characterize the morphological and functional impairments resulting from a TTN truncation mutation identified in DCM patients, knock-in mice recapitulating this mutation were generated [79]. In addition, two MYH7 missense mutations causing DCM in humans, S532P and F746L, have also been investigated in knock-in mice, and have provided important pathogenic insights [80]. Similarly, a causal TNNT2 deletion mutation, ΔK210, was introduced in mice, resulting in adverse cardiac alterations closely resembling the clinical phenotype in patients [81].

In some instances, reports on animal models of DCM preceded the recognition of a mutation in the same gene in humans and contributed to the discovery of genetic forms of human DCM. In fact, one of the first engineered murine models of DCM was the muscle LIM protein knock-out mouse, which exhibits many phenotypic features of human DCM [82]. It is also noteworthy that some animal models of DCM, such as desmin-deficient mice, are associated with skeletal myopathy [83].

Finally, the cardiomyopathic Syrian hamster is an established animal model of naturally occurring hypertrophic and DCM. Hamster strains Bio 14.6, UM-X7.1, CHF 146, and CHF 147 are characterized by significant cardiac hypertrophy with progressing ventricular dilatation, whereas strains 53.58 and Bio TO-2 display dilatation without hypertrophy. Genetic studies have shown that a mutation in the δ-sarcoglycan gene is responsible for both hypertrophic and DCM [84].

6.4 IMPROVEMENTS IN ANIMAL MODELS (TABLE 6.2)

When designing an experiment, many factors have to be considered, particularly, the balance between the incremental value of improving the model and the feasibility of the improvement. Aging animals with multiple cardiovascular risk factors may reflect human patients better, however, inducing cardiovascular risk factors (for example, insulin resistance using high-fat diet or pressure overload using aortic constriction [85]) increases the variability of the cardiac response, the length of the experiment, and the overall cost. Furthermore, these modifications are by essence imperfect reflections of the clinical development of human risk factors.

Recently, mice with polygenic abnormalities have been developed, in order to better mimic human pathologies. For example, mice with polygenic abnormalities that develop progressive obesity have been generated [86]. The advantage of these mice, however, may be limited by the decrease in mechanistic insights when multiple genes are involved in the development of a pathology. Recent advances in the generation of genetically modified mice have included inducible gene modifications and the generation of knock-in mice, in which the nonmouse gene of interest is inserted at a particular chosen locus as opposed to randomly.

Attempts should be encouraged to develop more physiological procedures. Progressive aortic constriction, which, for example, is extensively used in rats, better mimics the development of chronic hypertension or aortic stenosis than an acute banding and is being developed in mice. The disadvantage of this model, however, is that is it performed on very young mice as the progression of the constriction occurs as the mice are growing. Models of coronary clots formation in dogs thought to better mimic a thrombogenic event than coronary ligation have been reported [87] but have not been used recently due to their variability. Similarly, diets have been perfected to better reflect human diets, however, applying them in mice with different metabolic capabilities than humans still remain problematic.

Finally, in selected investigations, a process of high throughput triage of genes or treatments through the use of a crude model, such as the zebrafish, can be envisioned [88]. Once the selection is achieved, more sophisticated models are employed, bridging the gap toward the clinical environment.

TABLE 6.2 Possible Changes in Animal Models to Better Reflect Human Pathologies

Animals	Aging
	Sedentary Added CV risk factors
Mice	Polygenic (but may limit mechanistic insights) Knock-ins/inducible
Models	More physiological
Strategy	High throughput followed by more clinically relevant but fewer animals

References

[1] Mouse Genome Sequencing Consortium, Waterston RH, Lindblad-Toh K, Birney E, Rogers J, Abril JF, et al. Initial sequencing and comparative analysis of the mouse genome. Nature 2002;420:520–62.
[2] Gao E, Koch WJ. A novel and efficient model of coronary artery ligation in the mouse. Methods Mol Biol 2013;1037:299–311.
[3] Bers DM. Cardiac Na/Ca exchange function in rabbit, mouse and man: what's the difference? J Mol Cell Cardiol 2002;34:369–73.
[4] Endoh M. Force-frequency relationship in intact mammalian ventricular myocardium: physiological and pathophysiological relevance. Eur J Pharmacol 2004;500:73–86.
[5] Haghighi K, Kolokathis F, Pater L, Lynch RA, Asahi M, Gramolini AO, et al. Human phospholamban null results in lethal dilated cardiomyopathy revealing a critical difference between mouse and human. J Clin Invest 2003;111:869–76.
[6] Metzger D, Chambon P. Site- and time-specific gene targeting in the mouse. Methods 2001;24:71–80.
[7] Ho KK, Anderson KM, Kannel WB, Grossman W, Levy D. Survival after the onset of congestive heart failure in Framingham Heart Study subjects. Circulation 1993;88:107–15.
[8] Klocke R, Tian W, Kuhlmann MT, Nikol S. Surgical animal models of heart failure related to coronary heart disease. Cardiovasc Res 2007;74:29–38.
[9] Scherrer-Crosbie M, Ullrich R, Bloch KD, Nakajima H, Nasseri B, Aretz HT, et al. Endothelial Nitric oxide synthase limits left ventricular remodeling after myocardial infarction in mice. Circulation 2001;104:1286–91.
[10] Zolotareva AG, Kogan ME. Production of experimental occlusive myocardial infarction in mice. Cor Vasa 1978;20:308–14.
[11] Michael LH, Entman ML, Hartley CJ, Youker KA, Zhu J, Hall SR, et al. Myocardial ischemia and reperfusion: a murine model. Am J Physiol 1995;269:H2147–54.
[12] Ahn D, Cheng L, Moon C, Spurgeon H, Lakatta EG, Talan MI. Induction of myocardial infarcts of a predictable size and location by branch pattern probability-assisted coronary ligation in C57BL/6 mice. Am J Physiol Heart Circ Physiol 2004;286:H1201–7.
[13] Salto-Tellez M, Yung Lim S, El-Oakley RM, Tang TPL, ALmsherqi ZAM, Lim S-K. Myocardial infarction in the C57BL/6J mouse: a quantifiable and highly reproducible experimental model. Cardiovasc Pathol 2004;13:91–7.
[14] Gao E, Lei YH, Shang X, Huang ZM, Zuo L, Boucher M, et al. A novel and efficient model of coronary artery ligation and myocardial infarction in the mouse. Circ Res 2010;107:1445–53.
[15] Pfeffer MA, Pfeffer JM, Fishbein MC, Fletcher PJ, Spadaro J, Kloner RA, et al. Myocardial infarct size and ventricular function in rats. Circ Res 1979;44:503–12.
[16] Liu YH, Yang XP, Nass O, Sabbah HN, Peterson E, Carretero OA. Chronic heart failure induced by coronary artery ligation in Lewis inbred rats. Am J Physiol 1997;272:H722–7.
[17] Nossuli TO, Lakshminarayanan V, Baumgarten G, Taffet GE, Ballantyne CM, Michael LH, et al. A chronic mouse model of myocardial ischemia-reperfusion: essential in cytokine studies. Am J Physiol Heart Circ Physiol 2000;278:H1049–55.
[18] Vandervelde S, van Amerongen MJ, Tio RA, Petersen AH, van Luyn MJA, Harmsen MC. Increased inflammatory response and neovascularization in reperfused vs. non-reperfused murine myocardial infarction. Cardiovasc Pathol 2006;15:83–90.
[19] Reimer KA, Lowe JE, Rasmussen MM, Jennings RB. The wavefront phenomenon of ischemic cell death. 1. Myocardial infarct size vs duration of coronary occlusion in dogs. Circulation 1977;56:786–94.
[20] Przyklenk K, Vivaldi MT, Schoen FJ, Malcolm J, Arnold O, Kloner RA. Salvage of ischaemic myocardium by reperfusion: importance of collateral blood flow and myocardial oxygen demand during occlusion. Cardiovas Res 1986;20:403–14.
[21] Sabbah HN, Stein PD, Kono T, Gheorghiade M, Levine TB, Jafri S, et al. A canine model of chronic heart failure produced by multiple sequential coronary microembolizations. Am J Physiol 1991;260:H1379–84.
[22] Dörge H, Schulz R, Belosjorow S, Post H, van de Sand A, Konietzka I, et al. Coronary microembolization: the role of TNF-alpha in contractile dysfunction. J Mol Cell Cardiol 2002;34:51–62.
[23] Kukielka GL, Smith CW, Manning AM, Youker KA, Michael LH, Entman ML. Induction of interleukin-6 synthesis in the myocardium. Potential role in postreperfusion inflammatory injury. Circulation 1995;92:1866–75.
[24] Youker KA, Hawkins HK, Kukielka GL, Perrard JL, Michael LH, Ballantyne CM, et al. Molecular evidence for induction of intracellular adhesion molecule-1 in the viable border zone associated with ischemia-reperfusion injury of the dog heart. Circulation 1994;89:2736–46.
[25] Faxon DP, Gibbons RJ, Chronos NAF, Gurbel PA, Sheehan F. The effect of blockade of the CD11/CD18 integrin receptor on infarct size in patients with acute myocardial infarction treated with direct angioplasty: the results of the HALT-MI study. J Am Coll Cardiol 2002;40:1199–204.
[26] Baran KW, Nguyen M, McKendall GR, Lambrew CT, Dykstra G, Palmeri ST, et al. Double-blind, randomized trial of an anti-CD18 antibody in conjunction with recombinant tissue plasminogen activator for acute myocardial infarction: limitation of myocardial infarction following thrombolysis in acute myocardial infarction (LIMIT AMI) study. Circulation 2001;104:2778–83.
[27] Roberts R, DeMello V, Sobel BE. Deleterious effects of methylprednisolone in patients with myocardial infarction. Circulation 1976;53:I204–6.
[28] Weaver ME, Pantely GA, Bristow JD, Ladley HD. A quantitative study of the anatomy and distribution of coronary arteries in swine in comparison with other animals and man. Cardiovasc Res 1986;20:907–17.
[29] Gorman JH, Gorman RC, Plappert T, Jackson BM, Hiramatsu Y, St John-Sutton MG, et al. Infarct size and location determine development of mitral regurgitation in the sheep model. J Thorac Cardiovasc Surg 1998;115:615–22.
[30] Beeri R, Yosefy C, Guerrero JL, Nesta F, Abedat S, Chaput M, et al. Mitral regurgitation augments post-myocardial infarction remodeling failure of hypertrophic compensation. J Am Coll Cardiol 2008;51:476–86.
[31] Rockman HA, Ross RS, Harris AN, Knowlton KU, Steinhelper ME, Field LJ, et al. Segregation of atrial-specific and inducible expression of an atrial natriuretic factor transgene in an in vivo murine model of cardiac hypertrophy. Proc Natl Acad Sci U S A 1991;88:8277–81.
[32] Ichinose F. Pressure overload-induced LV hypertrophy and dysfunction in mice are exacerbated by congenital NOS3 deficiency. Am J Physiol Heart Circ Physiol 2003;286:1070H–5H.
[33] Takimoto E, Champion HC, Li M, Ren S, Rodriguez ER, Tavazzi B, et al. Oxidant stress from nitric oxide synthase-3 uncoupling stimulates cardiac pathologic remodeling from chronic pressure load. J Clin Invest 2005;115:1221–31.

[34] Feldman AM, Weinberg EO, Ray PE, Lorell BH. Selective changes in cardiac gene expression during compensated hypertrophy and the transition to cardiac decompensation in rats with chronic aortic banding. Circ Res 1993;73:184–92.
[35] Weinberg EO, Schoen FJ, George D, Kagaya Y, Douglas PS, Litwin SE, et al. Angiotensin-converting enzyme inhibition prolongs survival and modifies the transition to heart failure in rats with pressure overload hypertrophy due to ascending aortic stenosis. Circulation 1994;90:1410–22.
[36] Litwin SE, Katz SE, Weinberg EO, Lorell BH, Aurigemma GP, Douglas PS. Serial echocardiographic-Doppler assessment of left ventricular geometry and function in rats with pressure-overload hypertrophy. Chronic angiotensin-converting enzyme inhibition attenuates the transition to heart failure. Circulation 1995;91:2642–54.
[37] Mitchell GF, Pfeffer JM, Pfeffer MA. The transition to failure in the spontaneously hypertensive rat. Am J Hypertens 1997;10:120S–6S.
[38] Boluyt MO, O'Neill L, Meredith AL, Bing OH, Brooks WW, Conrad CH, et al. Alterations in cardiac gene expression during the transition from stable hypertrophy to heart failure. Marked upregulation of genes encoding extracellular matrix components. Circ Res 1994;75:23–32.
[39] Nagatomo Y, Carabello BA, Coker ML, McDermott PJ, Nemoto S, Hamawaki M, et al. Differential effects of pressure or volume overload on myocardial MMP levels and inhibitory control. Am J Physiol Heart Circ Physiol 2000;278:H151–61.
[40] Ferrario CM, Page IH, McCubbin JW. Increased cardiac output as a contributory factor in experimental renal hypertension in dogs. Circ Res 1970;27:799–810.
[41] Alyono D, Anderson RW, Parrish DG, Dai XZ, Bache RJ. Alterations of myocardial blood flow associated with experimental canine left ventricular hypertrophy secondary to valvular aortic stenosis. Circ Res 1986;58:47–57.
[42] Spinale FG, Ishihra K, Zile M, DeFryte G, Crawford FA, Carabello BA. Structural basis for changes in left ventricular function and geometry because of chronic mitral regurgitation and after correction of volume overload. J Thorac Cardiovasc Surg 1993;106:1147–57.
[43] Beeri R, Yosefy C, Guerrero JL, Abedat S, Handschumacher MD, Stroud RE, et al. Early repair of moderate ischemic mitral regurgitation reverses left ventricular remodeling: a functional and molecular study. Circulation 2007;116:I288–I1293.
[44] Pinsky WW, Lewis RM, Hartley CJ, Entman ML. Permanent changes of ventricular contractility and compliance in chronic volume overload. Am J Physiol 1979;237:H575–83.
[45] Beaudoin J, Levine RA, Guerrero JL, Yosefy C, Sullivan S, Abedat S, et al. Late repair of ischemic mitral regurgitation does not prevent left ventricular remodeling: importance of timing for beneficial repair. Circulation 2013;128:S248–52.
[46] Abassi Z, Goltsman I, Karram T, Winaver J, Hoffman A. Aortocaval fistula in rat: a unique model of volume-overload congestive heart failure and cardiac hypertrophy. J Biomed Biotechnol 2011;2011:729497.
[47] Yamamoto K, Protack CD, Tsuneki M, Hall MR, Wong DJ, Lu DY, et al. The mouse aortocaval fistula recapitulates human arteriovenous fistula maturation. Am J Physiol Heart Circ Physiol 2013;305:H1718–25.
[48] Spinale FG, Tempel GE, Mukherjee R, Eble DM, Brown R, Vacchiano CA, et al. Cellular and molecular alterations in the beta adrenergic system with cardiomyopathy induced by tachycardia. Cardiovasc Res 1994;28:1243–50.
[49] Nakamura R, Egashira K, Machida Y, Hayashidani S, Takeya M, Utsumi H, et al. Probucol attenuates left ventricular dysfunction and remodeling in tachycardia-induced heart failure: roles of oxidative stress and inflammation. Circulation 2002;106:362–7.
[50] Jaenke RS. An anthracycline antibiotic-induced cardiomyopathy in rabbits. Lab Invest 1974;30:292–304.
[51] Mettler FP, Young DM, Ward JM. Adriamycin-induced cardiotoxicity (cardiomyopathy and congestive heart failure) in rats. Cancer Res 1977;37:2705–13.
[52] Herman EH, el-Hage AN, Ferrans VJ, Ardalan B. Comparison of the severity of the chronic cardiotoxicity produced by doxorubicin in normotensive and hypertensive rats. Toxicol Appl Pharmacol 1985;78:202–14.
[53] Tan-Chiu E, Yothers G, Romond E, Geyer CE, Ewer M, Keefe D, et al. Assessment of cardiac dysfunction in a randomized trial comparing doxorubicin and cyclophosphamide followed by paclitaxel, with or without trastuzumab as adjuvant therapy in node-positive, human epidermal growth factor receptor 2-overexpressing breast cancer: NSABP B-31. J Clin Oncol 2005;23:7811–9.
[54] Neilan TG. Tissue Doppler imaging predicts left ventricular dysfunction and mortality in a murine model of cardiac injury. Eur Heart J 2006;27:1868–75.
[55] Van Vleet JF, Greenwood LA, Ferrans VJ. Pathologic features of adriamycin toxicosis in young pigs: nonskeletal lesions. Am J Vet Res 1979;40:1537–52.
[56] Gralla EJ, Fleischman RW, Luthra YK, Stadnicki SW. The dosing schedule dependent toxicities of adriamycin in beagle dogs and rhesus monkeys. Toxicology 1979;13:263–73.
[57] Sieber SM, Correa P, Young DM, Dalgard DW, Adamson RH. Cardiotoxic and possible leukemogenic effects of adriamycin in nonhuman primates. Pharmacology 1980;20:9–14.
[58] Buras JA, Holzmann B, Sitkovsky M. Animal models of sepsis: setting the stage. Nat Rev Drug Discov 2005;4:854–65.
[59] Szabó C. Role of nitric oxide in endotoxic shock. An overview of recent advances. Ann N Y Acad Sci 1998;851:422–5.
[60] Ichinose F, Hataishi R, Wu JC, Kawai N, Rodrigues ACT, Mallari C, et al. A selective inducible NOS dimerization inhibitor prevents systemic, cardiac, and pulmonary hemodynamic dysfunction in endotoxemic mice. Am J Physiol Heart Circ Physiol 2003;285:H2524–30.
[61] Ullrich R, Scherrer-Crosbie M, Bloch KD, Ichinose F, Nakajima H, Picard MH, et al. Congenital deficiency of nitric oxide synthase 2 protects against endotoxin-induced myocardial dysfunction in mice. Circulation 2000;102:1440–6.
[62] López A, Lorente JA, Steingrub J, Bakker J, McLuckie A, Willatts S, et al. Multiple-center, randomized, placebo-controlled, double-blind study of the nitric oxide synthase inhibitor 546C88: effect on survival in patients with septic shock. Crit Care Med 2004;32:21–30.
[63] Morita H, Seidman J, Seidman CE. Genetic causes of human heart failure. J Clin Invest 2005;115:518–26.
[64] Keren A, Syrris P, McKenna WJ. Hypertrophic cardiomyopathy: the genetic determinants of clinical disease expression. Nat Clin Pract Cardiovasc Med 2008;5:158–68.
[65] Geisterfer-Lowrance AA, Kass S, Tanigawa G, Vosberg HP, McKenna W, Seidman CE, et al. A molecular basis for familial hypertrophic cardiomyopathy: a beta cardiac myosin heavy chain gene missense mutation. Cell 1990;62:999–1006.
[66] Geisterfer-Lowrance AA, Christe M, Conner DA, Ingwall JS, Schoen FJ, Seidman CE, et al. A mouse model of familial hypertrophic cardiomyopathy. Science 1996;272:731–4.
[67] Marian AJ, Wu Y, Lim DS, McCluggage M, Youker K, Yu QT, et al. A transgenic rabbit model for human hypertrophic cardiomyopathy. J Clin Invest 1999;104:1683–92.

[68] Watkins H, McKenna WJ, Thierfelder L, Suk HJ, Anan R, O'Donoghue A, et al. Mutations in the genes for cardiac troponin T and alpha-tropomyosin in hypertrophic cardiomyopathy. N Engl J Med 1995;332:1058–64.
[69] Tardiff JC, Factor SM, Tompkins BD, Hewett TE, Palmer BM, Moore RL, et al. A truncated cardiac troponin T molecule in transgenic mice suggests multiple cellular mechanisms for familial hypertrophic cardiomyopathy. J Clin Invest 1998;101:2800–11.
[70] Frey N, Franz WM, Gloeckner K, Degenhardt M, Müller M, Müller O, et al. Transgenic rat hearts expressing a human cardiac troponin T deletion reveal diastolic dysfunction and ventricular arrhythmias. Cardiovasc Res 2000;47:254–64.
[71] Tardiff JC, Hewett TE, Palmer BM, Olsson C, Factor SM, Moore RL, et al. Cardiac troponin T mutations result in allele-specific phenotypes in a mouse model for hypertrophic cardiomyopathy. J Clin Invest 1999;104:469–81.
[72] Niimura H, Patton KK, McKenna WJ, Soults J, Maron BJ, Seidman JG, et al. Sarcomere protein gene mutations in hypertrophic cardiomyopathy of the elderly. Circulation 2002;105:446–51.
[73] McConnell BK, Jones KA, Fatkin D, Arroyo LH, Lee RT, Aristizabal O, et al. Dilated cardiomyopathy in homozygous myosin-binding protein-C mutant mice. J Clin Invest 1999;104:1235–44.
[74] Harris SP, Bartley CR, Hacker TA, McDonald KS, Douglas PS, Greaser ML, et al. Hypertrophic cardiomyopathy in cardiac myosin binding protein-C knockout mice. Circ Res 2002;90:594–601.
[75] James J, Zhang Y, Osinska H, Sanbe A, Klevitsky R, Hewett TE, et al. Transgenic modeling of a cardiac troponin I mutation linked to familial hypertrophic cardiomyopathy. Circ Res 2000;87:805–11.
[76] Rajan S, Ahmed RPH, Jagatheesan G, Petrashevskaya N, Boivin GP, Urboniene D, et al. Dilated cardiomyopathy mutant tropomyosin mice develop cardiac dysfunction with significantly decreased fractional shortening and myofilament calcium sensitivity. Circ Res 2007;101:205–14.
[77] Sanbe A, Nelson D, Gulick J, Setser E, Osinska H, Wang X, et al. In vivo analysis of an essential myosin light chain mutation linked to familial hypertrophic cardiomyopathy. Circ Res 2000;87:296–302.
[78] Kittleson MD, Meurs KM, Munro MJ, Kittleson JA, Liu SK, Pion PD, et al. Familial hypertrophic cardiomyopathy in maine coon cats: an animal model of human disease. Circulation 1999;99:3172–80.
[79] Gramlich M, Michely B, Krohne C, Heuser A, Erdmann B, Klaassen S, et al. Stress-induced dilated cardiomyopathy in a knock-in mouse model mimicking human titin-based disease. J Mol Cell Cardiol 2009;47:352–8.
[80] Schmitt JP, Debold EP, Ahmad F, Armstrong A, Frederico A, Conner DA, et al. Cardiac myosin missense mutations cause dilated cardiomyopathy in mouse models and depress molecular motor function. Proc Natl Acad Sci U S A 2006;103:14525–30.
[81] Du C-K, Morimoto S, Nishii K, Minakami R, Ohta M, Tadano N, et al. Knock-in mouse model of dilated cardiomyopathy caused by troponin mutation. Circ Res 2007;101:185–94.
[82] Arber S, Hunter JJ, Ross J, Hongo M, Sansig G, Borg J, et al. MLP-deficient mice exhibit a disruption of cardiac cytoarchitectural organization, dilated cardiomyopathy, and heart failure. Cell 1997;88:393–403.
[83] Milner DJ, Weitzer G, Tran D, Bradley A, Capetanaki Y. Disruption of muscle architecture and myocardial degeneration in mice lacking desmin. J Cell Biol 1996;134:1255–70.
[84] Sakamoto A, Ono K, Abe M, Jasmin G, Eki T, Murakami Y, et al. Both hypertrophic and dilated cardiomyopathies are caused by mutation of the same gene, delta-sarcoglycan, in hamster: an animal model of disrupted dystrophin-associated glycoprotein complex. Proc Natl Acad Sci U S A 1997;94:13873–8.
[85] Raher MJ, Thibault HB, Buys ES, Kuruppu D, Shimizu N, Brownell AL, et al. A short duration of high-fat diet induces insulin resistance and predisposes to adverse left ventricular remodeling after pressure overload. Am J Physiol Heart Circ Physiol 2008;295:H2495–502.
[86] Allan MF, Eisen EJ, Pomp D. The M16 mouse: an outbred animal model of early onset polygenic obesity and diabesity. Obes Res 2004;12:1397–407.
[87] Adrie C, Bloch KD, Moreno PR, Hurford WE, Guerrero JL, Holt R, et al. Inhaled nitric oxide increases coronary artery patency after thrombolysis. Circulation 1996;94:1919–26.
[88] Chico TJA, Ingham PW, Crossman DC. Modeling cardiovascular disease in the zebrafish. Trends Cardiovasc Med 2008;18:150–5.

S E C T I O N 2

INFLAMMATORY BIOMARKERS

CHAPTER

7

Inflammatory Biomarkers in Post-infarction Heart Failure and Cardiac Remodeling

Olga Frunza, Nikolaos G. Frangogiannis

Department of Medicine (Cardiology), The Wilf Family Cardiovascular Research Institute, Albert Einstein College of Medicine, Bronx, New York, USA

7.1 INTRODUCTION

What is a biomarker? In 2001, the "biomarkers definitions working group" defined a biomarker as a "characteristic that is objectively measured and evaluated as an indicator of normal biological processes, pathogenic processes or pharmacologic responses to a therapeutic intervention" [1]. According to this broad definition, physiologic measurements and quantitative imaging studies should also be considered biomarkers [2]; however, in most cases, the term is restricted to molecular or biochemical indicators [3]. Over the last few decades, the use of biomarkers has been expanded in all areas of clinical decision making. Biomarkers are potentially useful along the whole spectrum of the disease process. Prior to diagnosis, markers may be used for screening and risk assessment. Antecedent biomarkers determine the risk of developing a pathologic condition and screening biomarkers identify individuals with subclinical disease, whereas diagnostic biomarkers are important tools in diagnosing overt disease. Other types of biomarkers are useful for assessment and treatment of patients after the diagnosis has been established. Thus, staging biomarkers assess disease severity and prognostic biomarkers provide information on the course of a disease, and guide therapy by predicting responses to specific therapeutic interventions, or by monitoring efficacy of treatment [2,4].

Advances in basic and clinical research resulted in successful introduction of biomarkers to expedite diagnosis, predict mortality and morbidity, and guide therapy in patients with heart failure [5,6]. Biomarkers that reflect myocardial stretch, such as the natriuretic peptides, have been successfully introduced in clinical practice as diagnostic tools, prognostic indicators, and relevant guides for effective therapy of patients with heart failure [7]. Indicators reflecting cardiomyocyte injury (such as the cardiac troponins) [8] and neurohormonal biomarkers [5,6] may also prove useful independent prognostic tools for patients with heart failure. Recognition of the role of inflammatory cascades in the pathogenesis of heart failure generated intense interest in the potential role of inflammatory biomarkers as prognostic indicators and therapeutic guides [9]. Since the early findings by Levine and coworkers documenting elevated tumor necrosis factor (TNF) levels in patients with heart failure [10], numerous studies have suggested the potential usefulness of a wide range of inflammatory biomarkers, including several members of the cytokine and chemokine family, in heart failure [9]. In most cases, the role of these indicators in clinical practice has not been established. However, the significance of these biomarkers should not be limited to their potential associations with severity and progression of the disease. Perhaps the most promising potential role of biomarkers in patients with heart failure is to provide a window into the molecular and cellular environment of the failing heart, allowing identification of patient subpopulations with distinct pathophysiologic responses, and guiding therapeutic decisions on the basis of knowledge of the underlying biology. Such an approach could be most productive in the setting of post-infarction heart failure, where the balance between pro- and anti-inflammatory responses may drive morphologic and structural changes of the remodeling myocardium affecting the functional response. This chapter

http://dx.doi.org/10.1016/B978-0-12-800039-7.00007-4
© 2015 Elsevier Inc. All rights reserved.

discusses our knowledge on the role of inflammatory biomarkers in heart failure with an emphasis on their significance in predicting the development of heart failure following myocardial infarction. In addition to a systematic review of the clinical evidence, we highlight the links between the proposed biomarkers and the pathophysiology of post-infarction heart failure. Considering the pathophysiologic complexity of the response to myocardial infarction in human patients, we stress the need for development of biomarker-based approaches to identify subpopulations of patients with distinct responses in order to guide therapy.

7.2 THE ROLE OF THE INFLAMMATORY RESPONSE IN REPAIR AND REMODELING OF THE INFARCTED HEART

Reperfusion strategies have significantly improved survival in patients with acute myocardial infarction. However, improved survival of patients after the acute event results in an increased pool of individuals at risk of developing heart failure. The pathogenesis of heart failure following myocardial infarction is intricately linked with ventricular remodeling, a constellation of morphologic, molecular, and proteomic changes that lead to dilation and increased sphericity of the ventricle following myocardial infarction [11]. Adverse post-infarction remodeling is dependent on the size of the infarct and on the characteristics of the healing wound. Extensive evidence from experimental studies suggests that defective repair of the infarcted heart is associated with accentuated dilative remodeling following myocardial infarction [12]. Healing of the infarcted heart is dependent on a superbly orchestrated reparative response that can be divided into three distinct but overlapping phases. During the inflammatory phase (see Figure 7.1), activation of danger signals by dying cardiomyocytes triggers chemokine-driven inflammatory leukocyte recruitment. Inflammatory leukocytes serve to clear the wound from dead cells and matrix debris and set the stage for activation of mesenchymal reparative cells. As the infarct is debrided, inflammatory leukocytes undergo apoptosis and proinflammatory cytokine and chemokine synthesis is repressed. Timely resolution of the inflammatory reaction is critical for the transition to the proliferative phase of cardiac repair. During the proliferative phase, resident cardiac fibroblasts undergo transdifferentiation into myofibroblasts through interactions that involve transforming growth factor (TGF)-β [13] and specialized matrix proteins [14–16]. Inhibition of fibrogenic signaling marks the end of the proliferative phase and the transition to maturation of the scar. During the maturation phase, the matrix becomes cross-linked, while fibroblasts become quiescent and may undergo apoptosis [17].

Disruption of the sequence of events that leads to formation of a collagen-based scar is associated with dilative remodeling [18]. Overactive, prolonged, or spatially unrestrained proinflammatory activation triggers excessive matrix degradation, resulting in dilation of the chamber. In contrast, overactive matrix-preserving responses may be associated with uncontrolled fibrosis and development of diastolic dysfunction. Biomarkers mirroring the intensity of the inflammatory and fibrotic response following myocardial infarction may provide key information on the underlying pathophysiology in human patients, thus guiding therapeutic strategies [4]. Although the potential role of biomarker-based strategies in designing therapy for patients with acute myocardial infarction has not been tested, several studies have examined the prognostic implications of specific inflammatory mediators.

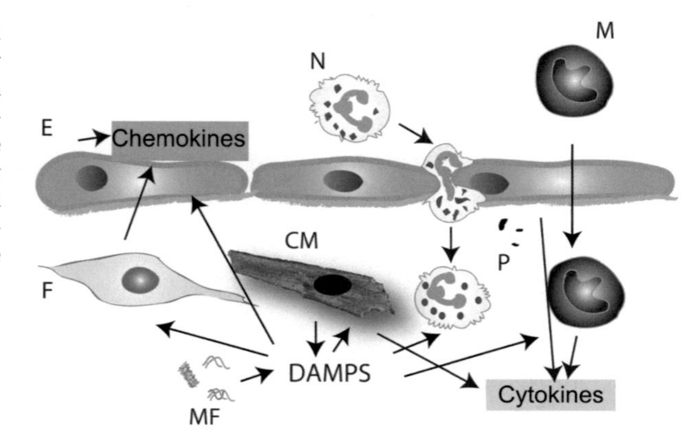

FIGURE 7.1 The inflammatory phase of cardiac repair. In the infarcted myocardium, release of danger-associated molecular patterns (DAMPS) by dying cardiomyocytes (CM) and matrix fragments (MF) induces activation of inflammatory cascades, leading to upregulation of chemokines and cytokines. Chemokines bind to glycosaminoglycans on the endothelial surface and activate rolling neutrophils (N) and mononuclear cells (M), ultimately leading to their extravasation and recruitment into the infarcted area. Both blood-derived cells (such as leukocytes and platelets/P) and resident myocardial cells (cardiomyocytes, fibroblasts/F, endothelial cells/E) contribute to the post-infarction inflammatory reaction.

7.3 SPECIFIC INFLAMMATORY BIOMARKERS AS PREDICTORS OF POST-INFARCTION REMODELING

7.3.1 General Markers of Inflammation

7.3.1.1 C-Reactive Protein

An acute phase protein produced by the liver in response to stimulation with proinflammatory cytokines, C-reactive protein (CRP) has been extensively studied as a biomarker in cardiovascular disease. A growing body of evidence suggests that CRP may be a useful biomarker for prediction of adverse outcome and development of heart failure in patients surviving from acute myocardial infarction. Elevated peak CRP levels independently predicted cardiac rupture in patients with acute myocardial infarction [19,20], and admission CRP levels predicted mortality and dysfunction in patients with acute myocardial infarction receiving thrombolytics [21]. In 112 patients with ST elevation myocardial infarction (STEMI) undergoing primary percutaneous coronary intervention (PCI), pre-discharge high-sensitivity CRP (hs-CRP) levels were independently associated with the development of dilative remodeling (defined as a >20% increase in left ventricular end-diastolic volume 6 months after the acute event) [22]. Several additional studies have suggested a relation between elevated hs-CRP levels and adverse dilative remodeling following myocardial infarction (see Table 7.1) [23–25,32,33]. Elevated peak CRP levels were associated with a greater increase in left ventricular end-diastolic volumes following an anterior infarction [26,27] suggesting that hs-CRP may be a useful tool in predicting chamber dilation. In a study enrolling 106 Chinese patients with acute

TABLE 7.1 Association Between Circulating CRP Levels and Post-infarction Remodeling or Development of Heart Failure in Patients with Acute Myocardial Infarction

Study	Number of Patients	Clinical Setting	Timing of Sampling	Method and Timing of Remodeling Assessment	Main Finding
Anzai *et al.* [19]	220	First STEMI	Within 24 h of onset of AMI	Echocardiography. Patients were followed for >12 months	Positive correlation between CRP elevation and cardiac rupture, LV aneurysm formation, 1-year cardiac death
Ueda *et al.* [20]	37	AMI with or without cardiac rupture	24 h after the onset of acute myocardial infarction	Echocardiography	persistently high serum CRP level of >20 mg/dl predicted cardiac rupture after MI
Urbano-Moral *et al.* [22]	112	STEMI	At discharge	6-month follow-up, echocardiography	CRP predicted dilative remodeling in STEMI patients treated with primary PCI
Fertin *et al.* [23]	246	STEMI	At hospital discharge and the 1-month, 3-month, and 1-year follow-up visits	Echocardiographic studies at hospital discharge, 3 months, and 1 year after MI	Absence of independent relation between CRP levels and dilative remodeling
Mather *et al.* [24]	48	STEMI	2 days, 1 week, 1 month, and 3 months after presentation	Cardiovascular magnetic resonance (CMR); 72 h, at 1 week, 1 month, and 3 months	CRP assessed 2 days after reperfusion was the strongest independent predictor of dilative remodeling at 3 months
Orn *et al.* [25]	42	First time STEMI	On admission to the hospital, and 2 days, 1 week, and 2 months following hospitalization	Cardiovascular magnetic resonance; 2 days, 1 week, 2 month	CRP predicts adverse remodeling

(*Continued*)

TABLE 7.1 Association Between Circulating CRP Levels and Post-infarction Remodeling or Development of Heart Failure in Patients with Acute Myocardial Infarction—Cont'd

Study	Number of Patients	Clinical Setting	Timing of Sampling	Method and Timing of Remodeling Assessment	Main Finding
Takahashi *et al.* [26]	31	STEMI	On admission, and 2 weeks and 6 months after AMI	Coronary angiography; on admission, and then 2 weeks and 6 months after infarction	Increased peak serum CRP level was associated with a greater increase in LV volume after anterior AMI
Uehara, *et al.* [27]	139	STEMI	Immediately after 1, 2, 3, and 7 days, and 1 month after the onset of AMI	Echocardiography, 1 month after infarction	CRP is a useful factor for predicting LV remodeling
Xiaozhou *et al.* [28]	106	First infarction	3 days after AMI	Echocardiography on the third day and third month after infarction	NT-proBNP was more effective than hs-CRP as a predictor of dilative remodeling
Berton *et al.* [29]	220	Myocardial infarction	On the first, third, and seventh day after admission	Echocardiogram between the third and the seventh day after admission and 1 year after recruitment	Peak CRP is a strong independent predictor of global and heart failure-related mortality following infarction
Bursi *et al.* [30]	329	STEMI and non-STEMI	On admission	Evaluation of medical records (1.0 ± 0.6 years after the event)	CRP is an independent predictor of heart failure and death
Hartford *et al.* [31]	1618	Acute coronary syndromes	Day 1 ($n = 757$) or day 4 ($n = 533$) after admission	Follow up 75 months	CRP is associated with long-term mortality and heart failure, but not reinfarction

myocardial infarction, both NT-proBNP and hs-CRP measured 3 days after the acute event correlated with increases in left ventricular end-diastolic volume (LVEDV) during the remodeling phase; however, the correlation coefficient was lower for hs-CRP [28]. A growing body of evidence suggests that elevated hs-CRP not only is associated with adverse remodeling, but also predicts the development of heart failure following acute infarction. In patients with acute myocardial infarction, high peak levels of CRP were independently associated with the development and progression of heart failure [29]. In a multimarker approach, baseline troponin, BNP, and CRP measurements provided independent unique prognostic information predicting the development of heart failure [34]. In patients surviving myocardial infarction, there was a strong positive graded association between CRP levels and the risk of developing heart failure; this relation was independent of the size of the infarct and of the occurrence of recurrent ischemic events [30]. In patients with ACS, CRP (and other more specific inflammatory mediators, including interleukin-6 (IL-6), sPLA(2)-IIA and intercellular adhesion molecule (ICAM)-1) assessed on the first day after the acute event, were associated with long-term mortality and development of heart failure, but not with reinfarction [31]. The usefulness of CRP as a biomarker providing relevant pathophysiologic information in patients with myocardial infarction is limited by its nonspecific role in the inflammatory process; use of CRP in this setting may be more informative when accompanied by measurement of other more specific inflammatory mediators.

7.3.1.2 *Myeloperoxidase*

Myeloperoxidase (MPO), an enzyme with potent oxidant effects that is abundantly produced and released by myeloid cells [35], is expressed in vulnerable plaques and is considered a marker for unstable coronary lesions. Increased serum MPO levels have adverse prognostic implications in healthy individuals, predicting risk of coronary heart disease [36]. In patients with established coronary disease, MPO levels provide important prognostic information. Baseline plasma levels of MPO were associated with the incidence of recurrent ischemic events in patients with acute coronary syndromes [37] and with increased mortality and nonfatal infarction in patients with STEMI [38]. Experimental evidence has suggested a crucial role for MPO in the pathogenesis of post-infarction cardiac remodeling mediated through activation of oxidative stress [39,40]. MPO was an independent predictor of mortality in STEMI patients presenting with cardiogenic shock [41]; however, diurnal variations in MPO levels [42] and the absence of systematic analysis of time-dependent changes following myocardial infarction makes interpretation of the findings quite challenging. Thus, the potential role of MPO as a biomarker predicting development of post-infarction heart failure remains poorly supported by clinical data.

7.3.2 Cytokines (Table 7.2)

7.3.2.1 The TNF-α System

TNF-α is the best-studied cytokine in heart failure. Elevated plasma TNF-α levels are consistently observed in patients with heart failure [10,50]; extensive evidence suggests that TNF levels correlate with disease severity [51]. Moreover, in patients with severe chronic heart failure, increased plasma levels of TNF-α predict adverse outcome and are associated with increased mortality [52]. The role of plasma TNF-α levels as a biomarker predicting adverse

TABLE 7.2 Associations Between Circulating Cytokines and Cytokine Receptor Levels and the Development of Post-infarction Remodeling or Heart Failure in Patients with Acute Myocardial Infarction

Cytokine/Cytokine Receptor	Study	Number of Patients	Clinical Setting	Timing of Sampling	Main Finding
IL-6	Hartford *et al.* [31]	1618	Acute coronary syndromes	Day 1 ($n=757$) or day 4 ($n=533$) after admission	IL-6 is associated with long-term mortality and heart failure, but not reinfarction
TNF-α, IL-6, sTNFR1, sTNFR2	Valgimigli *et al.* [43]	184	STEMI	0.5-34 h after symptom onset	sTNFR-1 is a major short- and long-term predictor of mortality and HF in patients with AMI
ST2	Shimpo *et al.* [44]	810 patients enrolled in TIMI- 14 (362 patients), or TIMI 23 (448 patients)	STEMI	At baseline and 1, 3, 12, and 24 h after enrollment (TIMI-14)	Serum levels of ST2 predict mortality and heart failure in patients with acute myocardial infarction
ST2	Sabatine *et al.* [45]	1239 patients enrolled in TIMI-28	STEMI	Time of enrollment	In STEMI, high baseline ST2 levels predict cardiovascular death and heart failure independently of baseline characteristics and NT-proBNP. The combination of ST2 and NT-proBNP significantly improves risk stratification
ST2	Kohli *et al.* [46]	4426 (TIMI-36)	Non-STEMI	Time of enrollment	ST2 correlates weakly with biomarkers of acute injury and hemodynamic stress but is strongly associated with the risk of heart failure after NSTE-ACS
ST2	Dhillon *et al.* [47]	667	STEMI	3-5 days after admission	Elevated ST2 and IL-33 were both associated with increased mortality. ST2 demonstrated incremental value over contemporary risk markers but IL-33 did not
GDF-15	Wollert *et al.* [48]	2081	Non-STEMI	On admission	GDF-15 levels are strongly associated with mortality
GDF-15	Kempf *et al.* [49]	741	STEMI	Upon enrollment	GDF-15 levels predicted mortality

remodeling and heart failure following myocardial infarction is less convincingly established. In experimental models of myocardial infarction, TNF-α expression is upregulated [53,54] and TNF bioactivity is increased in the cardiac interstitium [55]. In 184 human patients presenting with acute myocardial infarction, admission plasma levels of TNF-α and its type 1 and 2 soluble receptors (sTNFR1 and sTNFR2) were significantly increased [43]. Although in univariate analysis TNF, sTNFR-1, and sTNFR-2 levels predicted the composite endpoint of death and new onset heart failure, multivariate analysis suggested that only sTNFR-1 was an independent predictor of adverse outcome following acute infarction [43]. In a much smaller population of 33 patients undergoing PCI for acute myocardial infarction, TNF-α levels increased significantly 7-14 days after the acute event but were not associated with worse dilative remodeling [56]. In patients with heart failure, sTNFR levels may provide greater prognostic information than TNF-α levels, better reflecting global systemic inflammation, as several proinflammatory mediators promote sTNFR shedding. Moreover, elevated sTNFR levels may better identify patients with tissue activation of the TNF-α cascade.

7.3.2.2 IL-6

Several studies have suggested a role for IL-6 as a biomarker in patients with heart failure. In elderly subjects without prior myocardial infarction (and free of heart failure), baseline levels of IL-6 predicted the development of heart failure [57]. Extensive evidence from clinical investigations suggested that in patients with chronic heart failure, plasma IL-6 levels are elevated [58] and correlate with the severity of disease [59]. In patients with advanced heart failure, increased IL-6 levels predict adverse outcome and are associated with higher mortality [60,61]. Although IL-6 is upregulated in the infarcted heart and may play a role in the pathogenesis of post-infarction remodeling, the significance of elevated IL-6 levels in prediction of chamber dilation and post-infarction heart failure remains poorly documented. In a study examining plasma cytokine levels in 184 patients presenting with acute myocardial infarction, admission IL-6 levels were elevated [43]. In univariate analysis, IL-6 levels predicted the occurrence of death or new onset of heart failure. However, multivariate analysis suggested that the prognostic information provided by IL-6 levels was not independent of left ventricular systolic function, Killip class, and peak CK-MB [43]. Moreover, in a small study using cardiac magnetic resonance to evaluate cardiac remodeling in 42 patients undergoing PCI for acute myocardial infarction, elevated IL-6 levels were associated with worse systolic function and adverse cardiac remodeling [25].

7.3.2.3 The Expanding Role of ST2

IL-1 signaling is centrally involved in post-infarction inflammatory response and in the pathogenesis of cardiac remodeling [62–64]. The IL-1 family is comprised of agonists, receptors, and antagonists; one of the receptor members of the family, ST2, has great potential as a prognostic biomarker for patients with heart failure and acute myocardial infarction [65,66]. ST2 is found in both soluble and transmembrane forms and is markedly induced in cardiomyocytes and fibroblasts in response to mechanical strain and to stimulation with proinflammatory cytokines [67,68]. ST2 levels carry important and independent prognostic information in patients with acute heart failure [69] and in individuals with chronic heart failure [70]. ST2 levels were significantly elevated in the first day after acute myocardial infarction peaking at 12 h; elevated ST2 levels were associated with 30-day mortality [44,71]. In patients with STEMI, high baseline ST2 was a strong independent predictor of cardiovascular death and heart failure [45]. Moreover, ST2 levels were strongly associated with the risk of heart failure in patients with non-ST elevation acute coronary syndromes despite a weak correlation with markers of acute injury [46]. Although ST2 may play a role in risk stratification of patients with acute myocardial infarction, its ligand IL-33 may have much less value as a biomarker; a recent investigation suggested that when measured 3-5 days after STEMI, IL-33 levels were independently associated with 30-day and 1-year mortality, but (in contrast to ST2 levels) did not add incremental value over GRACE-RS and natriuretic peptide assessment [47].

7.3.2.4 The Role of Growth Differentiation Factor-15, a TGF-β Family Member

Members of the TGF-β family, including the TGF-β isoforms TGFβ1, β2, and β3 [72], activin A, [73] and growth differentiation factor (GDF)-15 [74] are markedly upregulated in the infarcted myocardium. Most members of the TGF-β family may have a limited role as circulating biomarkers because they are bound to tissues and are locally activated to modulate cellular responses [75]. GDF-15 is a notable exception; extensive evidence suggests that GDF-15 may be an important prognostic biomarker for patients with heart failure and coronary artery disease [76]. GDF-15 is a predictor of mortality in both STEMI and non-STEMI populations [49,77]. In patients with acute coronary syndromes, GDF-15 levels are associated with recurrent events independent of clinical indicators, natriuretic peptide, and hs-CRP levels [48,78]. A growing body of evidence suggests that GDF-15 levels may also predict the development of heart failure and adverse remodeling following myocardial infarction. In 1142 patients with acute myocardial infarction (STEMI and non-STEMI), GDF-15 levels measured 3-5 days after the acute event predicted death and the development of heart failure [79]. Moreover, in 97 patients with STEMI, GDF-15 levels measured on the first day after myocardial infarction independently correlated with the development of adverse dilative remodeling [80].

7.3.2.5 Chemokines

Although chemokine induction is a hallmark of the post-infarction inflammatory response, and several members of the family are implicated in the pathogenesis of cardiac remodeling [81], the usefulness of chemokines as biomarkers in cardiovascular disease remains limited. Chemokine-mediated actions are dependent on their binding to endothelial glycosaminoglycans; this important biological property may limit their potential significance as circulating indicators of myocardial inflammation. In patients with acute coronary syndromes, persistent elevation of CCL2/monocyte chemoattractant protein (MCP)-1 levels (assessed 30 days after the acute event) was associated with increased mortality [82]. Adverse prognosis in patients with prolonged elevation of chemokine levels was not associated with recurrent coronary events; whether accentuation of inflammatory myocardial injury was responsible for worse remodeling in these patients is unknown [83]. Limited evidence suggests the role of MCP-1 as a biomarker predicting adverse remodeling following myocardial infarction [84]. Weir and coworkers [85] measured serum MCP-1 levels in 100 patients with acute myocardial infarction and systolic dysfunction at baseline and at 24 weeks after the acute event. Changes in MCP-1 levels correlated positively with indicators of adverse remodeling [85] assessed through magnetic resonance imaging. In a small study enrolling 35 patients with acute myocardial infarction, elevated baseline CCL3/macrophage inflammatory protein (MIP)-1α levels were associated with accentuated dilative remodeling [86]. Because of their marked induction in experimental models of myocardial infarction [87–89], several members of the CXC chemokine family have also attracted interest as potential biomarkers in patients with acute myocardial infarction. Serum CXCL10/interferon-γ-inducible protein (IP)-10 levels were found to be elevated following myocardial infarction and independently associated with the cumulative creatine kinase (CK) release [90]. Orn and coworkers measured levels of several chemokines over the first week following acute myocardial infarction and examined their relation with indicators of myocardial injury, dysfunction, and remodeling. CCL4/MIP-1β, CXCL16, CXCL10/IP-10, and CXCL8/IL-8 levels correlated with infarct size and with impaired function 2 months after the acute event [91]; the findings may simply reflect a more intense inflammatory response in patients with larger infarcts. Despite the direct involvement of members of the chemokine family in post-infarction remodeling, several relatively small studies failed to provide significant support for the potential role of chemokines as biomarkers.

7.3.3 The Matrix

Activation of inflammatory signaling in the infarcted heart induces dynamic changes in the matrix network; these alterations are critically implicated in post-infarction remodeling and in the pathogenesis of heart failure following myocardial infarction [92]. Thus, serum biomarkers that reflect matrix activity could serve as useful windows to the remodeling heart and may reflect activation of inflammatory signaling. Several matrix-related biomarkers have been tested as prognostic indicators following myocardial infarction.

7.3.3.1 Matrix Metalloproteinases

Induction and activation of matrix metalloproteinases (MMPs) promote matrix degradation, contributing to clearance of the wound following myocardial infarction. In the healing infarct, unrestrained inflammatory signaling activates MMPs and accentuates chamber dilation, playing an important role in the pathogenesis of cardiac remodeling [93,94]. Several small investigations have suggested relations between serum MMP levels and the development of adverse remodeling following myocardial infarction [95]. Most studies have focused on measurement of serum levels of the gelatinases MMP-2 and MMP-9; results on the predictive value of these proteases have been conflicting. Squire and coworkers measured serum MMP-9 and MMP-2 levels in 60 patients with acute myocardial infarction and found no relation between MMP levels and the development of dilative remodeling 6 weeks after the acute event [96]. In a study measuring MMP levels in 52 clinically stable patients surviving acute infarction, Orn *et al.* found no association between gelatinase levels and dilative remodeling [97]. On the other hand, Webb and coworkers found that persistent elevations of MMP-9 levels 5 days after acute infarction had important prognostic implications, predicting chamber dilation 28 days after the acute event [98]. In a larger study of 404 patients with acute infarction, MMP-9 and tissue inhibitor of metalloproteinases (TIMP)-1 levels predicted chamber dilation. Differences in timing of MMP measurements in relation to the acute event, the specificity and sensitivity of various assays, and the use of various strategies examining temporal changes in protease expression may explain the conflicting results.

7.3.3.2 Matricellular Proteins

Induction of matricellular proteins is a crucial event in the healing infarct and critically regulates the inflammatory and reparative response. Because several matricellular proteins are critically implicated in post-infarction remodeling, biomarkers reflecting myocardial expression and activity of matricellular proteins could provide important information with prognostic implications. However, by definition, matricellular interactions involve immobilization of the matricellular macromolecules on the interstitial matrix and may not be mirrored by increases in circulating levels. Some members of the matricellular family can also be secreted and act as cytokines; these mediators may have potential as biomarkers reflecting the extent of cardiac remodeling.

7.3.3.2.1 GALECTIN-3

Galectin-3 is an inflammatory β-galactoside-binding lectin secreted by activated macrophages that, when bound to the matrix, exerts matricellular functions. Extensive evidence suggests a potential role for galectin-3 as a prognostic biomarker in patients with heart failure. Galectin-3 is a strong independent predictor of mortality in patients with chronic heart failure [99]; its predictive value may be higher in individuals with preserved ejection fraction [100]. Galectin-3 is upregulated in the infarcted heart and is implicated in the pathogenesis of cardiac remodeling and fibrosis [101]; however, its potential role as a biomarker predicting post-infarction remodeling and heart failure is poorly supported by data. In 100 patients admitted with acute myocardial infarction and left-ventricular dysfunction, baseline galectin-3 levels correlated significantly with other indicators of matrix turnover (such as MMP-3) and with inflammatory markers (such as MCP-1), but had no significant relation with remodeling-associated parameters [102].

7.3.3.2.2 TENASCIN-C

Tenascin-C deposition is a useful histological marker of interstitial remodeling in animal models [103] and in human patients with ischemic cardiomyopathy [104]. Although tenascin-C is critically involved in the pathogenesis of post-infarction remodeling [105], evidence of its potential role as a biomarker for patients with myocardial infarction is limited. In 239 patients with myocardial infarction, tenascin-C levels independently predicted adverse outcome [106]. However, the potential role of tenascin-C as a circulating marker of cardiac remodeling requires confirmation by larger studies.

7.3.4 Indicators of Cellular Activation

Myocardial infarction is associated with mobilization and activation of leukocyte subsets that play an important role in the pathogenesis of cardiac remodeling. Several studies have tested the potential use of simple markers of cellular activation as predictors of cardiac remodeling following myocardial infarction. In 107 patients with anterior myocardial infarction, pre-discharge white blood cell counts independently predicted adverse remodeling [107]. Elevated peripheral blood mononuclear cell counts were also associated with adverse post-infarction remodeling [108,109]. In a study of 131 patients with acute myocardial infarction, a peak mononuclear cell count higher than 3600/mm^3 independently predicted adverse remodeling [109]. The phenotypic characteristics of circulating mononuclear cells may also provide important prognostic information. Peak levels of CD14+CD16− monocytes were negatively correlated with recovery of function following myocardial infarction [110]. A proinflammatory monocyte response is associated with adverse outcome in patients with STEMI [111].

7.4 IMPLEMENTATION OF BIOMARKER-BASED STRATEGIES IN PATIENTS WITH MYOCARDIAL INFARCTION

In human patients, the reparative response following myocardial infarction is characterized by pathophysiologic complexity. Age, gender, genetic diversity, the presence or absence of comorbidities (such as hypertension, diabetes, and dyslipidemias), treatment with medications (including statins, angiotensin converting enzyme inhibitors, or β-blockers) greatly affect inflammatory signaling pathways following myocardial infarction. Thus, the success of therapeutic approaches targeting inflammatory signals in the infarcted heart is dependent on identification of patient subpopulations with distinct pathophysiologic defects. Patients with accentuated dilative remodeling following infarction may exhibit prolonged or overactive inflammatory reactions in the myocardium; biomarker-based approaches for identification of these individuals could greatly contribute to design of specific pathophysiologically driven therapies targeting proinflammatory cytokine signaling. On the other hand, individuals with overactive matrix-preserving signaling may exhibit diastolic heart failure due to excessive matrix deposition; these patients may benefit from approaches targeting the profibrotic growth factors. Although serum biomarkers may contribute to identification of patients with distinct pathophysiologic responses, systemic levels of inflammatory cytokines are affected by many factors that may be unrelated to the myocardial response. Thus, molecular imaging may hold great promise by providing key insights into myocardial structural, molecular, and proteomic alterations with important therapeutic implications [112].

References

[1] Biomarkers Definitions Working Group. Biomarkers and surrogate endpoints: preferred definitions and conceptual framework. Clin Pharmacol Ther 2001;69:89–95.
[2] Vasan RS. Biomarkers of cardiovascular disease: molecular basis and practical considerations. Circulation 2006;113:2335–62.
[3] Rifai N, Gillette MA, Carr SA. Protein biomarker discovery and validation: the long and uncertain path to clinical utility. Nat Biotechnol 2006;24:971–83.
[4] Frangogiannis NG. Biomarkers: hopes and challenges in the path from discovery to clinical practice. Transl Res 2012;159:197–204.
[5] Chowdhury P, Kehl D, Choudhary R, Maisel A. The use of biomarkers in the patient with heart failure. Curr Cardiol Rep 2013;15:372.
[6] Braunwald E. Biomarkers in heart failure. N Engl J Med 2008;358:2148–59.
[7] Troughton R, Michael Felker G, Januzzi Jr. JL. Natriuretic peptide-guided heart failure management. Eur Heart J 2014;35:16–24.
[8] de Lemos JA. Increasingly sensitive assays for cardiac troponins: a review. Jama 2013;309:2262–9.
[9] Bozkurt B, Mann DL, Deswal A. Biomarkers of inflammation in heart failure. Heart Fail Rev 2010;15:331–41.
[10] Levine B, Kalman J, Mayer L, Fillit HM, Packer M. Elevated circulating levels of tumor necrosis factor in severe chronic heart failure. N Engl J Med 1990;323:236–41.
[11] Pfeffer MA, Braunwald E. Ventricular remodeling after myocardial infarction. Experimental observations and clinical implications. Circulation 1990;81:1161–72.
[12] Frangogiannis NG. Regulation of the inflammatory response in cardiac repair. Circ Res 2012;110:159–73.
[13] Dobaczewski M, Chen W, Frangogiannis NG. Transforming growth factor (TGF)-beta signaling in cardiac remodeling. J Mol Cell Cardiol 2011;51:600–6.
[14] Dobaczewski M, de Haan JJ, Frangogiannis NG. The extracellular matrix modulates fibroblast phenotype and function in the infarcted myocardium. J Cardiovasc Transl Res 2012;5:837–47.
[15] Frangogiannis NG. Matricellular proteins in cardiac adaptation and disease. Physiol Rev 2012;92:635–88.
[16] Frangogiannis NG, Ren G, Dewald O, Zymek P, Haudek S, Koerting A, et al. The critical role of endogenous Thrombospondin (TSP)-1 in preventing expansion of healing myocardial infarcts. Circulation 2005;111:2935–42.
[17] Christia P, Bujak M, Gonzalez-Quesada C, Chen W, Dobaczewski M, Reddy A, et al. Systematic characterization of myocardial inflammation, repair, and remodeling in a mouse model of reperfused myocardial infarction. J Histochem Cytochem 2013;61:555–70.
[18] Frangogiannis NG. The immune system and the remodeling infarcted heart: cell biological insights and therapeutic opportunities. J Cardiovasc Pharmacol 2014;63:185–95.
[19] Anzai T, Yoshikawa T, Shiraki H, Asakura Y, Akaishi M, Mitamura H, et al. C-reactive protein as a predictor of infarct expansion and cardiac rupture after a first Q-wave acute myocardial infarction. Circulation 1997;96:778–84.
[20] Ueda S, Ikeda U, Yamamoto K, Takahashi M, Nishinaga M, Nago N, et al. C-reactive protein as a predictor of cardiac rupture after acute myocardial infarction. Am Heart J 1996;131:857–60.
[21] Zairis MN, Manousakis SJ, Stefanidis AS, Papadaki OA, Andrikopoulos GK, Olympios CD, et al. C-reactive protein levels on admission are associated with response to thrombolysis and prognosis after ST-segment elevation acute myocardial infarction. Am Heart J 2002;144:782–9.
[22] Urbano-Moral JA, Lopez-Haldon JE, Fernandez M, Mancha F, Sanchez A, Rodriguez-Puras MJ, et al. Prognostic value of different serum biomarkers for left ventricular remodelling after ST-elevation myocardial infarction treated with primary percutaneous coronary intervention. Heart 2012;98:1153–9.
[23] Fertin M, Hennache B, Hamon M, Ennezat PV, Biausque F, Elkohen M, et al. Usefulness of serial assessment of B-type natriuretic peptide, troponin I, and C-reactive protein to predict left ventricular remodeling after acute myocardial infarction (from the REVE-2 study). Am J Cardiol 2010;106:1410–6.
[24] Mather AN, Fairbairn TA, Artis NJ, Greenwood JP, Plein S. Relationship of cardiac biomarkers and reversible and irreversible myocardial injury following acute myocardial infarction as determined by cardiovascular magnetic resonance. Int J Cardiol 2013;166:458–64.
[25] Orn S, Manhenke C, Ueland T, Damas JK, Mollnes TE, Edvardsen T, et al. C-reactive protein, infarct size, microvascular obstruction, and left-ventricular remodelling following acute myocardial infarction. Eur Heart J 2009;30:1180–6.
[26] Takahashi T, Anzai T, Yoshikawa T, Maekawa Y, Asakura Y, Satoh T, et al. Serum C-reactive protein elevation in left ventricular remodeling after acute myocardial infarction–role of neurohormones and cytokines. Int J Cardiol 2003;88:257–65.
[27] Uehara K, Nomura M, Ozaki Y, Fujinaga H, Ikefuji H, Kimura M, et al. High-sensitivity C-reactive protein and left ventricular remodeling in patients with acute myocardial infarction. Heart Vessels 2003;18:67–74.
[28] Xiaozhou H, Jie Z, Li Z, Liyan C. Predictive value of the serum level of N-terminal pro-brain natriuretic peptide and high-sensitivity C-reactive protein in left ventricular remodeling after acute myocardial infarction. J Clin Lab Anal 2006;20:19–22.
[29] Berton G, Cordiano R, Palmieri R, Pianca S, Pagliara V, Palatini P. C-reactive protein in acute myocardial infarction: association with heart failure. Am Heart J 2003;145:1094–101.
[30] Bursi F, Weston SA, Killian JM, Gabriel SE, Jacobsen SJ, Roger VL. C-reactive protein and heart failure after myocardial infarction in the community. Am J Med 2007;120:616–22.

[31] Hartford M, Wiklund O, Mattsson Hulten L, Persson A, Karlsson T, Herlitz J, et al. C-reactive protein, interleukin-6, secretory phospholipase A2 group IIA and intercellular adhesion molecule-1 in the prediction of late outcome events after acute coronary syndromes. J Intern Med 2007;262:526–36.
[32] Piestrzeniewicz K, Luczak K, Maciejewski M, Drozdz J. Low adiponectin blood concentration predicts left ventricular remodeling after ST-segment elevation myocardial infarction treated with primary percutaneous coronary intervention. Cardiol J 2010;17:49–56.
[33] Fertin M, Dubois E, Belliard A, Amouyel P, Pinet F, Bauters C. Usefulness of circulating biomarkers for the prediction of left ventricular remodeling after myocardial infarction. Am J Cardiol 2012;110:277–83.
[34] Sabatine MS, Morrow DA, de Lemos JA, Gibson CM, Murphy SA, Rifai N, et al. Multimarker approach to risk stratification in non-ST elevation acute coronary syndromes: simultaneous assessment of troponin I, C-reactive protein, and B-type natriuretic peptide. Circulation 2002;105:1760–3.
[35] Penn MS. The role of leukocyte-generated oxidants in left ventricular remodeling. Am J Cardiol 2008;101:30D–3D.
[36] Meuwese MC, Stroes ES, Hazen SL, van Miert JN, Kuivenhoven JA, Schaub RG, et al. Serum myeloperoxidase levels are associated with the future risk of coronary artery disease in apparently healthy individuals: the EPIC-Norfolk Prospective Population Study. J Am Coll Cardiol 2007;50:159–65.
[37] Morrow DA, Sabatine MS, Brennan ML, de Lemos JA, Murphy SA, Ruff CT, et al. Concurrent evaluation of novel cardiac biomarkers in acute coronary syndrome: myeloperoxidase and soluble CD40 ligand and the risk of recurrent ischaemic events in TACTICS-TIMI 18. Eur Heart J 2008;29:1096–102.
[38] Khan SQ, Kelly D, Quinn P, Davies JE, Ng LL. Myeloperoxidase aids prognostication together with N-terminal pro-B-type natriuretic peptide in high-risk patients with acute ST elevation myocardial infarction. Heart 2007;93:826–31.
[39] Vasilyev N, Williams T, Brennan ML, Unzek S, Zhou X, Heinecke JW, et al. Myeloperoxidase-generated oxidants modulate left ventricular remodeling but not infarct size after myocardial infarction. Circulation 2005;112:2812–20.
[40] Askari AT, Brennan ML, Zhou X, Drinko J, Morehead A, Thomas JD, et al. Myeloperoxidase and plasminogen activator inhibitor 1 play a central role in ventricular remodeling after myocardial infarction. J Exp Med 2003;197:615–24.
[41] Dominguez-Rodriguez A, Samimi-Fard S, Abreu-Gonzalez P, Garcia-Gonzalez MJ, Kaski JC. Prognostic value of admission myeloperoxidase levels in patients with ST-segment elevation myocardial infarction and cardiogenic shock. Am J Cardiol 2008;101:1537–40.
[42] Dominguez-Rodriguez A, Abreu-Gonzalez P, Kaski JC. Diurnal variation of circulating myeloperoxidase levels in patients with ST-segment elevation myocardial infarction. Int J Cardiol 2010;144:407–9.
[43] Valgimigli M, Ceconi C, Malagutti P, Merli E, Soukhomovskaia O, Francolini G, et al. Tumor necrosis factor-alpha receptor 1 is a major predictor of mortality and new-onset heart failure in patients with acute myocardial infarction: the Cytokine-Activation and Long-Term Prognosis in Myocardial Infarction (C-ALPHA) study. Circulation 2005;111:863–70.
[44] Shimpo M, Morrow DA, Weinberg EO, Sabatine MS, Murphy SA, Antman EM, et al. Serum levels of the interleukin-1 receptor family member ST2 predict mortality and clinical outcome in acute myocardial infarction. Circulation 2004;109:2186–90.
[45] Sabatine MS, Morrow DA, Higgins LJ, MacGillivray C, Guo W, Bode C, et al. Complementary roles for biomarkers of biomechanical strain ST2 and N-terminal prohormone B-type natriuretic peptide in patients with ST-elevation myocardial infarction. Circulation 2008;117:1936–44.
[46] Kohli P, Bonaca MP, Kakkar R, Kudinova AY, Scirica BM, Sabatine MS, et al. Role of ST2 in non-ST-elevation acute coronary syndrome in the MERLIN-TIMI 36 trial. Clin Chem 2012;58:257–66.
[47] Dhillon OS, Narayan HK, Khan SQ, Kelly D, Quinn PA, Squire IB, et al. Pre-discharge risk stratification in unselected STEMI: is there a role for ST2 or its natural ligand IL-33 when compared with contemporary risk markers? Int J Cardiol 2013;167:2182–8.
[48] Wollert KC, Kempf T, Peter T, Olofsson S, James S, Johnston N, et al. Prognostic value of growth-differentiation factor-15 in patients with non-ST-elevation acute coronary syndrome. Circulation 2007;115:962–71.
[49] Kempf T, Bjorklund E, Olofsson S, Lindahl B, Allhoff T, Peter T, et al. Growth-differentiation factor-15 improves risk stratification in ST-segment elevation myocardial infarction. Eur Heart J 2007;28:2858–65.
[50] Katz SD, Rao R, Berman JW, Schwarz M, Demopoulos L, Bijou R, et al. Pathophysiological correlates of increased serum tumor necrosis factor in patients with congestive heart failure. Relation to nitric oxide-dependent vasodilation in the forearm circulation. Circulation 1994;90:12–6.
[51] Testa M, Yeh M, Lee P, Fanelli R, Loperfido F, Berman JW, et al. Circulating levels of cytokines and their endogenous modulators in patients with mild to severe congestive heart failure due to coronary artery disease or hypertension. J Am Coll Cardiol 1996;28:964–71.
[52] Deswal A, Petersen NJ, Feldman AM, Young JB, White BG, Mann DL. Cytokines and cytokine receptors in advanced heart failure: an analysis of the cytokine database from the Vesnarinone trial (VEST). Circulation 2001;103:2055–9.
[53] Herskowitz A, Choi S, Ansari AA, Wesselingh S. Cytokine mRNA expression in postischemic/reperfused myocardium. Am J Pathol 1995;146:419–28.
[54] Dewald O, Ren G, Duerr GD, Zoerlein M, Klemm C, Gersch C, et al. Of mice and dogs: species-specific differences in the inflammatory response following myocardial infarction. Am J Pathol 2004;164:665–77.
[55] Frangogiannis NG, Lindsey ML, Michael LH, Youker KA, Bressler RB, Mendoza LH, et al. Resident cardiac mast cells degranulate and release preformed TNF-alpha, initiating the cytokine cascade in experimental canine myocardial ischemia/reperfusion. Circulation 1998;98:699–710.
[56] Kondo H, Hojo Y, Tsuru R, Nishimura Y, Shimizu H, Takahashi N, et al. Elevation of plasma granzyme B levels after acute myocardial infarction. Circ J 2009;73:503–7.
[57] Vasan RS, Sullivan LM, Roubenoff R, Dinarello CA, Harris T, Benjamin EJ, et al. Inflammatory markers and risk of heart failure in elderly subjects without prior myocardial infarction: the Framingham Heart Study. Circulation 2003;107:1486–91.
[58] Torre-Amione G, Kapadia S, Benedict C, Oral H, Young JB, Mann DL. Proinflammatory cytokine levels in patients with depressed left ventricular ejection fraction: a report from the Studies of Left Ventricular Dysfunction (SOLVD). J Am Coll Cardiol 1996;27:1201–6.

[59] Tsutamoto T, Hisanaga T, Wada A, Maeda K, Ohnishi M, Fukai D, et al. Interleukin-6 spillover in the peripheral circulation increases with the severity of heart failure, and the high plasma level of interleukin-6 is an important prognostic predictor in patients with congestive heart failure. J Am Coll Cardiol 1998;31:391–8.
[60] Mohler 3rd ER, Sorensen LC, Ghali JK, Schocken DD, Willis PW, Bowers JA, et al. Role of cytokines in the mechanism of action of amlodipine: the PRAISE Heart Failure Trial. Prospective Randomized Amlodipine Survival Evaluation. J Am Coll Cardiol 1997;30:35–41.
[61] Kell R, Haunstetter A, Dengler TJ, Zugck C, Kubler W, Haass M. Do cytokines enable risk stratification to be improved in NYHA functional class III patients? Comparison with other potential predictors of prognosis. Eur Heart J 2002;23:70–8.
[62] Bujak M, Frangogiannis NG. The role of IL-1 in the pathogenesis of heart disease. Arch Immunol Ther Exp (Warsz) 2009;57:165–76.
[63] Bujak M, Dobaczewski M, Chatila K, Mendoza LH, Li N, Reddy A, et al. Interleukin-1 receptor type I signaling critically regulates infarct healing and cardiac remodeling. Am J Pathol 2008;173:57–67.
[64] Saxena A, Chen W, Su Y, Rai V, Uche OU, Li N, et al. IL-1 induces proinflammatory leukocyte infiltration and regulates fibroblast phenotype in the infarcted myocardium. J Immunol 2013;191:4838–48.
[65] Kakkar R, Lee RT. The IL-33/ST2 pathway: therapeutic target and novel biomarker. Nat Rev Drug Discov 2008;7:827–40.
[66] Shah RV, Januzzi Jr JL. ST2: a novel remodeling biomarker in acute and chronic heart failure. Curr Heart Fail Rep 2010;7:9–14.
[67] Weinberg EO, Shimpo M, De Keulenaer GW, MacGillivray C, Tominaga S, Solomon SD, et al. Expression and regulation of ST2, an interleukin-1 receptor family member, in cardiomyocytes and myocardial infarction. Circulation 2002;106:2961–6.
[68] Sanada S, Hakuno D, Higgins LJ, Schreiter ER, McKenzie AN, Lee RT. IL-33 and ST2 comprise a critical biomechanically induced and cardioprotective signaling system. J Clin Invest 2007;117:1538–49.
[69] Rehman SU, Mueller T, Januzzi Jr JL. Characteristics of the novel interleukin family biomarker ST2 in patients with acute heart failure. J Am Coll Cardiol 2008;52:1458–65.
[70] Ky B, French B, McCloskey K, Rame JE, McIntosh E, Shahi P, et al. High-sensitivity ST2 for prediction of adverse outcomes in chronic heart failure. Circ Heart Fail 2011;4:180–7.
[71] Weinberg EO, Shimpo M, Hurwitz S, Tominaga S, Rouleau JL, Lee RT. Identification of serum soluble ST2 receptor as a novel heart failure biomarker. Circulation 2003;107:721–6.
[72] Deten A, Holzl A, Leicht M, Barth W, Zimmer HG. Changes in extracellular matrix and in transforming growth factor beta isoforms after coronary artery ligation in rats. J Mol Cell Cardiol 2001;33:1191–207.
[73] Yndestad A, Ueland T, Oie E, Florholmen G, Halvorsen B, Attramadal H, et al. Elevated levels of activin A in heart failure: potential role in myocardial remodeling. Circulation 2004;109:1379–85.
[74] Kempf T, Eden M, Strelau J, Naguib M, Willenbockel C, Tongers J, et al. The transforming growth factor-beta superfamily member growth-differentiation factor-15 protects the heart from ischemia/reperfusion injury. Circ Res 2006;98:351–60.
[75] Bujak M, Frangogiannis NG. The role of TGF-beta signaling in myocardial infarction and cardiac remodeling. Cardiovasc Res 2007;74:184–95.
[76] Xu X, Li Z, Gao W. Growth differentiation factor 15 in cardiovascular diseases: from bench to bedside. Biomarkers 2011;16:466–75.
[77] Wollert KC, Kempf T, Lagerqvist B, Lindahl B, Olofsson S, Allhoff T, et al. Growth differentiation factor 15 for risk stratification and selection of an invasive treatment strategy in non ST-elevation acute coronary syndrome. Circulation 2007;116:1540–8.
[78] Bonaca MP, Morrow DA, Braunwald E, Cannon CP, Jiang S, Breher S, et al. Growth differentiation factor-15 and risk of recurrent events in patients stabilized after acute coronary syndrome: observations from PROVE IT-TIMI 22. Arterioscler Thromb Vasc Biol 2011;31:203–10.
[79] Khan SQ, Ng K, Dhillon O, Kelly D, Quinn P, Squire IB, et al. Growth differentiation factor-15 as a prognostic marker in patients with acute myocardial infarction. Eur Heart J 2009;30:1057–65.
[80] Dominguez-Rodriguez A, Abreu-Gonzalez P, Avanzas P. Relation of growth-differentiation factor 15 to left ventricular remodeling in ST-segment elevation myocardial infarction. Am J Cardiol 2011;108:955–8.
[81] Frangogiannis NG. Chemokines in ischemia and reperfusion. Thromb Haemost 2007;97:738–47.
[82] de Lemos JA, Morrow DA, Blazing MA, Jarolim P, Wiviott SD, Sabatine MS, et al. Serial measurement of monocyte chemoattractant protein-1 after acute coronary syndromes: results from the A to Z trial. J Am Coll Cardiol 2007;50:2117–24.
[83] Frangogiannis NG. The prognostic value of monocyte chemoattractant protein-1/CCL2 in acute coronary syndromes. J Am Coll Cardiol 2007;50:2125–7.
[84] Gonzalez-Quesada C, Frangogiannis NG. Monocyte chemoattractant protein-1/CCL2 as a biomarker in acute coronary syndromes. Current Atherosclerosis Reports 2009;11:131–8.
[85] Weir RA, Murphy CA, Petrie CJ, Martin TN, Clements S, Steedman T, et al. Monocyte chemoattractant protein-1: a dichotomous role in cardiac remodeling following acute myocardial infarction in man? Cytokine 2010;50:158–62.
[86] Parissis JT, Adamopoulos S, Venetsanou KF, Mentzikof DG, Karas SM, Kremastinos DT. Serum profiles of C-C chemokines in acute myocardial infarction: possible implication in postinfarction left ventricular remodeling. J Interferon Cytokine Res 2002;22:223–9.
[87] Bujak M, Dobaczewski M, Gonzalez-Quesada C, Xia Y, Leucker T, Zymek P, et al. Induction of the CXC chemokine interferon-gamma-inducible protein 10 regulates the reparative response following myocardial infarction. Circ Res 2009;105:973–83.
[88] Kukielka GL, Smith CW, LaRosa GJ, Manning AM, Mendoza LH, Daly TJ, et al. Interleukin-8 gene induction in the myocardium after ischemia and reperfusion in vivo. J Clin Invest 1995;95:89–103.
[89] Frangogiannis NG, Mendoza LH, Lewallen M, Michael LH, Smith CW, Entman ML. Induction and suppression of interferon-inducible protein 10 in reperfused myocardial infarcts may regulate angiogenesis. FASEB J 2001;15:1428–30.
[90] Koten K, Hirohata S, Miyoshi T, Ogawa H, Usui S, Shinohata R, et al. Serum interferon-gamma-inducible protein 10 level was increased in myocardial infarction patients, and negatively correlated with infarct size. Clin Biochem 2008;41:30–7.
[91] Orn S, Breland UM, Mollnes TE, Manhenke C, Dickstein K, Aukrust P, et al. The chemokine network in relation to infarct size and left ventricular remodeling following acute myocardial infarction. Am J Cardiol 2009;104:1179–83.
[92] Dobaczewski M, Gonzalez-Quesada C, Frangogiannis NG. The extracellular matrix as a modulator of the inflammatory and reparative response following myocardial infarction. J Mol Cell Cardiol 2010;48:504–11.

[93] Spinale FG. Myocardial matrix remodeling and the matrix metalloproteinases: influence on cardiac form and function. Physiol Rev 2007;87:1285–342.

[94] Lindsey ML. MMP induction and inhibition in myocardial infarction. Heart Fail Rev 2004;9:7–19.

[95] Halade GV, Jin YF, Lindsey ML. Matrix metalloproteinase (MMP)-9: a proximal biomarker for cardiac remodeling and a distal biomarker for inflammation. Pharmacol Ther 2013;139:32–40.

[96] Squire IB, Evans J, Ng LL, Loftus IM, Thompson MM. Plasma MMP-9 and MMP-2 following acute myocardial infarction in man: correlation with echocardiographic and neurohumoral parameters of left ventricular dysfunction. J Card Fail 2004;10:328–33.

[97] Orn S, Manhenke C, Squire IB, Ng L, Anand I, Dickstein K. Plasma MMP-2, MMP-9 and N-BNP in long-term survivors following complicated myocardial infarction: relation to cardiac magnetic resonance imaging measures of left ventricular structure and function. J Card Fail 2007;13:843–9.

[98] Webb CS, Bonnema DD, Ahmed SH, Leonardi AH, McClure CD, Clark LL, et al. Specific temporal profile of matrix metalloproteinase release occurs in patients after myocardial infarction: relation to left ventricular remodeling. Circulation 2006;114:1020–7.

[99] Gullestad L, Ueland T, Kjekshus J, Nymo SH, Hulthe J, Muntendam P, et al. Galectin-3 predicts response to statin therapy in the Controlled Rosuvastatin Multinational Trial in Heart Failure (CORONA). Eur Heart J 2012;33:2290–6.

[100] de Boer RA, Lok DJ, Jaarsma T, van der Meer P, Voors AA, Hillege HL, et al. Predictive value of plasma galectin-3 levels in heart failure with reduced and preserved ejection fraction. Ann Med 2011;43:60–8.

[101] Yu L, Ruifrok WP, Meissner M, Bos EM, van Goor H, Sanjabi B, et al. Genetic and pharmacological inhibition of galectin-3 prevents cardiac remodeling by interfering with myocardial fibrogenesis. Circ Heart Fail 2013;6:107–17.

[102] Weir RA, Petrie CJ, Murphy CA, Clements S, Steedman T, Miller AM, et al. Galectin-3 and cardiac function in survivors of acute myocardial infarction. Circ Heart Fail 2013;6:492–8.

[103] Bujak M, Ren G, Kweon HJ, Dobaczewski M, Reddy A, Taffet G, et al. Essential role of Smad3 in infarct healing and in the pathogenesis of cardiac remodeling. Circulation 2007;116:2127–38.

[104] Frangogiannis NG, Shimoni S, Chang SM, Ren G, Dewald O, Gersch C, et al. Active interstitial remodeling: an important process in the hibernating human myocardium. J Am Coll Cardiol 2002;39:1468–74.

[105] Nishioka T, Onishi K, Shimojo N, Nagano Y, Matsusaka H, Ikeuchi M, et al. Tenascin-C may aggravate left ventricular remodeling and function after myocardial infarction in mice. Am J Physiol Heart Circ Physiol 2010;298:H1072–8.

[106] Sato A, Hiroe M, Akiyama D, Hikita H, Nozato T, Hoshi T, et al. Prognostic value of serum tenascin-C levels on long-term outcome after acute myocardial infarction. J Card Fail 2012;18:480–6.

[107] Bauters A, Ennezat PV, Tricot O, Lallemant R, Aumegeat V, Segrestin B, et al. Relation of admission white blood cell count to left ventricular remodeling after anterior wall acute myocardial infarction. Am J Cardiol 2007;100:182–4.

[108] Bauters A, Fertin M, Lamblin N, Pinet F, Bauters C. White blood cell and peripheral blood mononuclear cell counts for the prediction of left ventricular remodeling after myocardial infarction. J Cardiol 2011;58:197–8, author reply 198.

[109] Aoki S, Nakagomi A, Asai K, Takano H, Yasutake M, Seino Y, et al. Elevated peripheral blood mononuclear cell count is an independent predictor of left ventricular remodeling in patients with acute myocardial infarction. J Cardiol 2011;57:202–7.

[110] Tsujioka H, Imanishi T, Ikejima H, Kuroi A, Takarada S, Tanimoto T, et al. Impact of heterogeneity of human peripheral blood monocyte subsets on myocardial salvage in patients with primary acute myocardial infarction. J Am Coll Cardiol 2009;54:130–8.

[111] van der Laan AM, Hirsch A, Robbers LF, Nijveldt R, Lommerse I, Delewi R, et al. A proinflammatory monocyte response is associated with myocardial injury and impaired functional outcome in patients with ST-segment elevation myocardial infarction: monocytes and myocardial infarction. Am Heart J 2012;163: 57-65.e52.

[112] Frangogiannis NG. The inflammatory response in myocardial injury, repair and remodeling. Nat Rev Cardiol 2014;11:255–65.

CHAPTER

8

Technological Aspects of Measuring Inflammatory Markers

Raffaele Altara[1,2], *W. Matthijs Blankesteijn*[1]

[1]Department of Pharmacology, Cardiovascular Research Institute Maastricht, Maastricht University, Maastricht, The Netherlands

[2]Department of Physiology and Biophysics, University of Mississippi Medical Center, Jackson, MS, USA

8.1 IMMUNOASSAYS DEVELOPMENT AND NEW DIRECTIONS

Immunochemical techniques have an important role in the identification of inflammatory markers. Using the specificity provided by the antigen-antibody reaction, where the antigen is the analyte of interest, the immunochemical techniques capitalize on the sensitivity of the measurable label (i.e., radioactive substances, enzymes, fluorescence substances, etc.), called the reporter.

Introduced in the 1960s, the enzyme immunoassays (EIAs) together with the primitive form of the enzyme-linked immunosorbent assay (ELISA) differed from the original radioimmunoassay (RIA). Both techniques followed the principle of the immunoassay, but enzymes rather than the radioactive substances were adopted as reporter labels [1]. The ELISA technique was conceptualized and developed by Peter Perlmann, and Eva Engvall at Stockholm University, Sweden, and the EIA technique was developed by Anton Schuurs and Bauke van Weemen at the Research Laboratories of NV Organon, Oss, The Netherlands [1].

The impact of the introduction of the conjugated enzymes replacing the radioactive reporters revolutionized the use of immunochemical assays. The final effect was evident by the number of publications that came out in the following years (see Figure 8.1). The RIA use dropped, while the popularity of the ELISA consistently increased over the years.

The standard approach for detecting and measuring these inflammatory mediators today is clearly the ELISA, however, since the beginning of the new millennium, novel techniques have been proposed to detect multiple cytokines/chemokines in a single run. The principle of these assays still resembles that of classical ELISA, however, new detection techniques have been introduced, as will be explained later in this chapter. Today, this kind of assay is known as multiplex immunoassay (MIA).

8.2 METHODOLOGY AND INSTRUMENTATION

As mentioned above, the current methodology in clinical chemistry for the identification of inflammatory mediators is changing. In this chapter, we aim to introduce the reader into an area that is quite common nowadays in basic research: identification/quantification of cytokines/chemokines in serum, plasma, tissue lysates, and other fluid samples (i.e., urine, cerebrospinal fluid). Therefore, the reader should not expect a detailed description of the biochemical methodology or functionality of the instruments, which will be available though the cited references, but can expect to be informed about various aspects behind the validation of a chosen assay. Those aspects are often neglected before using such assays because the manufacturer ensures a proper validation before releasing the kit to the market. However, the robustness and suitability of the assay cannot be tested in every condition, and this is the classic situation when issues may appear.

Inflammation in Heart Failure
http://dx.doi.org/10.1016/B978-0-12-800039-7.00008-6

© 2015 Elsevier Inc. All rights reserved.

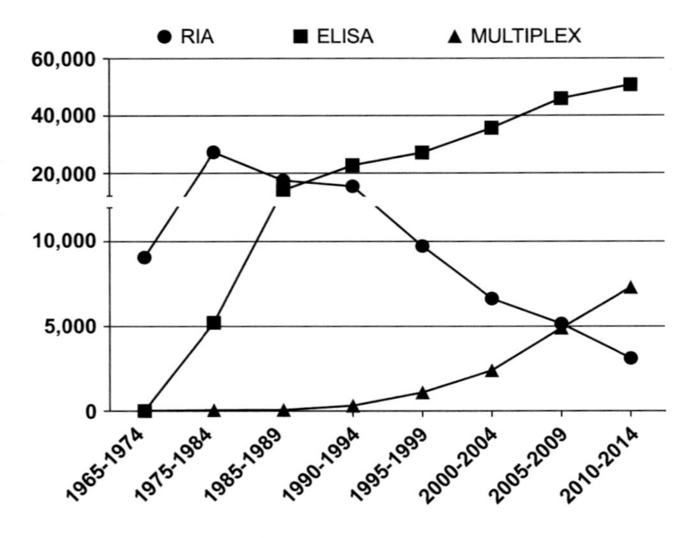

FIGURE 8.1 Estimates of published articles mentioning RIA, ELISA, and MULTIPLEX immunoassays between 1965 and 2014. The query was done in June 2014 in PubMed/National Library of Medicine of the NIH (www.ncbi.nlm.nih.gov/pubmed). The entry terms were radioimmunoassay (RIA), enzyme-linked immunosorbent assay (ELISA), and multiplex immunoassay. On the left side of the abscissa there are two periods of 10 years, followed by 5-year ranges (Nota bene: the purpose of this graph is to show the trends of use for these three assays, hence, the numbers of publications are not necessarily precise).

8.2.1 Solid Phase/Planar Assays

The ancestor of the solid-phase immunoassays is ELISA. The principle is very simple: a primary or capture antibody is immobilized to the bottom of solid carrier, typically a 96-well plate, and recognizes the antigen. The capture antibody is subsequently recognized by a secondary antibody that is often referred to as the detection antibody. Detection antibodies are typically labeled or conjugated to an enzyme that is responsible of the conversion of a substrate into a readable signal. Technically, newer ELISAs are not "enzyme-linked" but are instead linked to some nonenzymatic reporters [2,3] (see Figure 8.2). Although different variations may apply to the procedure, that is, the analyte of interest is immobilized to the plate or there is an addition of a tertiary antibody, the critical point of the assay is the choice of proper antibodies. To ensure their flawless performance in a sandwich ELISA antibody pair, it

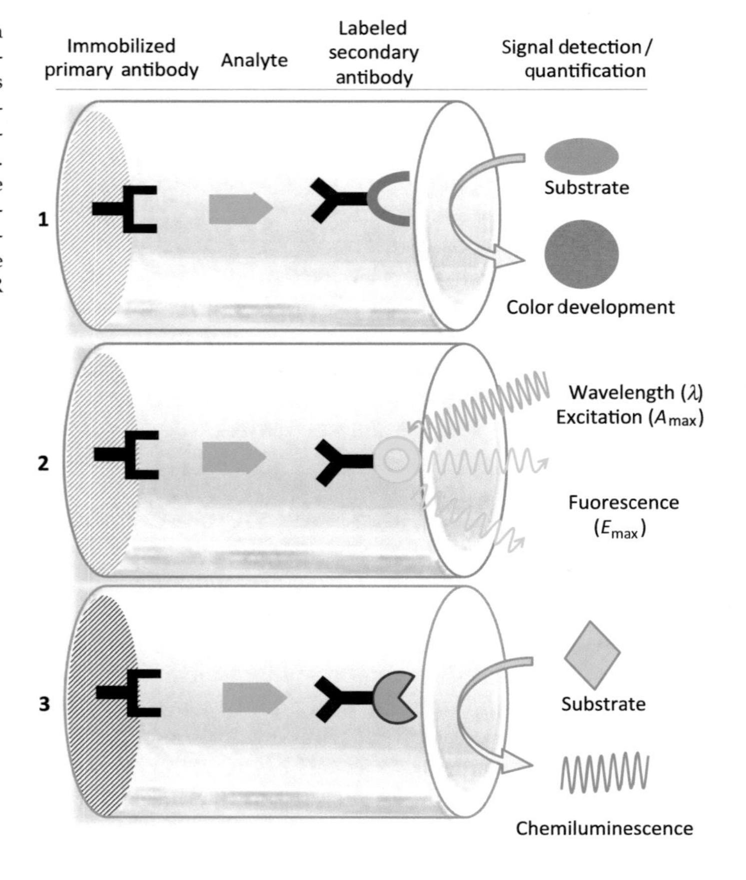

FIGURE 8.2 Main ELISA reporter formats. The first step of an ELISA is the immobilization of an antigen on a solid support (usually a polystyrene microtiter plate, also known as 96-well plate). This mechanism can be either nonspecific (printing the antigen on the bottom of the well) or specific (a previously immobilized capture antibody specific for the antigen, then generating the "sandwich" ELISA). Subsequently, a secondary antibody that is conjugated to an enzyme detects the antigen. Traditional ELISA-involved chromogenic reporters and substrates that produced an observable change of color to indicate the analyte presence (1) modern ELISA-like techniques make use of fluorogenic, (2) chemiluminescent, and (3) even real-time PCR reporters (not shown) to develop quantifiable signals.

is necessary to select antibodies with specific high-affinity recognition of the target analytes with as little nonspecific analyte detection as possible.

The type of antibody that is most likely to fulfill these criteria belongs to the monoclonal antibody class. Thanks to its monovalent affinity, it binds to a single epitope on the antigen. This characteristic is not always certain for the polyclonal antibody, which can bind to different epitopes on the same antigen. The manufacturer does not always provide detailed information about the epitope that is recognized by the antibody, although this information is quite relevant. It is therefore best practice to always inquire about the nature of the antibodies included in the kit/assay purchased to be aware of the possible limitations of the test.

In the last decades the antibody portfolio has been expanded extensively and new concepts have risen around the ELISA test. Moreover, the interest of researchers is nowadays shifting from information on single antigens toward the levels of panels of different antigens. This, together with the desire to obtain as much information as possible from a small amount of sample, gave rise to the development of multiplex assays. The main purpose is to have an assay for simultaneous detection and quantification of multiple target analytes in qualified complex sample types.

8.2.1.1 Multiplexed ELISA

The traditional ELISA test is far from being abandoned (see Figure 8.1), and this methodology is still considered the gold standard for quantification. To date, various methods have been developed and introduced to measure multiple inflammatory mediators in a quantitative way. For this purpose, novel concepts have been added to the classical assay to create a multiplexed ELISA. This is the case of the multiplex ELISA from Quansys Bioscences (www.quansysbio.com). With the aid of a robotic liquid handler, the capture antibodies are printed in a multiplex array within the same well on a 96-well microplate (Figure 8.3). In such a manner, it is possible to perform up to 25 ELISA tests in the same amount of time it takes to complete a single ELISA. Adopting this assay sounds attractive, as many other multiplex assays that will be described later in this chapter, however, there are many facets that have to be considered in advance. A schematic summary of the pros and cons of the major aspects of each MIA is presented in Table 8.1 of this chapter.

8.2.1.2 Electrochemical Multiplexed ELISA

An alternative multiplexed ELISA is the MULTI-ARRAY® from Meso Scale Discovery (www.mesoscale.com). As shown in Figure 8.3, multiple analytes can be detected within the same well (96-well microplate: 4-10 spots; 384-well microplate: 1-4 spots). The distinctive innovation that characterizes this assay is in the detection technology. Primary antibodies are immobilized on a carbon electrode plate surface and upon electrochemical stimulation a chain reaction of chemiluminescence is triggered [4]. Ultimately, light intensity is measured (see Figure 8.4 for more details). This method, referred to as ElectroChemiLuminescence or ECL, requires special equipment to read the plates as other more sophisticated MIAs. Nevertheless, the instruments needed do not require special attention making the entire procedure user-friendly.

The ECL technology has been applied to profile inflammatory mediators in different cardiovascular diseases. A strong relationship between heart failure and pulmonary arterial hypertension (PAH) has been reported in the literature [5]. In reviewing the published studies related to this topic, there is substantial evidence supporting a role for inflammatory mediators in the advancement of idiopathic PAH [6]. Using the ECL technology, Soon *et al.* [7] determined the circulating inflammatory profile of patients with idiopathic and hereditable PAH in order to examine the relationship between an 11 cytokines panel, the hemodynamics, functional capacity, and survival of these patients. The conclusion of their study was intriguing as they could find a clear relationship between the circulating levels of

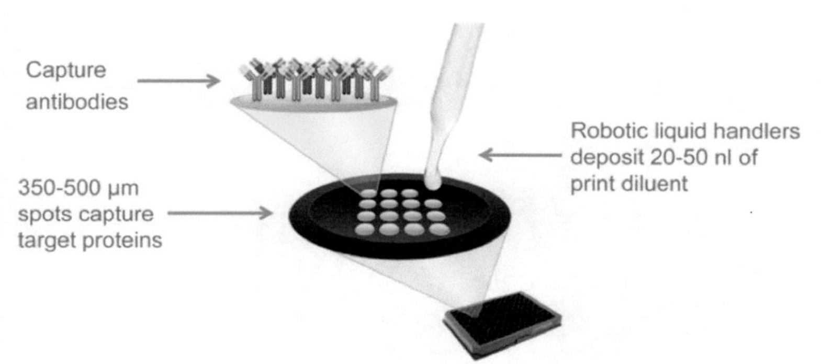

FIGURE 8.3 Configuration of a multiplex array. With the aid of a robotic liquid handler, up to 25 different capture antibodies are printed in a multiplex array within the same well on a 96-well microplate, in order to capture specific target proteins in a defined configuration.

TABLE 8.1 Differences Among the Main Commercially Available Immunoassay Platforms

	Solid Phase			Suspension Technology	
	ELISA	ECL	Membrane-Based Assay	CBA	Beads-Based Assay
Number of analytes determined	1/well	4-10/well	Up to 40/membrane	Up to 30/tube	Up to 500/well
	One analyte per plate	All analytes in one well Relatively small number for a multiplexed assay	All analytes on one membrane	All analytes in a single tube	All analytes in one well The panel has to be combined: not all analytes can be measured together
Sample size needed (μL)	50-100/well	15-50/well	500-1000/membrane	50/tube	25-50/well
	High volume requested for multiple analytes	Low sample size considering the number of analytes determined	Low sample size considering the number of analytes determined	Low sample size considering the number of analytes determined	Very low sample size
Specific devices	Plate reader	ECL reader	Infrared imaging system	Flow cytometer	Luminex™ plate reader + magnetic or standard washer
	Basic, simple to use, low budget	Expensive, dedicated to ECL assay only	Easy to use, multitasking (i.e. WB imaging) Intermediate cost	Routinely used in diagnostics, multiple use	Sophisticated, expensive, dedicated to beads-based assay only
Range of concentrations	Intermediate	Large dynamic range	Large dynamic range	Large dynamic range	Dynamic range
Handling skills	Intermediate	Intermediate	Novice/Intermediate	Expert/Intermediate	Expert
Cost of the commercial kit	Various costs Expensive when requested for multiple analytes determination	Low budget when compared to a single ELISA kit	Low budget considering the number of analytes	Expensive, especially when standards must be run each time.	Low budget when compared to a single ELISA kit Expensive
Time (hrs)	5 - O.N.	3	5 - O.N.	3 - 5	6 - O.N.
Other	May need multiple dilutions	Ability to perform repeated measurements. Reduces matrix effect	Ability to perform repeated measurements. Reduces matrix effect	Ability to perform repeated measurements	Ability to perform repeated measurements Multiple dilutions may apply. Matrix effect probability
Assessment	Quantitative	Quantitative	Semi-quantitative	Quantitative	Quantitative and qualitative

This table summarizes the prominent advantages (green) and disadvantages (red) that distinguish, to the personal opinion of the authors, the technology of the immunoassays for the identification of inflammatory mediators in qualified samples which are currently available on the market.

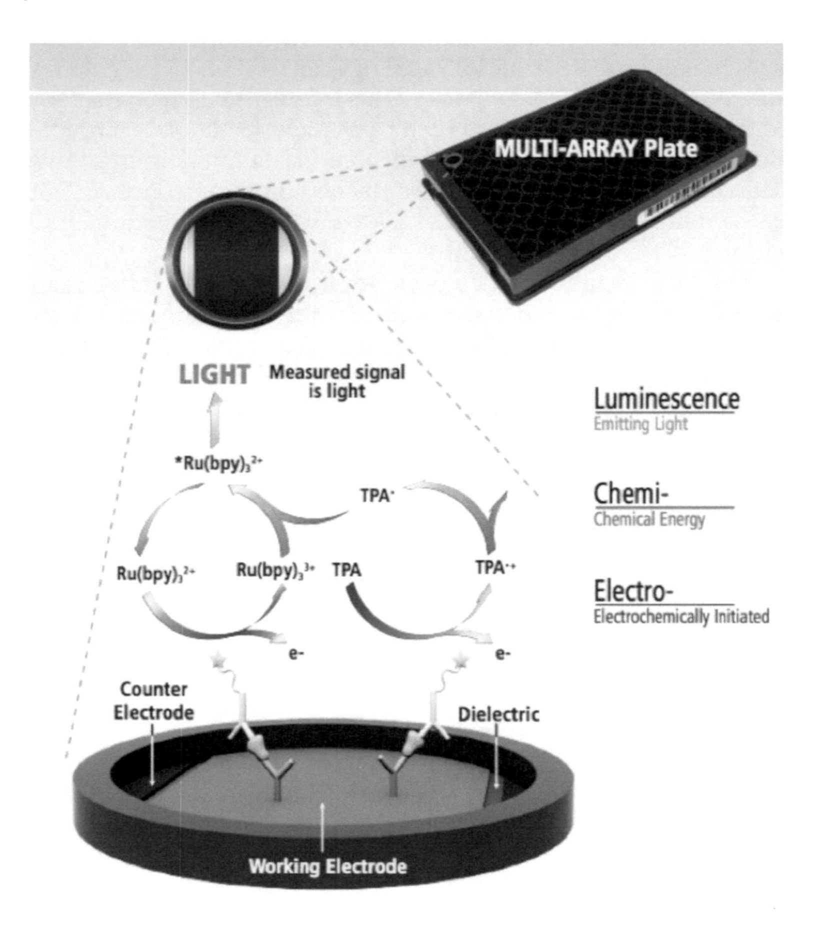

FIGURE 8.4 Principle of Electrochemical multiplexed ELISA. Microplates contain integrated carbon electrodes at the bottom of the plate; primary antibodies are immobilized on the carbon electrode plate surface and labeled with substances emitting light at ~620nm after a chemiluminescence chain reaction, when electrochemically stimulated. The detection process of the final light intensity is initiated at the electrodes located at the bottom of the microplates. Only labels near the electrode are excited and detected, enabling a wash-free assay. Multiple excitation cycles of each label amplify the signal to enhance light levels and improve sensitivity.

IL-2, IL-6, IL-8, IL-10, and IL-12p70 and the survival rate proving the importance of inflammatory biomarkers in the risk stratification of PAH patients. Looking at the results of their study, it is clear that the ECL technique offered a major advantage over more traditional approaches: Indeed, all cytokines measurements were present in low concentration ranges but the combination of high sensitivity and specificity of the assay with the small variation within the groups allowed the detection of significant differences among some of the cytokines levels determined in patients versus controls.

Data from several sources have identified that endothelial dysfunction and low-grade inflammation are common markers occurring together during cardiovascular disease [8]. Van Bussel and coworkers [9] used the ECL technique to assess the association between biomarkers of endothelial dysfunction and low-grade inflammation in a 6-year longitudinal study. Conducting their investigations on 293 apparently healthy individuals, they evaluated an extensive array of biomarkers of endothelial dysfunction and low-grade inflammation in relation to arterial stiffness. Although the cytokine concentration levels were low, the ECL assay gave precise measurements with little variation, allowing the authors to propose the hypothesis that endothelial dysfunction and low-grade inflammation lead to cardiovascular disease.

Another study that benefited from the use of the ECL multiplex technology was the one published by Collier *et al.* [10]. Studying two different cohorts—one of asymptomatic hypertensive patients ($n=94$) and one of HF patients with preserved ejection fraction ($n=181$)—they observed that varying fibro-inflammatory profiles can be detected throughout different stages of hypertensive heart disease. In contrast to the previously presented studies, this one presented larger variation between the cytokine values of the two groups, showing that ECL technology can be accurately performed on a wider range of analyte concentrations.

8.2.1.3 Membrane-Based Assay

Recently, paper-based microfluidics has emerged as a multiplexable point-of-care platform showing the opportunity to transform existing assays into resource-limited settings at or near the site of patient care [11]. Recently, an innovative membrane-based MIA called Proteome Profiler from R&D System (www.rndsystems.com) has been introduced. In addition to the multiplex features, it differs from the classical ELISA in that its primary antibodies are immobilized on nitrocellulose membranes rather than on polystyrene plates. In general, nitrocellulose membranes are high-quality membranes ideal for irreversible binding of proteins and nucleic acids. This characteristic provides the opportunity to immobilize antibodies in spots and to pour the entire sample on the same membrane (see Figure 8.5).

In principle, this assay has been designed by the manufacturer to yield qualitative data. However, in our laboratory, we developed a normalization system, using a fluorescence reporter system and a near-infrared imaging device to acquire the signal and showed that the assay could be upgraded to yield semiquantitative data [12]. Moreover, in the same study a proof of concept for this methodology was shown, and a strong relationship between the membrane-based assay and the ELISA outcome could be demonstrated (see Figure 8.6). Even with the upgraded analysis protocol, this assay does not provide quantitative results, but it is a valid alternative to more expensive MIA assays.

To date, there are no publications about this assay in the context of heart failure. Nevertheless, we observed that the membrane-based technique allowed the detection of a robust increase of at least 4 analytes (MIG, IP-10, MIP-2, and IL-17) out of the 30 detectable inflammatory mediators present in pooled serum samples of five

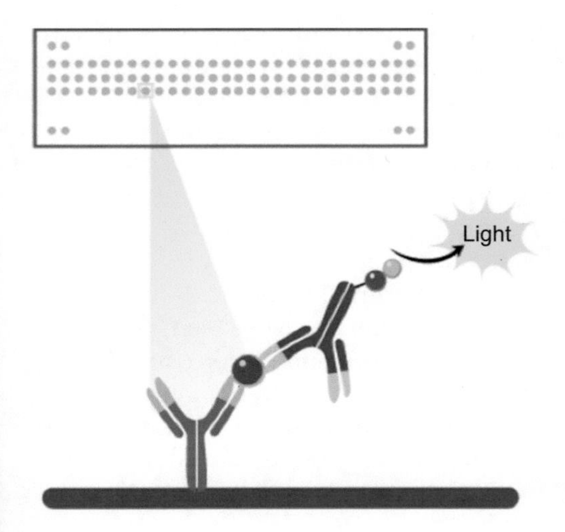

FIGURE 8.5 Principle of the Proteome Profiler membrane-based multiplex immune assay. In the Proteome Profiler technology, typically 35-40 different capture antibodies are immobilized in distinct spots on a nitrocellulose membrane in duplicate. The membranes are incubated with 0.5-1 ml of the sample of interest, washed, and further processed for either chemiluminescent of fluorescent detection.

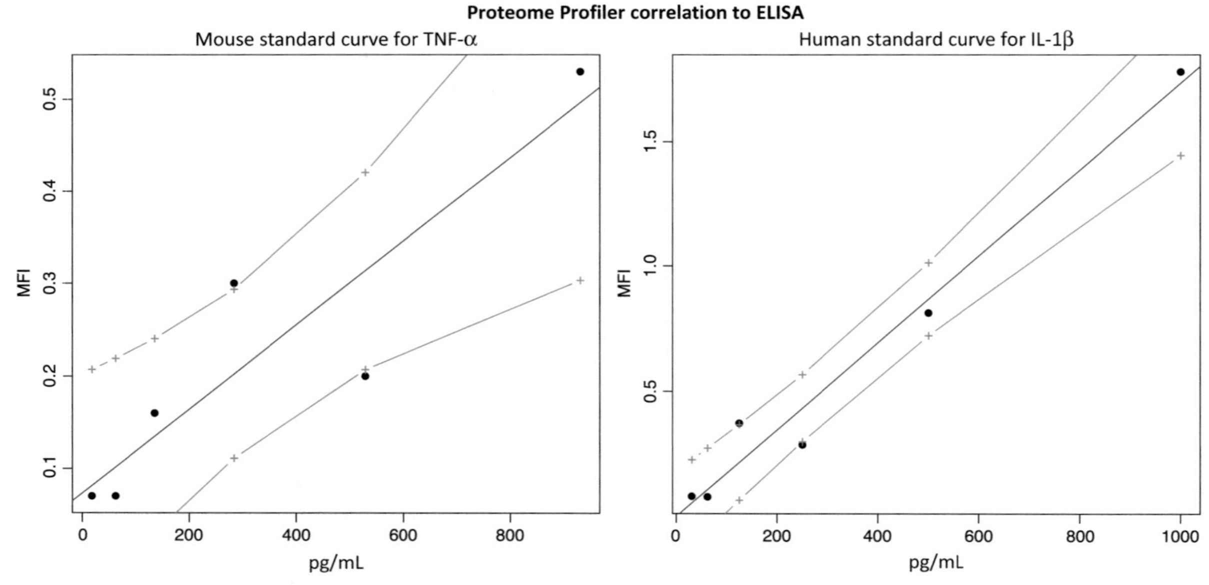

FIGURE 8.6 Proteome Profiler calibration curve for TNF-α and IL-1β. Visualization of the regression line (blue) fitted for the concentrations, measured by ELISA (abscissa), and membrane's fluorescence readouts (ordinate) of TNF-α (a) and IL-1β (b). The dots are compactly distributed around the predicted line, and the unexplained error is small. The red lines delimited the 95% of CI. The equations of the curves are (a) $y = 561.73x + 8.97$ with Pearson's $r = 0.97$; (b) $y = 1830.6x - 79.4$ with Pearson's $r = 0.91$. The red dashed lines delimited the 95% confidence interval.

mice exposed to pressure overload-induced cardiac hypertrophy. However, these results need to be interpreted with caution, as the current study was purely methodological; indeed, those observations came from a process in which the membranes underwent chemiluminescent detection, stripping, and finally fluorescent detection (see Figure 8.7).

8.2.2 Suspension Array Technology/Bead-Based Immunoassays

The suspension array technology is nowadays the first choice for a multiplexed high-throughput screening platform for investigations concerning inflammatory cytokines. The methodology is based on a marriage between the principles of ELISA and flow cytometry. Despite some variations among the commercially available assays, the principle remains the same: the primary antibodies are immobilized on microsphere beads, which are suspended in solution (i.e., serum, plasma, etc.) and the detection system is homologous to the flow cytometry in that it can detect signals from the individual beads. These kinds of assays are therefore named multiplex bead array assays (MBAA).

8.2.2.1 Cytometric Bead Assay

Cytometric bead array (CBA) from BD™ (www.bdbiosciences.com) uses a combination of different antibody-coated beads to capture the analytes of interest. The beads are labeled per analyte, allowing the simultaneous identification and signal intensity detection in clinical flow cytometer (see Figure 8.8 for more details). Although a limited list of kits are provided by BD Biosciences, it is possible to customize the assay ordering unconjugated beads from the company. This feature allows researchers to conjugate their own antibody or protein of interest using sulfo-SMCC chemistry and enhancing the flexibility of the assay. Moreover, a flow cytometer is a device that is usually present in a laboratory environment; hence the assay can be performed in practically every clinic.

In the study published by Tarnok *et al.* [13], the CBA assay displayed several advantages over the classic ELISA. In their work, 6 cytokines were measured in serum from 19 children (ranging in age from 2 weeks to 16 years) operated for a cardiopulmonary bypass implant. The first observation of the study was that by using the CBA, about 80% of sample was saved over ELISAs. Second, the calibration range of the CBA could be extended from the original, reaching about 10 times higher ranges compared to the corresponding ELISA. Interestingly, with both assay forms the serum cytokine concentrations could be determined at the reported cutoff concentrations [14–16]. Hence, this test provided a major advantage in addition to the multiplexing: samples could be measured without special manipulations because the concentration of cytokines to be determined was well within the calibration curve.

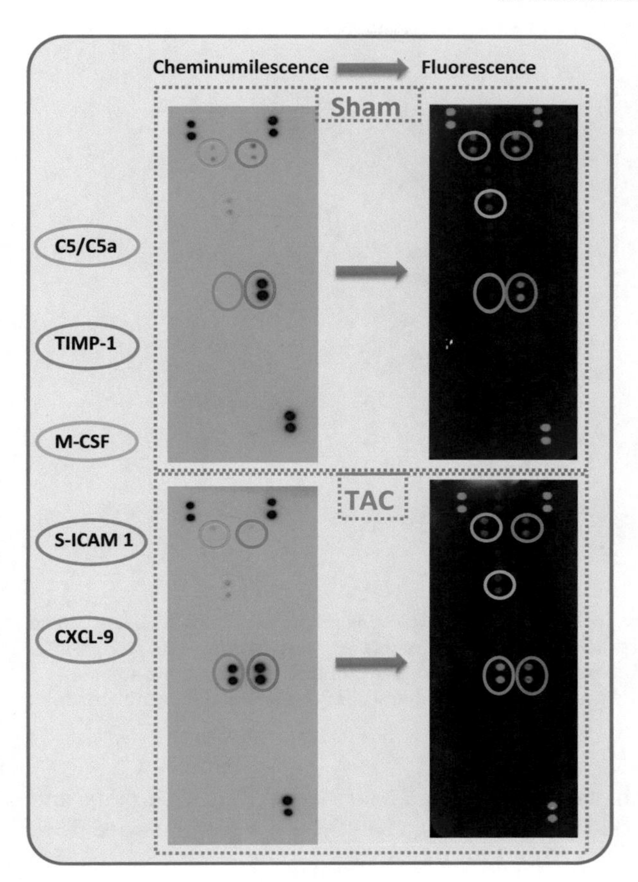

FIGURE 8.7 Comparison of the membrane development using chemiluminescence and fluorescence. Chemiluminescence development (left membranes) and fluorescence images after stripping (right membranes) are represented. The latter method yielded a higher number of detectable analytes that can be visualized (green spots). The fluorescence readout allowed us to compare the circulating inflammatory profile of pooled samples of TAC-operated mice to the relative shams and calculate the relative ratio TAC/Sham.

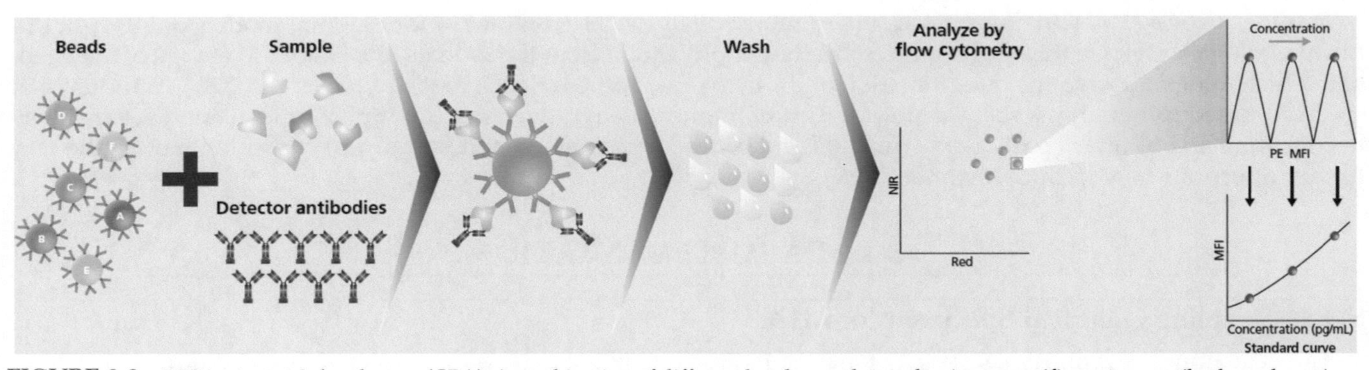

FIGURE 8.8 BD™ cytometric bead array (CBA). A combination of different beads—each type having a specific capture antibody and a unique corresponding fluorescent tag—is mixed with a sample or standard. Subsequently, a mixture of detection antibodies that are conjugated to a reporter molecule (PE) are added. After incubation and washing, the samples are acquired on a flow cytometer, where the individual beads are separated and identified based on their characteristic fluorescent signature. Finally, the FCAP Array analysis software determines the median fluorescence intensity (MFI) for each analyte in the array. It performs interpolation of sample concentrations by comparison to the generated standard curve and enables viewing analysis reports in graphical or tabular format.

The CBA found its early application in heart failure-linked investigations through the work of George *et al.* [17]. During that period, an emerging role of adiponectin acting as an anti-inflammatory and antiatherogenic cytokine during vascular remodeling was suggested by Haluzik *et al.* [18]. To determine whether there was a correlation between the concentrations of adiponectin and circulating inflammatory cytokines in a cohort of patients with congestive heart failure (CHF) the CBA assay was used. Despite the negative correlation found, this technique allowed George and coworkers to measure several other circulating analytes as they needed very little serum volume for the test.

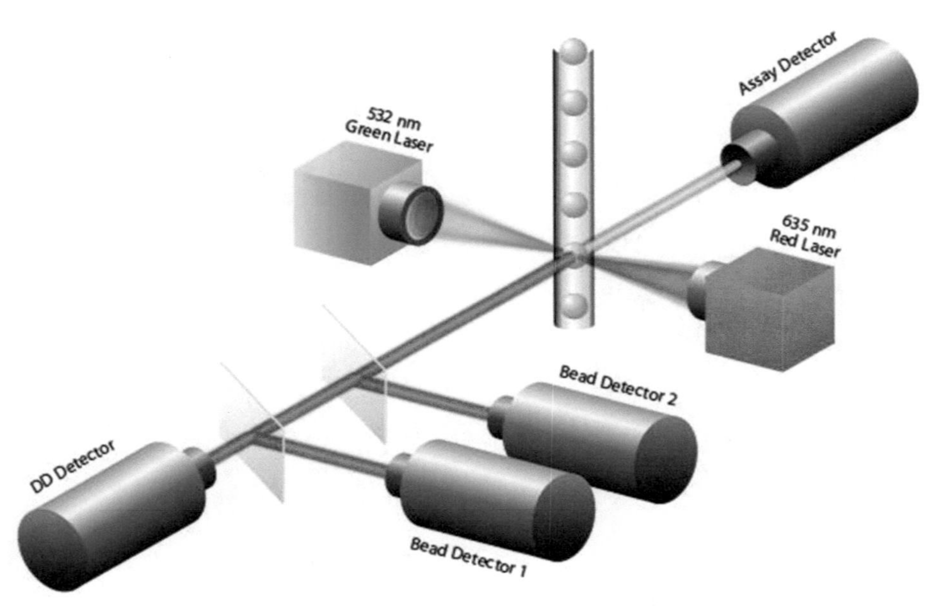

FIGURE 8.9 Principle of xMAP technology. The method uses up to 500 distinct color-coded tiny microsphere sets, each bead set being coated with a reagent specific to a particular bioassay. The beads flow inside a channel where a light source (two lasers: one green and one red) excites the internal dyes of each microsphere particle and also any reporter dye captured during the assay. Inside the Luminex analyzer there are multiple detectors in addition to the assay detector. These revealers are necessary to identify beads-aggregation that would compromise the entire readout.

8.2.2.2 Luminex Screening and Performance Assays

In the past 15 years, a new technology has emerged offering the benefits of ELISA, advanced fluidics, optics, and digital signal processing benefits all in one assay. Using color-coded microspheres, coated with a reagent specific to a particular bioassay, the xMAP (Multi-Analyte Profiling) technology from Luminex Corp. (www.luminexcorp.com) offers several advantages over the flow cytometry assay. The main advantage of this assay is given by the property of the beads in conjunction with the high-tech reader-detecting instrument—namely, the Luminex™ reader (Figure 8.9).

In 2010, Vistnes *et al.* [19] determined, via Luminex analysis, 25 cytokines in four animal models of HF with different etiologies. Despite the small volume size they could obtain from the rodents, the major advantage of the bead-based technology allowed the determination of the cytokine profiles in each model. Vistnes *et al.* [20], confident that the bead-based technology was the state of the art of immunoassays, declared "multiplex protein analyses are likely to constitute an important part of experimental and clinical research on heart failure and cytokines, paving the way for more accurate heart failure treatment" [19].

8.3 MIA IMPLEMENTATION

8.3.1 Sampling Qualified Specimen for MIA

The availability of multiple assay forms does not guarantee that every sample can be measured. Indeed, there are factors that influence the possibility to use the sample for an assay or hamper the performance of the test. Hence, conceiving an immunochemical assay for inflammatory mediator detection without a proper sample collection may have deleterious results.

Additional factors like type of blood sampled (i.e., venous, capillary, or arterials), use of medication, diet and fasting, physical condition, and circadian rhythms should be taken into account before a correct sampling protocol is achieved. These and many more factors are extensively delineated in the immunoassay handbook, *Theory and Applications of Ligand Binding, ELISA and Related Techniques*, edited by Wild [21]. We therefore invite the reader to consult this book in view of sampling qualified specimens for immunoassays.

8.3.2 Analytical Challenges and Clinical Utility

8.3.2.1 Cytokines/Biomarkers Stability

In preclinical studies it may happen that measurements of inflammatory cytokines are not performed directly but at the end of the study. This is a common situation in clinical studies, too, for which the lag time between sample collection and the performance of the actual measurement is dictated by the inclusion, intervention, and follow-up of patients and may span over several years.

In 2009, de Jager *et al.* [22] published the prerequisites for cytokine measurements in clinical trials with bead-based MIAs. They assessed the influence of various processes, such as blood-collecting tubes, duration of storage, and number of freeze-thawing cycles to study the impact of these parameters on the measurement of the cytokine concentration. One of their main conclusions was that cytokine stability is critical in time. In fact, reduced cytokine concentrations could be observed already within 1 h after the sampling was performed. Moreover, the samples, rigorously stored at −80 °C, have to be analyzed preferably within 2-4 years. A valid suggestion from de Jager and co-workers is stated in their conclusion: an internal control sample should be stored together with the sample collection and kept under similar conditions as the other clinical samples until the analysis. This method will allow calculating the recovery for each analyte that will be determined.

In our lab, we were also interested in determining the stability of a few inflammatory mediators over a 2-month period. Therefore, we performed a stability test using serum obtained from a healthy subject[1] to build up a concentration-curve over time. The concentrations used mimicked a disease situation where the levels of cytokines are usually high. Hence, we spiked the plasma with additional recombinant cytokines, and we measured the cytokine concentrations by ELISA. Here, we report the case of IP-10, IL-17, and IL-1β. As shown in Figure 8.10, IL-1β recovery was around 100% at 1 month, but then decreased to 20% at 2 months. Conversely, the stability of IL-17 was compromised at day 0, but it remained stable, with a slight increase in recovery along the 2-month period. The third analyte measured, IP-10, showed a clear variability between day 0 and day 1, although the recovery turned to be stable at 1 and 2 months. This trend suggested that an experimental error occurred on day 1, highlighting how sensitive the measurement might be to changes in the experimental setting.

8.3.2.2 The Importance of Validation

In our department, we evaluated the feasibility of several bead-based multiplex assay kits for the measurement of inflammatory biomarkers in animal models for HF, and we compared the results with those obtained by the gold standard immunoassay, namely, ELISA. To do that we made use of samples obtained from TAC mice and MI rats. LPS/treated rats were used as positive controls as they have high circulating levels of various inflammatory mediators (i.e, TNF-α). Our results are reported herewith in two cases.

8.3.2.2.1 CASE 1

The Bio-Plex Pro™ (Bio-Rad Laboratories) mouse assay offered incoherent results for 13 out of the 23 cytokines that were measured. A representative example is given in Figure 8.11a where the values determined in serum for IL-1β were variable. In both the Sham and TAC3 sample, IL-1β concentrations remained unchanged in the 4× and 8× diluted samples, while readings for the 16× dilution were below the detection limit. Also in the TAC1 sample, about the same concentration (≈150 pg/ml) was found as in Sham and TAC1 in the 4× dilution sample, but values increased paradoxically at every dilution step, reaching a threefold higher value (519 pg/ml) after 16× dilution. In fact, we observed a minimum detection threshold ranging from 131 to 10,666 pg/ml for the Bio-Rad assay and from

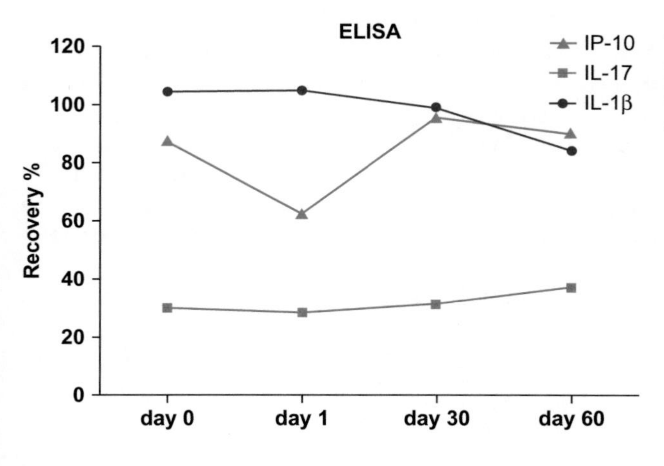

FIGURE 8.10 Recovery of spiked human recombinant IP-10, IL-17, and IL-1β in serum obtained from a healthy subject. All measurements were performed by ELISA (R&D System™). Recovery was calculated as: (real concentration/theoretical concentration) × 100.

[1] The recruitment of healthy volunteers was performed according to the Dutch Medical Ethical Committee (protocol: METC 11-3-056) and in respect of the Declaration of Helsinki.

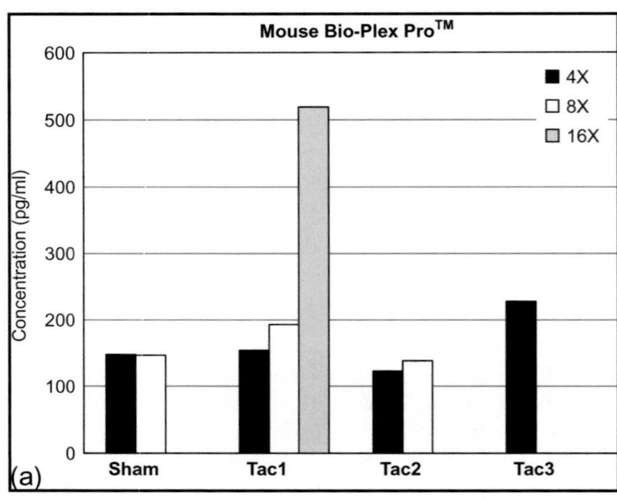

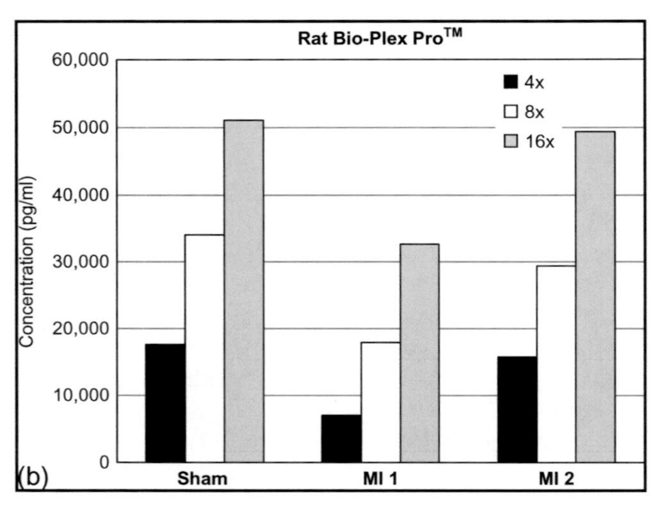

FIGURE 8.11 (a) IL-1β concentrations measured with Mouse Bio-Plex Pro™ in three TAC samples. TAC1 showed increased levels with the increasing of dilution. Conversely, IL-1β in the samples Sham, TAC2, and TAC3 were undetectable at higher dilutions. (b) With Rat Bio-Plex Pro™ rats serum showed an increase in concentration progressive with the dilutions.

3 to 117 pg/ml for the Affimetrix assay, whereas detection thresholds in the order of 1-10 pg/ml were described by the supplier for both assays.

In standard buffer diluents, the detection limit and sensitivity, in fact, do meet the indicated values. However, in "real" plasma or serum samples, the matrix components compromise a reliable measurement [23]. This phenomenon is known as the "hook effect" [24,25], and occurs in MIA whenever the level of the analyte of interest is close to the background; hence, the signal-to-noise ratio is too low. It could be suggested that a "matrix effect" [26] is interfering with the measurements, but very few studies have been carried out on how, and to what extent, the particles that compose the samples may affect the cytokine measurements. Kocbach *et al.* [27] showed that serum proteins contained in the sample solution could reduce partly or completely the detection of certain cytokines. This work could help put our results in the right perspective. Due to the fact that our study is limited to rodents, we would recommend to test and validate the performance of the beads-based assay ahead of the experiment.

The hook effect was even more profound in a second multiplex assay provided by a different supplier—namely, the Bio-Plex Pro™ rat assay (Bio-Rad). As shown in Figure 8.11b, this assay detected very high concentrations even at the first dilution, which is not an adequate representation of the physiological levels [28]. The mathematical increase of concentration was a pure reflection of the multiplication of the fold of dilution by the same intensity values measured for every dilution.

8.3.2.2.2 CASE 2

To further evaluate the multiplex assay and check whether we observed the classic hook effect, we spiked the blood of mouse and rat with the standards provided by the supplier. Starting from high concentrations, we created a dilution series of cytokines in mouse blood that guaranteed us to cover the entire concentration range of the standard curve (see Figure 8.12). The test, performed with the Mouse Procarta® assay (Affimetrix), showed no hook effect ($R=0.98$), although the recovery was low (recovery $=52\pm12\%$).

Conversely, when we used the spikes to build up a standard curve in rat serum that reflected the one of the supplier, the reproducibility was poor for all the brands tested. More specifically, the main issue of the Bio-Rad assay was a high threshold; that is, in the case of TNF-α, the threshold was about 400 pg/ml. Hence, the detection of the analyte of interest was impossible even in the linear part of the standard curve. Similar results were obtained for the nine other analytes investigated.

The Affymetrix assay showed a different issue compared to the one from Bio-Rad. In this case, the standard curves in rodent plasma were sufficiently linear, and the threshold for the detection was varying between 3 and 117 pg/ml. However, the recovery of the spikes was really poor. In Figure 8.13, the example of TNF-α is shown where the recovery ranged from 16% to 46%.

8.3.2.3 The IL-6 Case (Systematic Review)

In several chapters of this book there is evidence that the application of an inflammatory biomarker as a diagnostic tool could be used to identify heart failure at an earlier stage, as the inflammatory events that contribute

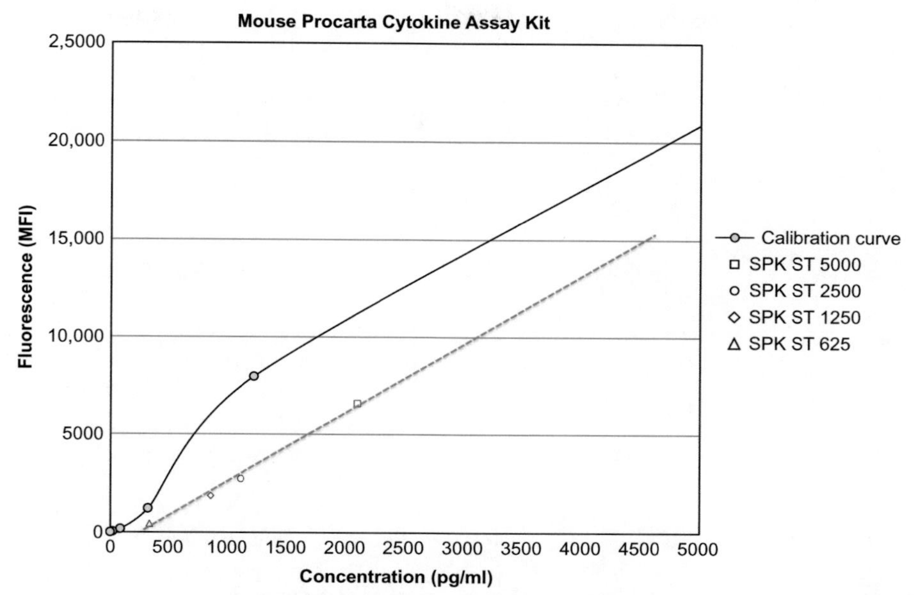

FIGURE 8.12 IL-1β concentration measured with Mouse Procarta® assay. The calibration curve is represented here as a continuous line analyzed through the five-parameter logistic (5-PL) nonlinear regression curve-fitting model. The spike concentrations measured in serum resulted in a linear curve (dashed line, $R=0.97$). However, the recovery of the cytokines is about half of the original concentration spiked into the blood: recovery $=52\pm12\%$.

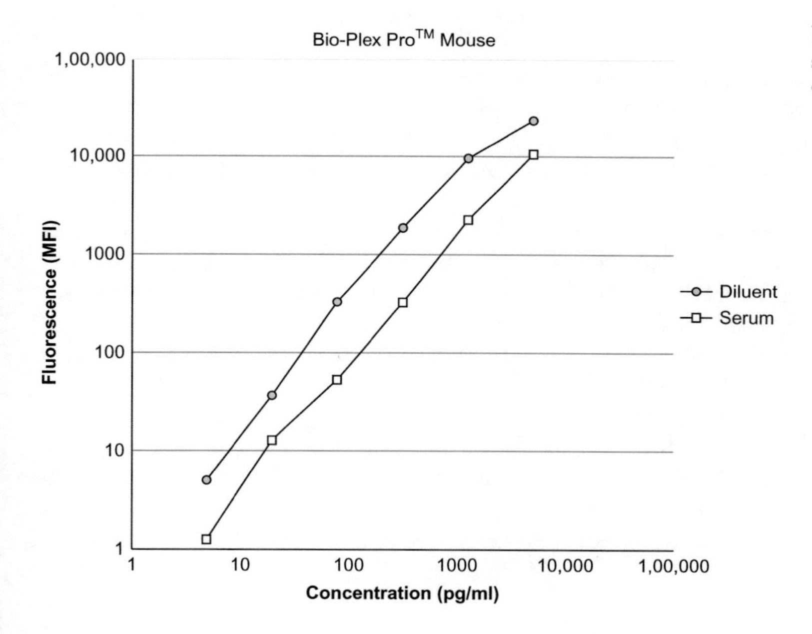

FIGURE 8.13 TNF-α in Bio-Plex Pro™ Mouse cytokine assay showed relatively good linearity for the spikes in serum, despite the low recovery: 16-46%.

to the development of heart failure are already active when no structural adaptions or symptoms can be found in patients [29,30]. Therefore, the immunoassay used to determine the inflammatory mediators should be accurate and reproducible in every study before it can be implemented to the clinic. However, despite a continuous progression in the development of immunoassays, this is not yet the case. To give a pragmatic example on what would be the causes of inter-assays and inter-studies variations, we give a single example on interleukin-6 herein.

In a considerable amount of literature, it was reported that IL-6 blood levels are not only increased in patients suffering from heart failure but blood concentrations are also predictive for the severity of heart failure [31,32]. Besides this, IL-6 signaling in the heart is directly responsible for a hypertrophic response of myocytes and enhances thereby adverse cardiac remodeling [31]. This suggests that IL-6 levels are increased in HF patients in an early phase and therefore would be a suitable biomarker for early disease detection.

TABLE 8.2 Results of the Individual Studies

	IL-6 (pg/ml)			
	Patients	**Healthy**	***p*-Value**	**Additional**
[35] Mean (±SD)	10.9 ± 18.2	<6.0	0.0243	Highest IL-6 levels in NYHA IV (*F*-value 5.284, $p < 0.0005$)
[36] Median (95% CI)	23 (10-65)	15 (11-28)	<0.0001	
[37] Mean (±SEM)	18.5 ± 11.8	1.7 ± 1.6	<0.01	NYHA II 3.8 ± 2.0 ($p < 0.05$). NYHA III/IV significant higher compared to NYH II ($p < 0.05$)
[38] Mean (±SD)	6.4 ± 3.5	1.7 ± 0.6	<0.0001	NYHA II 5.5 ± 2.3 pg/ml in comparison with control NS. NYHA III/IV 7.1 ± 4.2 pg/ml significant higher compared to control ($p < 0.001$) and to NYHA II ($p < 0.05$)
[39] Mean (95% CI)	8.9 (0.9-51.5)	3.2 (1.2-5.2)	<0.01	NYHA II ($n = 55$): 5.1 (4.1-6.4) NYHA III ($n = 97$): 6.2 (5.2-7.4) NYHA IV ($n = 62$): 9.1 (7.4-11.3) <0.001 compared NYHA IV with NYHA II <0.01 compared NYHA IV with NYHA III
[40] Mean (±SD)	18 ± 19	<5.0	<0.0001	IL-6 elevation significant higher in NYHA III/IV compared to NYHA I/II ($p < 0.001$)

We studied the literature for existing evidence concerning elevations of interleukin-6 in blood of patients suffering from heart failure. Furthermore, we analyzed whether interleukin-6 blood levels are useful as a biomarker in identifying different stages of heart failure[2] [33]. Table 8.2 summarizes the results of the included studies.

With this review, a solid ground of evidence for statistically significant elevations of IL-6 in patients suffering from HF compared to healthy subjects is shown. However, we observed a large variability among the results that might be attributed to the different sample collection and processing and storage before the samples have been measured. Overall, the samples have been measured via ELISA, but the antibodies set used for each study was different as their source were from different suppliers.

Taken together, IL-6 has characteristics that are interesting for its application as a potential biomarker in HF, but the mentioned limitations suggest that IL-6 alone is not sufficient. This is also indicated by the varying ratios of IL-6 levels of patients versus controls (see supplementary Table 8.1). In addition, these studies used different assays and protocols for assessing IL-6 levels, which may partly explain the variation of the ratios. The optimal procedure for measuring IL-6 blood levels in patients is not yet found, claiming for the establishment of a standardized and reliable approach for assessing IL-6 before using IL-6 as a biomarker in the clinic.

Revealing distinct clusters of pro- and anti-inflammatory in HF groups may also be of importance for new treatment strategies (see Chapter 14 on "The role of cytokines in clinical heart failure"), given that approaches targeting only TNF-α in patients with HF were not effective. In conclusion, it might be possible to identify inflammatory markers in patients who are at risk of developing HF in time or with adverse cardiac events, but it is vital that studies based on immunochemical assay for inflammatory mediator detection rely on a proper sample collection and assay validation.

8.4 THE IMMUNOASSAY MARKET: OPPORTUNITIES AND ISSUES

At the beginning of this chapter, we described the introduction of the immunoassay in clinical laboratories since the 1960s. As depicted in Figure 8.1, the immunoassays prospered and developed into different forms. Novel methodologies have been introduced, and today these biochemical tests are indispensable tools in clinical diagnostics. Even though the field is now mature, companies have continued to develop new immunoassays and immunoassay instrument platforms to further improve the sensitivity of the assays, to make multiplexing possible, to miniaturize

[2] The methodological quality of the included studies was assessed using an adapted scheme of the Cochrane Collaboration's tool for assessing risk of bias. Furthermore, the STrengthening the Reporting of OBservational studies in Epidemiology – Molecular Epidemiology (STROBE-ME) [24] guideline was used for assessing the quality and risks of the reviewed studies.

the platforms for point-of-care, and to identify and develop assays for novel biomarkers, thus further expanding the potential of immunoassays for the future [34].

The Kalorama Information report on the World Market of Immunoassays reported that the competitiveness within this market is "stubbornly large and long established." By the end of 2013, more than 275 companies were active in this market and the number appears to be increasing. This matter is not trivial though. Indeed, one major drawback of this approach is that the competition among the suppliers is pushing the R&D to faster develop novel and better technology, while the quality of the assays does not receive the highest priority. The pressure to rapidly bring an assay to the market might have been the underlying cause of the issues described in the previous sections.

For all these reasons, we suggest to the reader who is planning to perform measurements of inflammatory mediators through an immunoassay to consider the various aspects mentioned in this chapter that might strongly influence the outcomes of studies on inflammatory mediators.

Acknowledgments

The authors wish to thank Matthias Bush for his assistance in preparing a systematic review on IL-6.

References

[1] Lequin RM. Enzyme immunoassay (EIA)/enzyme-linked immunosorbent assay (ELISA). Clin Chem 2005;51:2415–8. doi:10.1373/clinchem.2005.051532.

[2] Matsukuma E, Kato Z, Omoya K, Hashimoto K, Li A, Yamamoto Y, et al. Development of fluorescence-linked immunosorbent assay for high throughput screening of interferon-gamma. Allergol Int 2006;55:49–54. doi:10.2332/allergolint.55.49.

[3] Swartzman EE, Miraglia SJ, Mellentin-Michelotti J, Evangelista L, Yuan PM. A homogeneous and multiplexed immunoassay for high-throughput screening using fluorometric microvolume assay technology. Anal Biochem 1999;271:143–51. doi:10.1006/abio.1999.4128.

[4] Forster RJ, Bertoncello P, Keyes TE. Electrogenerated chemiluminescence. Annu Rev Anal Chem (Palo Alto Calif) 2009;2:359–85. doi:10.1146/annurev-anchem-060908-155305.

[5] Vonk Noordegraaf A, Galie N. The role of the right ventricle in pulmonary arterial hypertension. Eur Respir Rev 2011;20:243–53. doi:10.1183/09059180.00006511.

[6] Cohen-Kaminsky S, Hautefort A, Price L, Humbert M, Perros F. Inflammation in pulmonary hypertension: what we know and what we could logically and safely target first. Drug Discov Today 2014;19(8):1251–6. doi:10.1016/j.drudis.2014.04.007.

[7] Soon E, Holmes AM, Treacy CM, Doughty NJ, Southgate L, Machado RD, et al. Elevated levels of inflammatory cytokines predict survival in idiopathic and familial pulmonary arterial hypertension. Circulation 2010;122:920–7. doi:10.1161/CIRCULATIONAHA.109.933762.

[8] de Jager J, Dekker JM, Kooy A, Kostense PJ, Nijpels G, Heine RJ, et al. Endothelial dysfunction and low-grade inflammation explain much of the excess cardiovascular mortality in individuals with type 2 diabetes: the Hoorn Study. Arterioscler Thromb Vasc Biol 2006;26:1086–93. doi:10.1161/01.ATV.0000215951.36219.a4.

[9] van Bussel BC, Schouten F, Henry RM, Schalkwijk CG, de Boer MR, Ferreira I, et al. Endothelial dysfunction and low-grade inflammation are associated with greater arterial stiffness over a 6-year period. Hypertension 2011;58:588–95. doi:10.1161/HYPERTENSIONAHA.111.174557.

[10] Collier P, Watson CJ, Voon V, Phelan D, Jan A, Mak G, et al. Can emerging biomarkers of myocardial remodelling identify asymptomatic hypertensive patients at risk for diastolic dysfunction and diastolic heart failure? Eur J Heart Fail 2011;13:1087–95. doi:10.1093/eurjhf/hfr079.

[11] Yetisen AK, Akram MS, Lowe CR. Paper-based microfluidic point-of-care diagnostic devices. Lab Chip 2013;13:2210–51. doi:10.1039/c3lc50169h.

[12] Altara R, Manca M, Hessel MH, Janssen BJ, Struijker-Boudier HH, Hermans RJ, et al. Improving membrane based multiplex immunoassays for semi-quantitative detection of multiple cytokines in a single sample. BMC Biotechnol 2014;14:63. doi:10.1186/1472-6750-14-63.

[13] Tarnok A, Hambsch J, Chen R, Varro R. Cytometric bead array to measure six cytokines in twenty-five microliters of serum. Clin Chem 2003;49:1000–2.

[14] Mehr SS, Doyle LW, Rice GE, Vervaart P, Henschke P. Interleukin-6 and interleukin-8 in newborn bacterial infection. Am J Perinatol 2001;18:313–24. doi:10.1055/s-2001-17857.

[15] Franz AR, Steinbach G, Kron M, Pohlandt F. Interleukin-8: a valuable tool to restrict antibiotic therapy in newborn infants. Acta Paediatr 2001;90:1025–32.

[16] Romagnoli C, Frezza S, Cingolani A, De Luca A, Puopolo M, De Carolis MP, et al. Plasma levels of interleukin-6 and interleukin-10 in preterm neonates evaluated for sepsis. Eur J Pediatr 2001;160:345–50.

[17] George J, et al. Circulating adiponectin concentrations in patients with congestive heart failure. Heart 2006;92:1420–4. doi:10.1136/hrt.2005.083345.

[18] Haluzik M, Parizkova J, Haluzik MM. Adiponectin and its role in the obesity-induced insulin resistance and related complications. Physiol Res 2004;53:123–9.

[19] Vistnes M, Waehre A, Nygard S, Sjaastad I, Andersson KB, Husberg C, et al. Circulating cytokine levels in mice with heart failure are etiology dependent. J Appl Physiol 1985;108:1357–64. doi:10.1152/japplphysiol.01084.2009.

[20] Vistnes M, Christensen G, Omland T. Multiple cytokine biomarkers in heart failure. Expert Rev Mol Diagn 2010;10:147–57. doi:10.1586/erm.10.3.

[21] Wild D. The Immunoassay Handbook: "Theory and applications of ligand binding, ELISA and related techniques". Amsterdam: Elsevier; 2013.

[22] de Jager W, Bourcier K, Rijkers GT, Prakken BJ, Seyfert-Margolis V. Prerequisites for cytokine measurements in clinical trials with multiplex immunoassays. BMC Immunol 2009;10:52. doi:10.1186/1471-2172-10-52.
[23] Hoadley ME, Hopkins SJ. Overcoming matrix matching problems in multiplex cytokine assays. J Immunol Methods 2013;396:157–62. doi:10.1016/j.jim.2013.07.005.
[24] Ryall RG, Story CJ, Turner DR. Reappraisal of the causes of the "hook effect" in two-site immunoradiometric assays. Anal Biochem 1982;127:308–15.
[25] Tate J, Ward G. Interferences in immunoassay. Clin Biochem Rev 2004;25:105–20.
[26] Rosenberg-Hasson Y, Hansmann L, Liedtke M, Herschmann I, Maecker HT. Effects of serum and plasma matrices on multiplex immunoassays. Immunol Res 2014;58:224–33. doi:10.1007/s12026-014-8491-6.
[27] Kocbach A, Totlandsdal AI, Lag M, Refsnes M, Schwarze PE. Differential binding of cytokines to environmentally relevant particles: a possible source for misinterpretation of in vitro results? Toxicol Lett 2008;176:131–7. doi:10.1016/j.toxlet.2007.10.014.
[28] Frangogiannis NG. Regulation of the inflammatory response in cardiac repair. Circ Res 2012;110:159–73. doi:10.1161/CIRCRESAHA.111.243162.
[29] Vasan RS, Sullivan LM, Roubenoff R, Dinarello CA, Harris T, Benjamin EJ, et al. Inflammatory markers and risk of heart failure in elderly subjects without prior myocardial infarction: the Framingham Heart Study. Circulation 2003;107:1486–91.
[30] Murray DR, Freeman GL. Proinflammatory cytokines: predictors of a failing heart? Circulation 2003;107:1460–2.
[31] Petersen JW, Felker GM. Inflammatory biomarkers in heart failure. Congest Heart Fail 2006;12:324–8.
[32] Parissis JT, Nikolaou M, Farmakis D, Paraskevaidis IA, Bistola V, Venetsanou K, et al. Self-assessment of health status is associated with inflammatory activation and predicts long-term outcomes in chronic heart failure. Eur J Heart Fail 2009;11:163–9. doi:10.1093/eurjhf/hfn032.
[33] Gallo V, Egger M, McCormack V, Farmer PB, Ioannidis JP, Kirsch-Volders M, et al. STrengthening the Reporting of OBservational studies in Epidemiology—Molecular Epidemiology (STROBE-ME): an extension of the STROBE statement. PLoS Med 2011;8:e1001117. doi:10.1371/journal.pmed.1001117.
[34] Sannes A. The world market for immunoassays, http://www.kaloramainformation.com; 2013.
[35] Emdin M, Passino C, Prontera C, Iervasi A, Ripoli A, Masini S, et al. Cardiac natriuretic hormones, neuro-hormones, thyroid hormones and cytokines in normal subjects and patients with heart failure. Clin Chem Lab Med 2004;42(6):627–36.
[36] Chin BS, Conway DS, Chung NA, Blann AD, Gibbs CR, Lip GY. Interleukin-6, tissue factor and von Willebrand factor in acute decompensated heart failure: relationship to treatment and prognosis. Blood Coagul Fibrin 2003;14(6):515–21.
[37] Hirota H, Izumi M, Hamaguchi T, Sugiyama S, Murakami E, Kunisada K, et al. Circulating interleukin-6 family cytokines and their receptors in patients with congestive heart failure. Heart Vessels 2004;19(5):237–41.
[38] Wykretowicz A, Furmaniuk J, Smielecki J, Deskur-Smielecka E, Szczepanik A, Banaszak A, et al. The oxygen stress index and levels of circulating interleukin-10 and interleukin-6 in patients with chronic heart failure. Int J Cardiol 2004;94(2–3):283–7.
[39] Marcucci R, Gori AM, Giannotti F, Baldi M, Verdiani V, Del Pace S, et al. Markers of hypercoagulability and inflammation predict mortality in patients with heart failure. J Thromb Haemost 2006;4(5):1017–22.
[40] Orus J, Roig E, Perez-Villa F, Pare C, Azqueta M, Filella X, et al. Prognostic value of serum cytokines in patients with congestive heart failure. J Heart Lung Transplant 2000;19(5):419–25.

CHAPTER

9

Molecular Imaging to Identify the Vulnerable Plaque: From Basic Research to Clinical Practice*

Dennis H.M. Kusters, Jan Tegtmeier, Leon J. Schurgers, Chris P.M. Reutelingsperger

Department of Biochemistry, Cardiovascular Research Institute Maastricht (CARIM), Maastricht University, Maastricht, The Netherlands

9.1 INTRODUCTION

Despite great advances in diagnosis and therapy over the past decades, cardiovascular disease (CVD) remains the leading cause of death in developed countries [1]. Globalization of Western lifestyle is associated with a sharp increase in cardiovascular-related deaths in various emerging economies. Atherosclerosis is considered to be the major cause of CVD [2]. Atherosclerosis is a systemic inflammatory disease of the arteries with focal manifestations of heterogeneous nature that develop over long periods of time [3,4]. The complete process of atherogenesis is described to great extent by Ross [2], and will therefore not be the subject of this review. Briefly, atherosclerosis is characterized by lesions of the intimal vessel wall that develop progressively. The immune and hemostatic systems are key players in the progression toward an advanced lesion [5]. The major complication of atherosclerosis is the formation of an occlusive thrombus after rupture of a plaque [3,6].

In order to identify high-risk patients prone to develop cardiovascular events, risk profiling such as the Framingham risk score is applied [7]. This score is based on information gathered by epidemiological studies and estimates the risk to develop acute myocardial infarction over a period of 10 years. Although this helps to visualize the risk of individual patients, it only offers a rough estimation of a possible event over a long time frame. Moreover, even for high-risk patients, it is currently impossible to discriminate between stable and unstable plaques, and thus to identify patients in need for immediate intervention. Different characteristics of atherosclerotic plaques can be taken into account to assess the severity of the disease. Contrary to common beliefs, plaque size itself does not determine plaque vulnerability. In fact, it has been shown that most atherosclerotic plaques that undergo abrupt rupture showed only negligible levels of vessel stenosis and these patients are most often asymptomatic [8,9].

Several diagnostic approaches have been employed to identify the patients at high risk for acute arterial events in the presymptomatic phase. One such approach comprises the quantitative measurement of circulating biomarkers. Despite great advances in understanding atherogenic processes at the molecular level, circulating biomarkers are still of limited value [10]. Other approaches comprise imaging techniques to visualize anatomy of atherosclerotic plaques including computed tomography (CT), intravascular ultrasound, optical coherence tomography (OCT), and magnetic resonance imaging (MRI). As stipulated above, these techniques fail to identify the vulnerable plaque because no unambiguous association exists between plaque morphology and stability. Therefore, recently, focus has shifted

* Reprinted from Kusters, Dennis H. M., Tegtmeier, Jan, Schurgers, Leon J., Reutelingsperger, Chris P. M., Molecular Imaging and Biology, Vol. 14, 523-533, Springer, 2012.

http://dx.doi.org/10.1016/B978-0-12-800039-7.00009-8

Published by Elsevier Inc.

toward molecular imaging in an attempt to identify the unstable plaque [11,12]. Several biological characteristics in humans have been associated with an unstable plaque phenotype, e.g., proinflammatory status [13–15], large (oxidized) lipid content [16], thin fibrous cap [17,18], presence of platelets/intraplaque thrombi or hemorrhage [19], low number of (viable) vascular smooth muscle cells (VSMCs), high content of activated matrix metalloproteases (MMPs) [20–25], neovascularization [26], (micro)calcification [27,28], high shear [29], large necrotic core [16], high levels of apoptosis [30], and impaired efferocytosis [31–35]. All these biological characteristics can be exploited for noninvasive molecular imaging of atherosclerosis in order to visualize the vulnerable plaque. This review focuses on existing and novel targets to image atherosclerosis noninvasively, and addresses which targets could possibly be translated from preclinic to clinic in order to filter out patients at risk for acute atherothrombotic events (Figure 9.1).

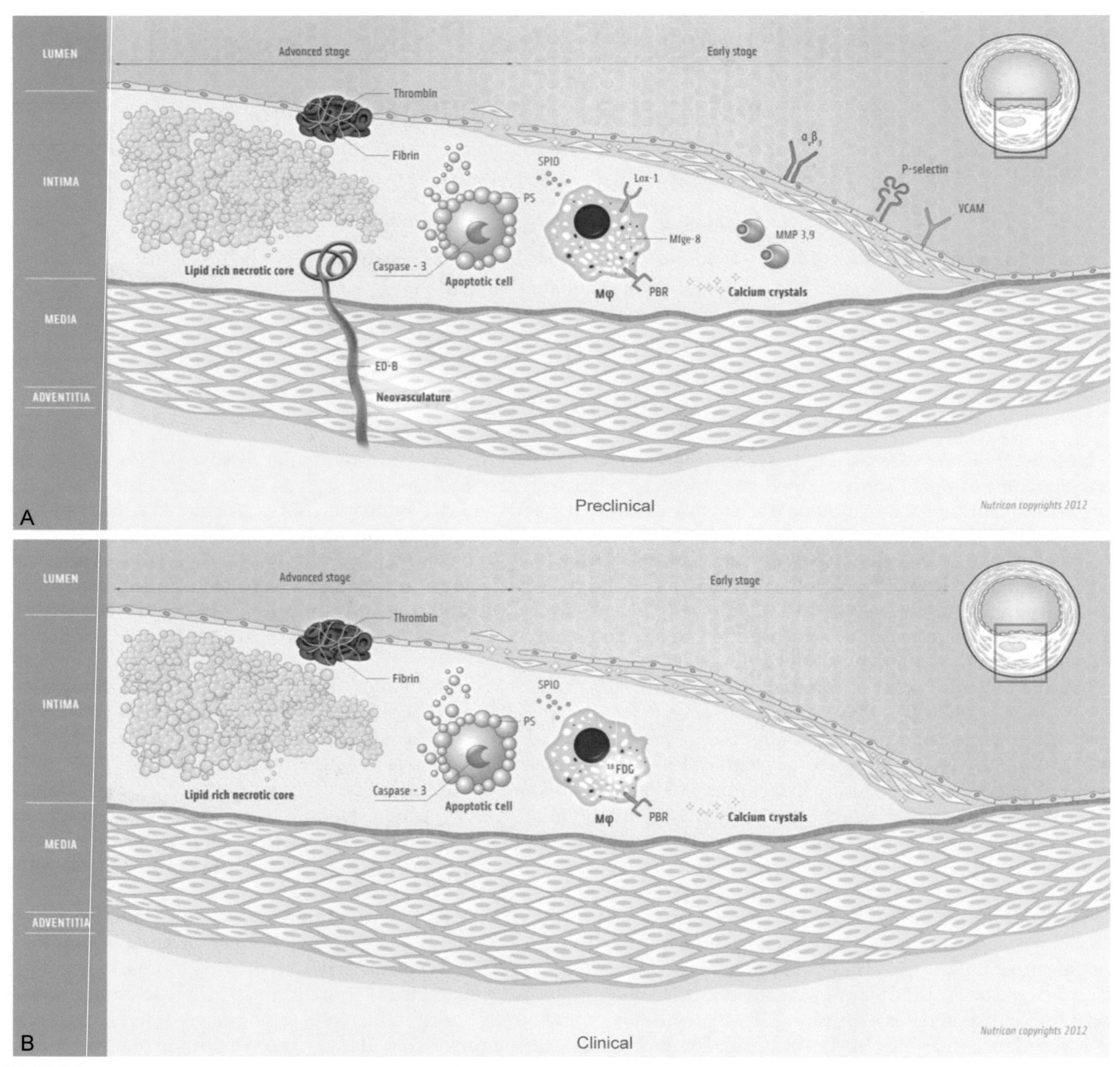

FIGURE 9.1 Molecular imaging targets for the detection of the vulnerable atherosclerotic plaque. (a) The process of atherogenesis is shown with preclinical and (b) clinical molecular imaging targets. Some targets have proven their potential in animal models but have not been validated in clinical trials. VCAM, vascular cell adhesion molecule 1; MMP, matrix metalloprotease; PBR, peripheral benzodiazepine receptor; Mφ, macrophage; SPIO, super paramagnetic iron-oxide particle; PS, phosphatidylserine; ED-B, fibronectin extra-domain B.

9.2 MOLECULAR IMAGING OF INFLAMMATION

9.2.1 Preclinical

The reader is referred to a recent review for details of inflammation and macrophages in atherosclerosis [36]. Cells participating in the inflammatory process use glucose as the main energy source. An analog of glucose, 2-deoxy-2-[^{18}F] fluoro-D-glucose (^{18}FDG), can be used to visualize these cells using positron emission tomography (PET) [37]. Although not the primary aim, it was first noted in cancer studies that even arterial walls showed uptake of ^{18}FDG and that this uptake was higher in older patients and patients with cardiovascular risk factors [38]. It was confirmed by several other studies that ^{18}FDG is taken up by the vessel wall, and that uptake correlates with atherosclerosis [39–41]. Uptake of ^{18}FDG was higher in more advanced and unstable plaques and correlated with the presence of macrophages inside the plaque [42,43].

Macrophages can also be targeted with antibodies against, e.g., LOX-1, peripheral benzodiazepine receptor (PBR), or lactadherin (mfg-E8) [15]. LOX-1 has been suggested as an interesting target since it is upregulated in response to high levels of oxLDL, proinflammatory cytokines, or mechanical stress [44,45]. For imaging purposes, an LOX-1 antibody was conjugated with ^{99m}Technetium (^{99m}Tc), which can be visualized by single photon emission computed tomography (SPECT). The efficacy of this antibody was first demonstrated in a rabbit model of atherosclerosis and evaluated by histological analysis [46]. Li *et al.* confirmed that this antibody reliably targets macrophages *in vivo* in ApoE$^{-/-}$ mice and LDLr$^{-/-}$ (LOX1$^{-/-}$) mice [47]. The PBR (also known as translocator protein) is minimally expressed in noninflamed tissue and highly expressed in activated macrophages, and is therefore exploited as molecular imaging target [48].

The most frequently used contrast agents for molecular imaging of macrophages in atherosclerotic plaques are (lipid-coated ultra) small superparamagnetic iron-oxide particles (LUSPIO/SPIO) or antibody coated micelles [49–52]. SPIOs are spontaneously engulfed by macrophages of animals and humans. Focal accumulation of SPIOs causes a signal decrease that can be detected by MRI. It has been shown that this decrease correlates with the level of intraplaque inflammation [53]. These nanoparticles can also be conjugated to specific ligands targeting macrophages with high sensitivity and selectivity [49]. Monocrystalline iron-oxide nanoparticles or cross-linked iron-oxide nanoparticles have been coupled to a variety of unstable plaque-specific ligands targeting adhesion molecules, lipoproteins, proteases and macrophages, and offer the possibility to add fluorescent dyes for optical verification *in vivo* and *in vitro* [11,54–57].

Another target for imaging of inflammation is P-selectin, which is expressed on activated endothelial cells and platelets. This molecule mediates tethering and rolling of leukocytes prior to migration into the plaque and is associated with increased plaque instability [58]. A recently described nuclear tracer (^{99m}Tc-fucoidan) can be used to successfully image P-selectin overexpression in arterial thrombosis and endothelial activation after ischemia [59]. Using different imaging probes, the feasibility of P-selectin imaging in atherosclerotic settings has already been shown [60,61]. Another adhesion molecule, vascular cell adhesion molecule 1 (VCAM-1), is also exploited as potential imaging target. In a recent publication, it is shown that a VCAM-1 targeted USPIO derivative (R832) shows specific plaque uptake in ApoE$^{-/-}$ mice, visualized by MRI within 30 min [62]. Specific uptake was confirmed by *ex vivo* histology. However, after 30 min, nonspecific uptake was observed, which was probably caused by the enhanced permeability and retention effect in the plaque. The authors show a possibility to identify features of unstable plaque phenotypes in these mice, which could potentially be translated to the clinic [62]. Besides adhesion molecules, activated endothelial cells also highly express the integrin $\alpha_v\beta_3$. Targeting this integrin has been done by a low molecular weight peptidomimetic of arginine-glycine-aspartic acid peptide conjugated to gadolinium-diethylenetriamine pentaacetate [63]. Specific integrin binding and imaging of aortic plaques was confirmed by competition experiments and subsequent histological analysis. Increased uptake of the contrast agent is directly correlated with an enlarged neovascular network and thus plaque instability.

9.2.2 Clinical

Imaging of vascular inflammation by ^{18}FDG uptake with PET has been performed in several clinical trials, which have recently been reviewed by Rudd *et al.* [64]. Menezes *et al.* performed a prospective study to determine the optimal time to image vascular inflammation using ^{18}FDG PET/CT among 17 asymptomatic patients undergoing routine surveillance for atherosclerotic abdominal aortic aneurysms [65]. There was no significant difference in signal to noise ratio (SNR), despite a significant difference between aortic wall and lumen uptake in time [65]. More recently, ^{18}FDG PET and CT angiography were performed in 21 patients undergoing endarterectomy in order to investigate plaque vulnerability. A positive correlation between plaque ^{18}FDG uptake and CD68/VEGF positivity was found. Furthermore, an inverse relationship between ^{18}FDG plaque uptake and plaque percentage calcium composition on CT was observed [66]. Meyers *et al.* conducted a multicenter clinical trial to evaluate arterial ^{18}FDG uptake in peripheral artery disease. The researchers included 21 patients and measured ^{18}FDG uptake in carotid, aortic, and femoral artery before atherectomy, and compared this uptake to immunohistochemical CD68 positivity. However, no significant correlation between lesion SNR and CD68 level was found [67].

A recent clinical trial (dal-PLAQUE) that evaluates the efficacy and safety of dalcetrapib, assigned 64 patients to dalcetrapib or placebo (66 patients) for 24 months. Total vessel area was determined by MRI and found to be reduced in patients who were given dalcetrapib. However, PET/CT showed no evidence of increased or decreased vascular inflammation [68,69].

Most of these clinical trials provide *in vivo* evidence that increased plaque metabolism is associated with increased biomarkers of inflammation. However, the greater part fails to show a correlation between ^{18}FDG uptake and plaque vulnerability, particularly in the coronary arteries. Imaging of coronary arteries is complicated because of the movement of the heart and the small size. Moreover, the myocardium has a high background uptake of ^{18}FDG which compromises the SNR. Dietary measures can suppress the background uptake of ^{18}FDG by the myocardium to some extent [70].

PET imaging of the PBR has also been performed. Although animal studies have yielded several promising tracers for PBR imaging (e.g., ^{11}C-PK11195, ^{11}C-DAA1106, ^{11}C-PBR28, ^{11}C-DPA713, ^{11}C-CLINME, and ^{18}F-FEDAA1106), hitherto only ^{11}C-PK11195 (1-(2-chloro-phenyl)-*N*-methyl-*N*-(1-methylpropyl)-3-isoquinoline-carboxamide) has been investigated in clinical studies [71]. Recently, ^{11}CPK11195 has been used in two small-scale clinical trials. Pugliese *et al.* imaged six patients with systemic inflammatory disorders and clinical suspicion of active vasculitis compared to nine controls. Focal ^{11}C-PK11195 uptake in the arterial wall of all symptomatic patients was noticed compared to no detectable uptake in the control group [72]. Gaemperli et *al.* imaged 32 patients with carotid stenosis (nine symptomatic/23 asymptomatic) and compared ^{11}C-PK11195 uptake into carotid plaques [73]. Both research groups concluded that ^{11}C-PK11195 PET allows noninvasive detection and quantification of intraplaque inflammation in patients. Moreover, the combination of ^{11}C-PK11195 PET with contrast-enhanced CT provides an integrated assessment of plaque structure, composition, and biological activity, and allows the distinction between recently symptomatic vulnerable plaques and asymptomatic plaques with a high positive predictive value. Whether ^{11}C-PK11195 is a molecular imaging agent to identify vulnerable plaque, also in coronary arteries, needs to be assessed by more clinical trials.

9.3 MOLECULAR IMAGING OF CELL DEATH

9.3.1 Preclinical

Apoptosis of macrophages and smooth muscle cells is associated with plaque instability, since this leads to a growing necrotic core, inflammation, the release of proteases, and subsequently thinning of the fibrous cap [17,31]. Apoptosis can be imaged using either intracellular- or membrane-bound targets.

During apoptosis, caspases are activated that degrade intracellular proteins and DNA in an orderly manner. Caspases are expressed as inactive zymogens in healthy cells, but cleave to their active form in response to apoptotic signals. This property makes targeting of activated caspases attractive. An inhibitor of caspase-3 and -7, radiolabeled isatin-5-sulfonamide (^{18}F-ICTM-11) and several of its analogs have been evaluated as imaging agent of apoptosis *in vitro* [74], *in vivo* [75], in a rat model of liver cell apoptosis [76], and in tumor-bearing mice [77]. All studies provide great evidence of the potential to use activated caspase-specific imaging agents to find culprit lesions. However, thus far, this has not been tested in animal models of atherosclerosis.

Another hallmark of apoptosis is externalization of phosphatidylserine (PS), which is normally predominantly confined to the inner leaflet of the plasma membrane [78]. The protein annexin A5 (anxA5) has been shown to bind PS with nanomolar affinity. This discovery has led to the use of PS as target for molecular imaging of apoptosis *in vitro* and *in vivo* in myocardial infarction [79], cancer [80,81], and atherosclerosis [42]. For this purpose, anxA5 was conjugated to radiotracers for nuclear imaging and fluorescent probes for histological analysis [82–84]. Pioneering studies of molecular imaging of apoptosis were performed with technetium-labeled anxA5 (^{99m}Tc-anxA5) and SPECT. One of the first approaches was to target apoptotic macrophages in animal models of atherosclerosis [42,50,85,86]. To verify the results of molecular imaging, histological analysis of the atherosclerotic plaques was performed. Atherosclerotic lesions showed increased uptake of ^{99m}Tc-AnxA5 compared to control animals. Furthermore, a gradient of ^{99m}Tc-anxA5 uptake was observed, which correlated to the grade of the atherosclerotic lesion. This level of ^{99m}Tc-anxA5 uptake could directly be linked to the level of macrophage infiltration [86]. These *in vivo* results confirmed that ^{99m}Tc-anxA5 can target apoptosis in living animals and that uptake correlates with the severity of atherosclerosis [42,86]. An alternative to anxA5 to target PS is lactadherin (mfg-E8). Falborg *et al.* compared the biodistribution of lactadherin and anxA5, both conjugated to ^{99m}Tc-labeled hydrazinonicotinamide (^{99m}Tc-HYNIC). Regarding biodistribution and blood clearance, lactadherin is comparable to anxA5. However, anxA5 showed higher uptake in the kidneys, whereas lactadherin showed higher uptake in the liver [87]. Another alternative for detection of cell death is the C2A domain of synaptotagmin-I, which also binds to PS [88]. The C2A domain has been labeled with paramagnetic and radionuclide labels and used to detect apoptosis *in vivo* successfully. However, a proper quantification of conjugated label could not be performed, and affinity for PS decreased after labeling [89]. A recent study compared a mutated form of C2A and anxA5 in their ability to quantitatively image cell death [90]. The mutant described was fluorescently labeled and showed lower binding to viable cells than a similarly labeled anxA5, thus increasing its specificity. This could partly be explained by a slightly lower affinity for PS, which might be important when radionuclide derivatives of this protein will be used to detect cell death *in vivo* [90]. Confirmation of its ability to detect apoptosis in atherosclerotic plaques still has to be performed.

9.3.2 Clinical

The first small-scale clinical trial to image apoptosis with labeled anxA5 was done in four patients with either recent or remote history of a transient ischemic attack (TIA) [91]. These patients were scheduled for carotid endarterectomy (CEA), and imaging was done before removal of the carotid lesions. The patients with a recent TIA showed lucid uptake of ^{99m}Tc-AnxA5, whereas those with a remote TIA did not show any significant uptake. After CEA, plaque analysis from patients with recent TIA showed that unstable plaques stained anxA5 positive. In contrast, lesions of patients with remote TIA, which had a stable phenotype, showed almost no anxA5 uptake. The researchers conclude that ^{99m}Tc-anxA5 imaging could help to identify unstable plaques in patients with carotid atherosclerosis [91]. Unfortunately, this clinical study involved only a very small number of patients; thus more and larger clinical cohort studies are needed to confirm these promising results. Although very promising, all other imaging tracers mentioned above for the imaging of cell death have not been tested in a clinical trial thus far.

9.4 MOLECULAR IMAGING OF REMODELING

9.4.1 Preclinical

As response to partial occlusion of a vessel, constrictive or expansive vascular remodeling can occur [92]. Vascular remodeling is associated with increasing plaque vulnerability and consists of several distinct processes. Two major processes are angiogenesis and hypertrophic compensation of the vessel [93].

Key players and potential imaging target of vascular remodeling are VSMCs. In general, VSMCs can exhibit two phenotypes, contractile or synthetic [94]. Synthetic VSMCs are predominantly present in the intima, are more prone to calcify and become apoptotic, and therefore seen as detrimental [95]. Several groups targeted vascular remodeling in atherosclerosis with a labeled antibody (Z2D3) specific against synthetic VSMCs. They verified binding *ex vivo* in human atherosclerotic endarterectomy specimens [96], *in vivo* by SPECT in a rabbit model [97,98] and a rat model [99], and by scintillation gamma counting in a swine model [100,101]. All studies show the possibility to selectively image synthetic VSMCs to great extent in animals, but this still has to be established in clinical trials.

Hypertrophic compensation is associated with the activation of several proteases (e.g., MMPs or cathepsin G). Besides their role in vascular remodeling and inflammation, these proteases have significant detrimental effects on cap thickness and therefore plaque stability [20–22,24]. Activated MMPs form an interesting target to identify unstable lesions, and have extensively been investigated using an antibody against activated MMPs (P947). This antibody was recently tested and validated in several *in vivo* studies [23,25,54,102]. Lancelot *et al.* showed excellent delinearization of the plaque with strong signal intensity enhancement using noninvasive MRI. However, a limitation of imaging with P947 is that it does not discriminate between different activated MMPs and Johnson *et al.* found that MMP-3 and 9 could even exert a beneficial effect on plaque phenotype [22]. Another strategy for imaging of activated MMPs is using MMP2/9-specific substrate. Deguchi *et al.* showed that labeling of a small substrate peptide with a near-infrared probe enables specific noninvasive optical molecular imaging of activated MMP2/9 in atherosclerotic plaques [103]. As an alternative for MMP imaging, Narula *et al.* used a ^{99m}Tc-labeled broad-spectrum MMP inhibitor and demonstrated feasibility of noninvasive detection of activated MMPs in atherosclerotic lesions [104].

TABLE 9.1 Molecular Imaging Targets for the Vulnerable Plaque

Process	Target	Agent	Imaging modality	Phase (Reference)
Inflammation	Macrophages	^{18}FDG	PET or PET/CT	Clinical [39–43,65–69]
		^{11}C-PK11195	PET	Clinical [71–73]
		^{99m}Tc-labeled anti-LOX-1	SPECT	Preclinical [46,47]
		Nanoparticles	MRI	Preclinical [51,53]
	P-selectin (EC and platelets)	^{99m}Tc-fucoidan	SPECT/MRI	Preclinical [59–61]
Apoptosis	VSMC, macrophages	^{99m}Tc-anxA5	SPECT	Clinical [42,83–85,91]
Vascular remodeling	VSMC	Z2D3	SPECT	Preclinical [96–101]
	MMP	P947	MRI	Preclinical [23,25,54,102]
	Integrin $\alpha_v\beta_3$	Nanoparticles	MRI	Preclinical [26]
	ED-B	L19	SPECT	Preclinical [107]

Another process in vascular remodeling is angiogenesis. Plaque and vasa vasorum neovascularization are associated with decreased plaque stability and offer several potential imaging targets that have been investigated—integrin $\alpha_A\beta_3$ [26,105,106], fibronectin extra-domain B [107], and vascular endothelial growth factor receptor (VEGFr). Winter *et al.* showed that *in vivo* imaging of neovascularization of atherosclerotic plaques using MRI targeting $\alpha_v\beta_3$ is feasible and might provide a method for defining the burden and evolution of atherosclerosis [26]. Matter et *al.* showed that targeting of fibronectin with a labeled human recombinant antibody (^{125}I-SIP[L19]) could selectively visualize atherosclerotic plaques in ApoE$^{-/-}$ mice with a high SNR and very high specificity [107]. Thus far, no molecular imaging studies targeting VEGFr in atherosclerosis have been performed, nevertheless molecular imaging of VEGFr in cancer has been investigated and validated extensively [108,109].

Taken together, these preclinical studies indicate high potential for noninvasive molecular imaging of vascular remodeling, even though none of the aforementioned probes has made it to clinical studies yet (Table 9.1).

9.5 MOLECULAR IMAGING OF THROMBOSIS

9.5.1 Preclinical

The hemostatic system is an important modulator of atherosclerosis and plays a key role in plaque rupture and subsequent thrombus formation [1]. Therefore, partakers of the coagulation cascade might be an interesting target for molecular imaging to identify the vulnerable plaque [110]. Inflamed endothelium and cells participating in the development of the necrotic core show high levels of fibrin and thrombin expression. Abundance of fibrin on the surface or within atherosclerotic plaques has thus been associated with vulnerable plaques [111,112]. The fibrin-targeted contrast agent (FTCA) uses gadolinium contrast and can be visualized by MRI. Besides its various applications in animal studies, FTCA has also been used in a clinical trial to detect thrombi in the heart and blood vessels [113–116]. A recent study showed that FTCA can be successfully applied to image endothelial and intraplaque fibrin in a mouse model of atherosclerosis [117]. Despite the ubiquitous distribution of fibrin, FTCA in atherosclerotic plaques could be visualized with a good SNR. FTCA uptake and signal enhancement was strongest in culprit lesions, as confirmed by immuno-histochemistry. This is in concordance with recent findings from autopsy studies that indicate that advanced lesions express high levels of fibrin [111]. Additionally, it was found that fibrin is a key component of plaque erosion [112,118] and can be applied to image eroding plaques, which can cause thrombosis without plaque rupture [117]. Although less common in general, plaque erosion is estimated to cause up to 25% of coronary thrombosis leading to MI and is especially important in women [119]. Even though molecular imaging of thrombosis has not yet been applied in clinical studies of atherosclerosis, it still represents a promising target. Fibrin is associated with two important characteristics of vulnerable plaques, and FTCA has been successfully used to detect thrombi in the heart and blood vessels of patients. Therefore, the future clinical use of thrombosis imaging to identify vulnerable plaques is highly probable.

9.6 MOLECULAR IMAGING OF (MICRO) CALCIFICATION

9.6.1 Preclinical

Calcification has long been associated with the process of atherosclerosis and acute cardiovascular events [118]. Although large calcifications inside the plaque have been shown to promote plaque stability [120], spotty calcifications—especially inside the fibrous cap—are suggested to cause plaque rupture and thrombotic events [28,121]. It is postulated that these microcalcifications, which are undetectable by conventional imaging, increase the local stress inside the thin fibrous cap by twofold and thus increase the likelihood of rupture [28]. It is assumed that calcification is a product of osteogenic action by osteocytic- and chondrocytic-like cells inside the plaque. The work of Aikawa and colleagues provided substantial insights in the course of microcalcification and possibilities to visualize this process [122]. In a mouse model of atherosclerosis, they showed that macrophage infiltration in early atherosclerotic plaques precedes calcification. Later, these events occur in temporal and spatial overlap. Conventional histological and CT imaging failed to visualize early microcalcifications. Using intravital fluorescence microscopy with a nanoparticle-based contrast agent to visualize macrophages and a bisphosphonate-based agent to detect calcifications, the authors were able to properly detect inflammation and microcalcifications in early atherosclerotic plaques. Due to the time at which these microcalcifications take place and their implications in destabilizing the plaque, molecular imaging would be able to detect rupture-prone plaques at a very early stage. Asymptomatic high-risk patients would benefit from this application. Additionally, molecular imaging of microcalcification provides an opportunity to monitor and evaluate the early success of therapeutic interventions.

9.6.2 Clinical

The association of vascular ^{18}FDG uptake and vascular calcification detected by contemporaneous CT was investigated. The researchers evaluated PET/CT images of 78 patients who were referred for tumor staging. These images were analyzed for the presence of vascular ^{18}FDG uptake and vascular calcification. They found that vascular calcification and vascular metabolic activity rarely (<2%) overlapped. This suggests that these findings represent different phases of atherosclerosis [123]. These findings were confirmed by Rudd *et al.* who investigated the relationship between inflammation and arterial calcification. The researchers imaged inflammation in 41 patients with ^{18}FDG PET/CT and scored calcium by CT analysis. Additionally, from 33 of these patients, a collection of biomarkers was determined. The authors suggested that ^{18}FDG PET imaging can be used as a surrogate marker of both atherosclerotic disease activity and drug effectiveness [40].

Calcification status can also be visualized by ^{18}F-sodium fluoride with PET. In a retrospective study, arterial ^{18}F-sodium fluoride uptake and calcification was evaluated. When spatial correlation between vascular ^{18}F-sodium fluoride uptake and calcification sites was analyzed per lesion, 12% of lesions with marked arterial wall ^{18}F-sodium fluoride uptake did not show concordant calcification. This study shows the feasibility of using ^{18}F-sodium fluoride for *in vivo* functional imaging of atherosclerotic lesions [124]. Hitherto, locating microcalcification in the vessel wall with ^{18}F-sodium fluoride has not been demonstrated; in order to find these minute alterations, OCT might be more feasible.

9.7 SOCIOECONOMIC IMPACT OF MOLECULAR IMAGING

As major cause of cardio- and cerebrovascular morbidity and mortality, atherosclerosis puts a significant strain on healthcare budgets. The following numbers serve as an example of the costs that originate from atherosclerosis and underline the need for new diagnostic tools and treatments. The total expenses for stroke in the Netherlands summed up to more than 1.6 billion Euros in the year 2007, making stroke one of the 10 most expensive conditions [125]. The costs for stroke made up 2.2% of the total healthcare costs in the Netherlands. A comparable situation is found in Germany (total costs in 2006: 1.2 billion Euros, 3.4% of total costs) [126].

Acute coronary syndromes (mostly unstable angina and acute MI) are also within the top 10 of the most expensive diseases in the Netherlands. The total expenses in 2007 summed up to 1.807 billion Euros [125]. In other European countries like the United Kingdom, France, Germany, Italy, and Spain, costs range from 1 to 3 billion Euros yearly [127]. These values represent between 1 and 3% of total healthcare costs in those countries. In the United States, total annual costs for CVD and stroke are estimated at 297.7 billion dollar, accounting for 16% of total healthcare expenditures in 2008 [128].

These numbers demonstrate the enormous expenses for the healthcare system. Therefore, any improvement in the diagnosis of atherosclerosis, and thus prevention of its adverse outcomes, would lower financial burden significantly.

Support for this approach comes from the SHAPE Guideline by Naghavi *et al.* [129]. In an extensive work, the SHAPE task force proposed protocols to identify the vulnerable patients with a high risk of MI or stroke. The report underlines the shortcomings of current risk stratification as the Framingham Risk Score and the System Coronary Risk Evaluation. These tools assign most people to a group with intermediate risk for adverse outcomes. The power to predict an acute event is especially low in this group; yet the incidence of MI is the highest. To improve risk profiling, the authors propose two noninvasive tests, namely, carotid intima-media thickness (CIMT) and coronary artery calcification (CACS) to be applied to the population at risk. Based on initial screening with CIMT or CACS, patients are categorized and treated if necessary. Large studies demonstrated predictive power and cost-effectiveness of this approach. All current nonmolecular imaging modalities are able to find vascular stenosis but fail to discriminate reliably between vulnerable and stable lesions in patients [130]. Newer imaging techniques can increase the accuracy of this risk stratification. In this regard, molecular imaging of the vulnerable plaque offers great potential. Once established, molecular imaging can be added to the SHAPE guideline to strengthen the power of diagnostics in preventive cardiology. Another beneficial aspect is that molecular imaging uses probes specific for targets on (unstable) atherosclerotic plaques, which could potentially be used as theranostic.

9.8 CONCLUSION AND FUTURE PERSPECTIVES

Molecular imaging of atherosclerosis bears potential as a powerful diagnostic tool to screen and select patients prone to develop CVD morbidity and mortality and, hence, to contribute to a personalized and more cost-effective treatment of atherosclerotic disease. This review has highlighted molecular imaging targets that have been studied *in vitro* and *in vivo* in experimental animal models of atherosclerosis and that have potential to be translated to clinical practice. Most promising targets are found in the process of apoptosis, vascular remodeling, and inflammation (Figure 9.1 and Table 9.1). Additionally, we suggest that efferocytosis of apoptotic cells might be added as an interesting molecular imaging target, since impairment of efferocytosis is associated with an increase of necrotic core and consequently with an increased vulnerability [31,32,34,131,132]. Ligands specific for noninvasive imaging of efferocytosis still need to be explored; potential targets might be C1q, MfgE-8, LRP-1, and Gas6.

Translation of molecular imaging ligands from preclinical research to clinical practice faces important scientific as well as financial challenges to be overcome. Although animal models can reflect human vulnerable plaque phenotype, results of preclinical research cannot simply be extrapolated to the human situation, since spontaneous rupture of a plaque without manipulation so far has not been observed in animal models. In addition, molecular imaging remains relatively expensive and time-consuming, both of which limit large clinical studies to validate targets and ligands.

Further scientific research into pathobiology of the vulnerable plaque, and technological development of imaging equipment that increases sensitivity and specificity are required if we wish to uphold the promise of molecular imaging to reduce health care costs. Successful translation of targets and ligands to clinical molecular imaging of the vulnerable plaque may also offer novel therapeutic avenues to treat the vulnerable plaque employing strategies such as targeted drug delivery.

Conflict of Interest: The authors declare that they have no conflict of interest.

References

[1] Borissoff JI, Spronk HM, ten Cate H. The hemostatic system as a modulator of atherosclerosis. N Engl J Med 2011;364(18):1746–60.
[2] Ross R. Atherosclerosis—an inflammatory disease. N Engl J Med 1999;340(2):115–26.
[3] Libby P. Molecular bases of the acute coronary syndromes. Circulation 1995;91(11):2844–50.
[4] Virmani R, Burke AP, Kolodgie FD, et al. Vulnerable plaque: the pathology of unstable coronary lesions. J Interv Cardiol 2002;15(6):439–46.
[5] Libby P, Ridker PM, Hansson GK. Progress and challenges in translating the biology of atherosclerosis. Nature 2011;473(7347):317–25.
[6] Libby P, Sukhova G, Lee RT, et al. Cytokines regulate vascular functions related to stability of the atherosclerotic plaque. J Cardiovasc Pharmacol 1995;25(Suppl 2):S9–S12.
[7] Tzoulaki I, Liberopoulos G, Ioannidis JP. Assessment of claims of improved prediction beyond the Framingham risk score. JAMA 2009;302(21):2345–52.
[8] Davies MJ. A macro and micro view of coronary vascular insult in ischemic heart disease. Circulation 1990;82(3 Suppl):II38–46.
[9] Hansson GK. Inflammation, atherosclerosis, and coronary artery disease. N Engl J Med 2005;352(16):1685–95.
[10] Wang TJ, Gona P, Larson MG, et al. Multiple biomarkers for the prediction of first major cardiovascular events and death. N Engl J Med 2006;355(25):2631–9.

[11] Matter CM, Stuber M, Nahrendorf M. Imaging of the unstable plaque: how far have we got? Eur Heart J 2009;30(21):2566–74.
[12] Yuan C, Hatsukami TS, Cai J. MRI plaque tissue characterization and assessment of plaque stability. Stud Health Technol Inform 2005;113:55–74.
[13] Bouki KP, Katsafados MG, Chatzopoulos DN, et al. Inflammatory markers and plaque morphology: an optical coherence tomography study. Int J Cardiol 2012;154(3):287–92.
[14] Kerwin WS, O'Brien KD, Ferguson MS, et al. Inflammation in carotid atherosclerotic plaque: a dynamic contrast-enhanced MR imaging study. Radiology 2006;241(2):459–68.
[15] Tabas I. Macrophage death and defective inflammation resolution in atherosclerosis. Nat Rev Immunol 2010;10(1):36–46.
[16] Gao T, Zhang Z, Yu W, et al. Atherosclerotic carotid vulnerable plaque and subsequent stroke: a high-resolution MRI study. Cerebrovasc Dis 2009;27(4):345–52.
[17] Kolodgie FD, Narula J, Haider N, et al. Apoptosis in atherosclerosis. Does it contribute to plaque instability? Cardiol Clin 2001;19(1):127–39, ix.
[18] Tabas I. Pulling down the plug on atherosclerosis: finding the culprit in your heart. Nat Med 2011;17(7):791–3.
[19] Derksen WJ, Peeters W, van Lammeren GW, et al. Different stages of intraplaque hemorrhage are associated with different plaque phenotypes: a large histopathological study in 794 carotid and 276 femoral endarterectomy specimens. Atherosclerosis 2011;218(2):369–77.
[20] Bazeli R, Coutard M, Duport BD, et al. *In vivo* evaluation of a new magnetic resonance imaging contrast agent (P947) to target matrix metalloproteinases in expanding experimental abdominal aortic aneurysms. Invest Radiol 2010;45(10):662–8.
[21] Galis ZS, Sukhova GK, Lark MW, et al. Increased expression of matrix metalloproteinases and matrix degrading activity in vulnerable regions of human atherosclerotic plaques. J Clin Invest 1994;94(6):2493–503.
[22] Johnson JL, George SJ, Newby AC, et al. Divergent effects of matrix metalloproteinases 3, 7, 9, and 12 on atherosclerotic plaque stability in mouse brachiocephalic arteries. Proc Natl Acad Sci U S A 2005;102(43):15575–80.
[23] Lancelot E, Amirbekian V, Brigger I, et al. Evaluation of matrix metalloproteinases in atherosclerosis using a novel noninvasive imaging approach. Arterioscler Thromb Vasc Biol 2008;28(3):425–32.
[24] Loftus IM, Naylor AR, Bell PR, et al. Matrix metalloproteinases and atherosclerotic plaque instability. Br J Surg 2002;89(6):680–94.
[25] Ouimet T, Lancelot E, Hyafil F, et al. Molecular and cellular targets of the MRI contrast agent p947 for atherosclerosis imaging. Mol Pharm 2012;9(4):850–61.
[26] Winter PM, Morawski AM, Caruthers SD, et al. Molecular imaging of angiogenesis in early-stage atherosclerosis with alpha(v)beta3-integrin-targeted nanoparticles. Circulation 2003;108(18):2270–4.
[27] Vengrenyuk Y, Cardoso L, Weinbaum S. Micro-CT based analysis of a new paradigm for vulnerable plaque rupture: cellular microcalcifications in fibrous caps. Mol Cell Biomech 2008;5(1):37–47.
[28] Vengrenyuk Y, Carlier S, Xanthos S, et al. A hypothesis for vulnerable plaque rupture due to stress-induced debonding around cellular microcalcifications in thin fibrous caps. Proc Natl Acad Sci U S A 2006;103(40):14678–83.
[29] Slager CJ, Wentzel JJ, Gijsen FJ, et al. The role of shear stress in the generation of rupture-prone vulnerable plaques. Nat Clin Pract Cardiovasc Med 2005;2(8):401–7.
[30] Bennett MR. Life and death in the atherosclerotic plaque. Curr Opin Lipidol 2010;21(5):422–6.
[31] Van Vre EA, Ait-Oufella H, Tedgui A, et al. Apoptotic cell death and efferocytosis in atherosclerosis. Arterioscler Thromb Vasc Biol 2012;32(4):887–93.
[32] Thorp E, Tabas I. Mechanisms and consequences of efferocytosis in advanced atherosclerosis. J Leukoc Biol 2009;86(5):1089–95.
[33] Tabas I. Macrophage apoptosis in atherosclerosis: consequences on plaque progression and the role of endoplasmic reticulum stress. Antioxid Redox Signal 2009;11(9):2333–9.
[34] Schrijvers DM, De Meyer GR, Kockx MM, et al. Phagocytosis of apoptotic cells by macrophages is impaired in atherosclerosis. Arterioscler Thromb Vasc Biol 2005;25(6):1256–61.
[35] Schoenenberger AW, Urbanek N, Bergner M, et al. Associations of reactive hyperemia index and intravascular ultrasound-assessed coronary plaque morphology in patients with coronary artery disease. Am J Cardiol 2012;109(12):1711–6.
[36] Moore KJ, Tabas I. Macrophages in the pathogenesis of atherosclerosis. Cell 2011;145(3):341–55.
[37] Visioni A, Kim J. Positron emission tomography for benign and malignant disease. Surg Clin North Am 2011;91(1):249–66.
[38] Yun M, Yeh D, Araujo LI, et al. F-18 FDG uptake in the large arteries: a new observation. Clin Nucl Med 2001;26(4):314–9.
[39] Rudd JH, Myers KS, Bansilal S, et al. (18)Fluorodeoxyglucose positron emission tomography imaging of atherosclerotic plaque inflammation is highly reproducible: implications for atherosclerosis therapy trials. J Am Coll Cardiol 2007;50(9):892–6.
[40] Rudd JH, Myers KS, Bansilal S, et al. Relationships among regional arterial inflammation, calcification, risk factors, and biomarkers: a prospective fluorodeoxyglucose positron-emission tomography/computed tomography imaging study. Circ Cardiovasc Imaging 2009;2(2):107–15.
[41] Tawakol A, Migrino RQ, Bashian GG, et al. *In vivo* 18 F-fluorodeoxyglucose positron emission tomography imaging provides a noninvasive measure of carotid plaque inflammation in patients. J Am Coll Cardiol 2006;48(9):1818–24.
[42] Laufer EM, Winkens HM, Corsten MF, et al. PET and SPECT imaging of apoptosis in vulnerable atherosclerotic plaques with radiolabeled Annexin A5. Q J Nucl Med Mol Imaging 2009;53(1):26–34.
[43] Paulmier B, Duet M, Khayat R, et al. Arterial wall uptake of fluorodeoxyglucose on PET imaging in stable cancer disease patients indicates higher risk for cardiovascular events. J Nucl Cardiol 2008;15(2):209–17.
[44] Goyal T, Mitra S, Khaidakov M, et al. Current concepts of the role of oxidized LDL receptors in atherosclerosis. Curr Atheroscler Rep 2012;14(2):150–9.
[45] Reiss AB, Anwar K, Wirkowski P. Lectin-like oxidized low density lipoprotein receptor 1 (LOX-1) in atherogenesis: a brief review. Curr Med Chem 2009;16(21):2641–52.
[46] Ishino S, Mukai T, Kuge Y, et al. Targeting of lectin-like oxidized low-density lipoprotein receptor 1 (LOX-1) with 99mTc-labeled anti-LOX-1 antibody: potential agent for imaging of vulnerable plaque. J Nucl Med 2008;49(10):1677–85.
[47] Li D, Patel AR, Klibanov AL, et al. Molecular imaging of atherosclerotic plaques targeted to oxidized LDL receptor LOX-1 by SPECT/CT and magnetic resonance. Circ Cardiovasc Imaging 2010;3(4):464–72.

[48] Canat X, Guillaumont A, Bouaboula M, et al. Peripheral benzodiazepine receptor modulation with phagocyte differentiation. Biochem Pharmacol 1993;46(3):551–4.
[49] Chen W, Cormode DP, Fayad ZA, et al. Nanoparticles as magnetic resonance imaging contrast agents for vascular and cardiac diseases. Wiley Interdiscip Rev Nanomed Nanobiotechnol 2010;3(2):146–61.
[50] Sadeghi MM, Glover DK, Lanza GM, et al. Imaging atherosclerosis and vulnerable plaque. J Nucl Med 2010;51(Suppl 1):51S–65S.
[51] Schmitz SA, Coupland SE, Gust R, et al. Superparamagnetic iron oxide-enhanced MRI of atherosclerotic plaques in Watanabe hereditable hyperlipidemic rabbits. Invest Radiol 2000;35(8):460–71.
[52] Sanz J, Fayad ZA. Imaging of atherosclerotic cardiovascular disease. Nature 2008;451(7181):953–7.
[53] Trivedi RA, Mallawarachi C, U-King-Im JM, et al. Identifying inflamed carotid plaques using *in vivo* USPIO-enhanced MR imaging to label plaque macrophages. Arterioscler Thromb Vasc Biol 2006;26(7):1601–6.
[54] Amirbekian V, Aguinaldo JG, Amirbekian S, et al. Atherosclerosis and matrix metalloproteinases: experimental molecular MR imaging in vivo. Radiology 2009;251(2):429–38.
[55] Frias JC, Williams KJ, Fisher EA, et al. Recombinant HDL-like nanoparticles: a specific contrast agent for MRI of atherosclerotic plaques. J Am Chem Soc 2004;126(50):16316–7.
[56] Kooi ME, Cappendijk VC, Cleutjens KB, et al. Accumulation of ultrasmall superparamagnetic particles of iron oxide in human atherosclerotic plaques can be detected by *in vivo* magnetic resonance imaging. Circulation 2003;107(19):2453–8.
[57] Lipinski MJ, Frias JC, Amirbekian V, et al. Macrophage-specific lipid-based nanoparticles improve cardiac magnetic resonance detection and characterization of human atherosclerosis. JACC Cardiovasc Imaging 2009;2(5):637–47.
[58] Blann AD, Nadar SK, Lip GY. The adhesion molecule P-selectin and cardiovascular disease. Eur Heart J 2003;24(24):2166–79.
[59] Rouzet F, Bachelet-Violette L, Alsac JM, et al. Radiolabeled fucoidan as a *p*-selectin targeting agent for in vivo imaging of platelet-rich thrombus and endothelial activation. J Nucl Med 2011;52(9):1433–40.
[60] Jacobin-Valat MJ, Deramchia K, Mornet S, et al. MRI of inducible *p*-selectin expression in human activated platelets involved in the early stages of atherosclerosis. NMR Biomed 2010;24(4):413–24.
[61] McAteer MA, Schneider JE, Ali ZA, et al. Magnetic resonance imaging of endothelial adhesion molecules in mouse atherosclerosis using dual-targeted microparticles of iron oxide. Arterioscler Thromb Vasc Biol 2008;28(1):77–83.
[62] Burtea C, Ballet S, Laurent S, et al. Development of a magnetic resonance imaging protocol for the characterization of atherosclerotic plaque by using vascular cell adhesion molecule-1 and apoptosis-targeted ultrasmall superparamagnetic iron oxide derivatives. Arterioscler Thromb Vasc Biol 2012;32(6):e36–48.
[63] Burtea C, Laurent S, Murariu O, et al. Molecular imaging of alpha v beta3 integrin expression in atherosclerotic plaques with a mimetic of RGD peptide grafted to Gd-DTPA. Cardiovasc Res 2008;78(1):148–57.
[64] Rudd JH, Narula J, Strauss HW, et al. Imaging atherosclerotic plaque inflammation by fluorodeoxyglucose with positron emission tomography: ready for prime time? J Am Coll Cardiol 2010;55(23):2527–35.
[65] Menezes LJ, Kotze CW, Hutton BF, et al. Vascular inflammation imaging with 18 F-FDG PET/CT: when to image? J Nucl Med 2009;50(6):854–7.
[66] Menezes LJ, Kotze CW, Agu O, et al. Investigating vulnerable atheroma using combined (18)F-FDG PET/CT angiography of carotid plaque with immunohistochemical validation. J Nucl Med 2011;52(11):1698–703.
[67] Myers KS, Rudd JH, Hailman EP, et al. Correlation between arterial FDG uptake and biomarkers in peripheral artery disease. JACC Cardiovasc Imaging 2012;5(1):38–45.
[68] Fayad ZA, Mani V, Woodward M, et al. Safety and efficacy of dalcetrapib on atherosclerotic disease using novel non-invasive multimodality imaging (dal-PLAQUE): a randomised clinical trial. Lancet 2011;378(9802):1547–59.
[69] Fayad ZA, Mani V, Woodward M, et al. Rationale and design of dal-PLAQUE: a study assessing efficacy and safety of dalcetrapib on progression or regression of atherosclerosis using magnetic resonance imaging and 18F-fluorodeoxyglucose positron emission tomography/computed tomography. Am Heart J 2011;162(2):214–21, e2.
[70] Wykrzykowska J, Lehman S, Williams G, et al. Imaging of inflamed and vulnerable plaque in coronary arteries with ^{18}F-FDG PET/CT in patients with suppression of myocardial uptake using a low-carbohydrate, high-fat preparation. J Nucl Med 2009;50(4):563–8.
[71] Debruyne JC, Versijpt J, Van Laere KJ, et al. PET visualization of microglia in multiple sclerosis patients using [11C]PK11195. Eur J Neurol 2003;10(3):257–64.
[72] Pugliese F, Gaemperli O, Kinderlerer AR, et al. Imaging of vascular inflammation with [11C]-PK11195 and positron emission tomography/computed tomography angiography. J Am Coll Cardiol 2010;56(8):653–61.
[73] Gaemperli O, Shalhoub J, Owen DR, et al. Imaging intraplaque inflammation in carotid atherosclerosis with 11C–PK11195 positron emission tomography/computed tomography. Eur Heart J 2012;33(15):1902–10.
[74] Kopka K, Faust A, Keul P, et al. 5-Pyrrolidinylsulfonyl isatins as a potential tool for the molecular imaging of caspases in apoptosis. J Med Chem 2006;49(23):6704–15.
[75] Chen DL, Zhou D, Chu W, et al. Radiolabeled isatin binding to caspase-3 activation induced by anti-Fas antibody. Nucl Med Biol 2012;39(1):137–44.
[76] Zhou D, Chu W, Rothfuss J, et al. Synthesis, radiolabeling, and *in vivo* evaluation of an 18F-labeled isatin analog for imaging caspase-3 activation in apoptosis. Bioorg Med Chem Lett 2006;16(19):5041–6.
[77] Nguyen QD, Smith G, Glaser M, et al. Positron emission tomography imaging of drug-induced tumor apoptosis with a caspase-3/7 specific [18F]-labeled isatin sulfonamide. Proc Natl Acad Sci U S A 2009;106(38):16375–80.
[78] Schutters K, Reutelingsperger C. Phosphatidylserine targeting for diagnosis and treatment of human diseases. Apoptosis 2010;15(9):1072–82.
[79] Taki J, Higuchi T, Kawashima A, et al. Detection of cardiomyocyte death in a rat model of ischemia and reperfusion using 99mTc-labeled annexin V. J Nucl Med 2004;45(9):1536–41.
[80] Takei T, Kuge Y, Zhao S, et al. Enhanced apoptotic reaction correlates with suppressed tumor glucose utilization after cytotoxic chemotherapy: use of 99mTc-Annexin V, 18F-FDG, and histologic evaluation. J Nucl Med 2005;46(5):794–9.

[81] Belhocine T, Steinmetz N, Hustinx R, et al. Increased uptake of the apoptosis-imaging agent (99 m)Tc recombinant human annexin V in human tumors after one course of chemotherapy as a predictor of tumor response and patient prognosis. Clin Cancer Res 2002;8(9):2766–74.
[82] Blankenberg FG, Katsikis PD, Tait JF, et al. *In vivo* detection and imaging of phosphatidylserine expression during programmed cell death. Proc Natl Acad Sci U S A 1998;95(11):6349–54.
[83] Dumont EA, Reutelingsperger CP, Smits JF, et al. Real-time imaging of apoptotic cell-membrane changes at the single-cell level in the beating murine heart. Nat Med 2001;7(12):1352–5.
[84] Hofstra L, Liem IH, Dumont EA, et al. Visualisation of cell death *in vivo* in patients with acute myocardial infarction. Lancet 2000;356(9225):209–12.
[85] Kolodgie FD, Petrov A, Virmani R, et al. Targeting of apoptotic macrophages and experimental atheroma with radiolabeled annexin V: a technique with potential for noninvasive imaging of vulnerable plaque. Circulation 2003;108(25):3134–9.
[86] Laufer EM, Reutelingsperger CP, Narula J, et al. Annexin A5: an imaging biomarker of cardiovascular risk. Basic Res Cardiol 2008;103(2):95–104.
[87] Falborg L, Waehrens LN, Alsner J, et al. Biodistribution of 99mTc-HYNIC-lactadherin in mice—a potential tracer for visualizing apoptosis *in vivo*. Scand J Clin Lab Invest 2010;70(3):209–16.
[88] Davletov BA, Sudhof TC. A single C2 domain from synapto-tagmin I is sufficient for high affinity Ca2+/phospholipid binding. J Biol Chem 1993;268(35):26386–90.
[89] Krishnan AS, Neves AA, De Backer MM, et al. Detection of cell death in tumors by using MR imaging and a gadolinium-based targeted contrast agent. Radiology 2008;246(3):854–62.
[90] Alam IS, Neves AA, Witney TH, et al. Comparison of the C2A domain of synaptotagmin-I and annexin-V as probes for detecting cell death. Bioconjug Chem 2010;21(5):884–91.
[91] Kietselaer BL, Reutelingsperger CP, Heidendal GA, et al. Noninvasive detection of plaque instability with use of radiolabeled annexin A5 in patients with carotid-artery atherosclerosis. N Engl J Med 2004;350(14):1472–3.
[92] Takeuchi H, Morino Y, Matsukage T, et al. Impact of vascular remodeling on the coronary plaque compositions: an investigation with *in vivo* tissue characterization using integrated backscatter-intravascular ultrasound. Atherosclerosis 2009;202(2):476–82.
[93] Pasterkamp G, Fitzgerald PF, De Kleijn DP. Atherosclerotic expansive remodeled plaques: a wolf in sheep's clothing. J Vasc Res 2002;39(6):514–23.
[94] Iyemere VP, Proudfoot D, Weissberg PL, et al. Vascular smooth muscle cell phenotypic plasticity and the regulation of vascular calcification. J Intern Med 2006;260(3):192–210.
[95] Shanahan CM, Weissberg PL. Smooth muscle cell phenotypes in atherosclerotic lesions. Curr Opin Lipidol 1999;10(6):507–13.
[96] Carrio I, Pieri PL, Narula J, et al. Noninvasive localization of human atherosclerotic lesions with indium 111-labeled monoclonal Z2D3 antibody specific for proliferating smooth muscle cells. J Nucl Cardiol 1998;5(6):551–7.
[97] Narula J, Petrov A, Bianchi C, et al. Noninvasive localization of experimental atherosclerotic lesions with mouse/human chimeric Z2D3 F(ab')2 specific for the proliferating smooth muscle cells of human atheroma. Imaging with conventional and negative charge-modified antibody fragments. Circulation 1995;92(3):474–84.
[98] Narula J, Petrov A, Pak KY, et al. Noninvasive detection of atherosclerotic lesions by 99mTc-based immunoscintigraphic targeting of proliferating smooth muscle cells. Chest 1997;111(6):1684–90.
[99] Tekabe Y, Einstein AJ, Johnson LL, et al. Targeting very small model lesions pretargeted with bispecific antibody with 99mTc-labeled high-specific radioactivity polymers. Nucl Med Commun 2010;31(4):320–7.
[100] Jimenez J, Donahay T, Schofield L, et al. Smooth muscle cell proliferation index correlates with 111In-labeled antibody Z2D3 uptake in a transplant vasculopathy swine model. J Nucl Med 2005;46(3):514–9.
[101] Johnson LL, Schofield LM, Verdesca SA, et al. *In vivo* uptake of radiolabeled antibody to proliferating smooth muscle cells in a swine model of coronary stent restenosis. J Nucl Med 2000;41(9):1535–40.
[102] Hyafil F, Vucic E, Cornily JC, et al. Monitoring of arterial wall remodelling in atherosclerotic rabbits with a magnetic resonance imaging contrast agent binding to matrix metalloproteinases. Eur Heart J 2011;32(12):1561–71.
[103] Deguchi JO, Aikawa M, Tung CH, et al. Inflammation in atherosclerosis: visualizing matrix metalloproteinase action in macrophages *in vivo*. Circulation 2006;114(1):55–62.
[104] Ohshima S, Petrov A, Fujimoto S, et al. Molecular imaging of matrix metalloproteinase expression in atherosclerotic plaques of mice deficient in apolipoprotein e or low-density-lipoprotein receptor. J Nucl Med 2009;50(4):612–7.
[105] ten Kate GL, Sijbrands EJ, Valkema R, et al. Molecular imaging of inflammation and intraplaque vasa vasorum: a step forward to identification of vulnerable plaques? J Nucl Cardiol 2010;17(5):897–912.
[106] Moulton KS. Angiogenesis in atherosclerosis: gathering evidence beyond speculation. Curr Opin Lipidol 2006;17(5):548–55.
[107] Matter CM, Schuler PK, Alessi P, et al. Molecular imaging of atherosclerotic plaques using a human antibody against the extradomain B of fibronectin. Circ Res 2004;95(12):1225–33.
[108] Cai W, Chen K, Mohamedali KA, et al. PET of vascular endothelial growth factor receptor expression. J Nucl Med 2006;47(12):2048–56.
[109] Rodriguez-Porcel M. Non-invasive monitoring of angiogenesis in cardiology. Curr Cardiovasc Imaging Rep 2009;2(1):59–66.
[110] Clofent-Sanchez G, Jacobin-Valat MJ, Laroche-Traineau J. The growing interest of fibrin imaging in atherosclerosis. Atherosclerosis 2012;222(1):22–5.
[111] Tavora F, Cresswell N, Li L, et al. Immunolocalisation of fibrin in coronary atherosclerosis: implications for necrotic core development. Pathology 2010;42(1):15–22.
[112] Sato Y, Hatakeyama K, Yamashita A, et al. Proportion of fibrin and platelets differs in thrombi on ruptured and eroded coronary atherosclerotic plaques in humans. Heart 2005;91(4):526–30.
[113] Botnar RM, Buecker A, Wiethoff AJ, et al. *In vivo* magnetic resonance imaging of coronary thrombosis using a fibrin-binding molecular magnetic resonance contrast agent. Circulation 2004;110(11):1463–6.
[114] Spuentrup E, Buecker A, Katoh M, et al. Molecular magnetic resonance imaging of coronary thrombosis and pulmonary emboli with a novel fibrin-targeted contrast agent. Circulation 2005;111(11):1377–82.

[115] Spuentrup E, Botnar RM, Wiethoff AJ, et al. MR imaging of thrombi using EP-2104R, a fibrin-specific contrast agent: initial results in patients. Eur Radiol 2008;18(9):1995–2005.
[116] Spuentrup E, Katoh M, Buecker A, et al. Molecular MR imaging of human thrombi in a swine model of pulmonary embolism using a fibrin-specific contrast agent. Invest Radiol 2007;42(8):586–95.
[117] Makowski MR, Forbes SC, Blume U, et al. *In vivo* assessment of intraplaque and endothelial fibrin in ApoE$^{(-/-)}$ mice by molecular MRI. Atherosclerosis 2012;222(1):43–9.
[118] Virmani R, Burke AP, Farb A, et al. Pathology of the vulnerable plaque. J Am Coll Cardiol 2006;47(8 Suppl):C13–8.
[119] Arbustini E, Dal Bello B, Morbini P, et al. Plaque erosion is a major substrate for coronary thrombosis in acute myocardial infarction. Heart 1999;82(3):269–72.
[120] Huang H, Virmani R, Younis H, et al. The impact of calcification on the biomechanical stability of atherosclerotic plaques. Circulation 2001;103(8):1051–6.
[121] Ehara S, Kobayashi Y, Yoshiyama M, et al. Spotty calcification typifies the culprit plaque in patients with acute myocardial infarction: an intravascular ultrasound study. Circulation 2004;110(22):3424–9.
[122] Aikawa E, Nahrendorf M, Figueiredo JL, et al. Osteogenesis associates with inflammation in early-stage atherosclerosis evaluated by molecular imaging *in vivo*. Circulation 2007;116(24):2841–50.
[123] Dunphy MP, Freiman A, Larson SM, et al. Association of vascular 18F-FDG uptake with vascular calcification. J Nucl Med 2005;46(8):1278–84.
[124] Derlin T, Richter U, Bannas P, et al. Feasibility of 18F-sodium fluoride PET/CT for imaging of atherosclerotic plaque. J Nucl Med 2010;51(6):862–5.
[125] Slobbe LCJ, Smit JM, Groen J, Poos MJJC, Kommer GJ. Cost of Illness in the Netherlands 2007: trends in healthcare expenditure 1999–2010. In RIVM rapport 2011, Rijksinstituut voor Volksgezondheid en Milieu RIVM Centraal Bureau voor de Statistiek CBS.
[126] Koch-Institut R. Krankheitskosten. Krankheitskosten [Gesundheitsberichterstattung-Themenhefte], Available from: http://www.gbe-bund.de/gbe10/ergebnisse.prc_tab?fid=12567&suchstring=&query_id=&sprache=D&fund_typ=TXT&methode=&vt=&verwandte=1&page_ret=0&seite=1&p_lfd_nr=42&p_news=&p_sprachkz=D&p_uid=gast&p_aid=78162345&hlp_nr=2&p_janein=J#Kap3.1; 2009.
[127] Taylor MJ, Scuffham PA, McCollam PL, et al. Acute coronary syndromes in Europe: 1-year costs and outcomes. Curr Med Res Opin 2007;23(3):495–503.
[128] Roger VL, Go AS, Lloyd-Jones DM, et al. Heart disease and stroke statistics—2012 update: a report from the American Heart Association. Circulation 2012;125(1):e2–e220.
[129] Naghavi M, Falk E, Hecht HS, et al. From vulnerable plaque to vulnerable patient—part III: executive summary of the Screening for Heart Attack Prevention and Education (SHAPE) Task Force report. Am J Cardiol 2006;98(2A):2H–15H.
[130] Chan KH, Ng MK. Is there a role for coronary angiography in the early detection of the vulnerable plaque? Int J Cardiol 2012;164(3):262–6.
[131] Ait-Oufella H, Kinugawa K, Zoll J, et al. Lactadherin deficiency leads to apoptotic cell accumulation and accelerated atherosclerosis in mice. Circulation 2007;115(16):2168–77.
[132] Aprahamian T, Rifkin I, Bonegio R, et al. Impaired clearance of apoptotic cells promotes synergy between atherogenesis and autoimmune disease. J Exp Med 2004;199(8):1121–31.

SECTION 3

TARGETING OF THE INFLAMMATORY RESPONSE

CHAPTER

10

Mineralcorticoid Receptor Antagonists

Federico Carbone[1,2,3], *Fabrizio Montecucco*[2,3,4]

[1]Division of Cardiology, Department of Medical Specialties, Foundation for Medical Researches, University of Geneva, Geneva, Switzerland

[2]Department of Internal Medicine, University of Genoa School of Medicine, Genoa, Italy

[3]IRCCS Azienda Ospedaliera Universitaria San Martino–IST Istituto Nazionale per la Ricerca sul Cancro, Genoa, Italy

[4]Division of Laboratory Medicine, Department of Genetics and Laboratory Medicine, Geneva University Hospitals, Geneva, Switzerland

10.1 INTRODUCTION

Molecular and cellular mechanisms of cardiac remodeling in hypertension, heart failure (HF), or ischemic heart diseases occur in the frame of upregulated neurohormonal stimuli, including adrenergic signaling and activation of the renin-angiotensin-aldosterone system (RAAS) [1]. A critical role in the pathogenesis of HF is played by aldosterone, the terminal effector of RAAS cascade that is not adequately suppressed by RAAS inhibitors [2]. Traditionally considered a salt-retaining hormone, aldosterone was isolated in 1953. However, it has been recognized for a long time only as regulator of blood pressure [3]. The discovery of the widespread distribution of mineralcorticoid receptor (MR) (as well as the local extra-adrenal synthesis of aldosterone) has changed this view. Accordingly, aldosterone is now known to directly promote cardiac remodeling by enhancing vascular and myocardial inflammation up to cardiac fibrosis [4]. Likewise, from a marginal role of potassium-sparing diuretics, MR antagonists have become a cornerstone for the treatment of HF, as demonstrated by several clinical trials [5]. Unfortunately, the relevance of preventing (or at least slowing) adverse cardiac remodeling is still poorly recognized by clinicians, and MR blockers are still under-prescribed. With this aim, we will update in this review the current knowledge about the antagonist of MR, focusing on clinical evidences and new insights about the selective actions of MR in myocardial tissue.

10.2 MOLECULAR BASIS FOR THE CLINICAL USE OF MR ANTAGONIST IN HF

Twenty years ago, Brilla and coworkers showed that chronic administration of aldosterone was associated with hypertension, cardiac hypertrophy, and fibrosis in uninephrectomized rats receiving a high-sodium diet [6]. This association was confirmed in the following experimental and clinical studies, including a substudy of RALES showing lower levels of serum markers of extracellular matrix turnover in patients randomized to spironolactone [7]. Consistent with these data, MR activation is now established to promote vascular inflammation, cardiac fibrosis, and hypertension independent on the blood pressure and, in large part, the systemic and local RAAS [8]. Aldosterone achieved these detrimental effects by regulating the transcription of several proinflammatory and profibrotic genes in vascular and immune cells, although the mechanisms translating MR signaling into cardiac remodeling are still being elucidated.

http://dx.doi.org/10.1016/B978-0-12-800039-7.00010-4

© 2015 Elsevier Inc. All rights reserved.

10.2.1 Generalities

Aldosterone is a steroid hormone synthesized from the cholesterol in the zona glomerulosa of the adrenal cortex. Into the mitochondria, cholesterol undergoes several reactions catalyzed by enzymes belonging to the cytochrome P450 family. Among these, the CYP11B2 is the rate-limiting enzyme subjected to a tight control by several factors. Angiotensin II and potassium levels are the best-known promoters of CYP11B2 transcription [9], but also adrenocorticotrophic hormone [10], adipose tissue factors [11], and atrial natriuretic peptide [12] has been described as regulators of aldosterone synthesis.

The discovery of MR dates back from three decades and now is well recognized as a pivotal transcription factor involved in many physiological processes and pathological disorders [13]. In the steady state, MR is located in cytoplasm and chaperone proteins maintain a conformational state of MR suitable for the ligand with hormone [14]. Upon hormone binding, the MR-hormone complex translocates into the nucleus where it acts as a transcription factor by binding specific elements (hormone responsive element (HRE)) in the target genes [15]. Furthermore, transcriptional coregulators address the MR activity on tissue-specific target genes in addition to controlling the expression of MR itself with feedback mechanism [16] (Figure 10.1).

Alongside classical genomic actions of aldosterone (requiring a lag time of 1-2h), rapid effects (occurring within 15min) have been recognized [17]. These rapid responses synergize with later transcriptional responses but through other independent pathways [18]. This alternative signaling pathways are shared with other steroid hormones, but the mechanisms involved in signal transduction are still a matter of debate. Unlike other steroid receptors, MR lacks of the palmitoylation motif required for both plasma-membrane binding and early signaling transduction (through the interaction with the scaffolding protein caveolin-1) [19]. However, a small fraction of MR has been recently discovered within the plasma membrane, and it might also be involved in the transactivation of EGFR, a pivotal step to activate the cascade of nongenomic effects of aldosterone [20]. Concerning the CV system, aldosterone promotes proliferation and hypertrophy of vascular smooth-muscle cells (VSMCs) [21] and cardiomyocyte [22,23] by activating the ERK1/2 pathway; whereas, through p38 MAPK aldosterone induces the shift of VSMCs toward a profibrotic phenotype. Furthermore, through the family of protein kinase C, aldosterone regulates a large amount of processes ranging from cell proliferation and apoptosis up to tight-junction formation. Finally, second messenger pathways also are involved in aldosterone signaling transduction, including intracellular Ca^{++} concentrations, synthesis and release of nitric oxide, and Na^+/H^+ exchanger.

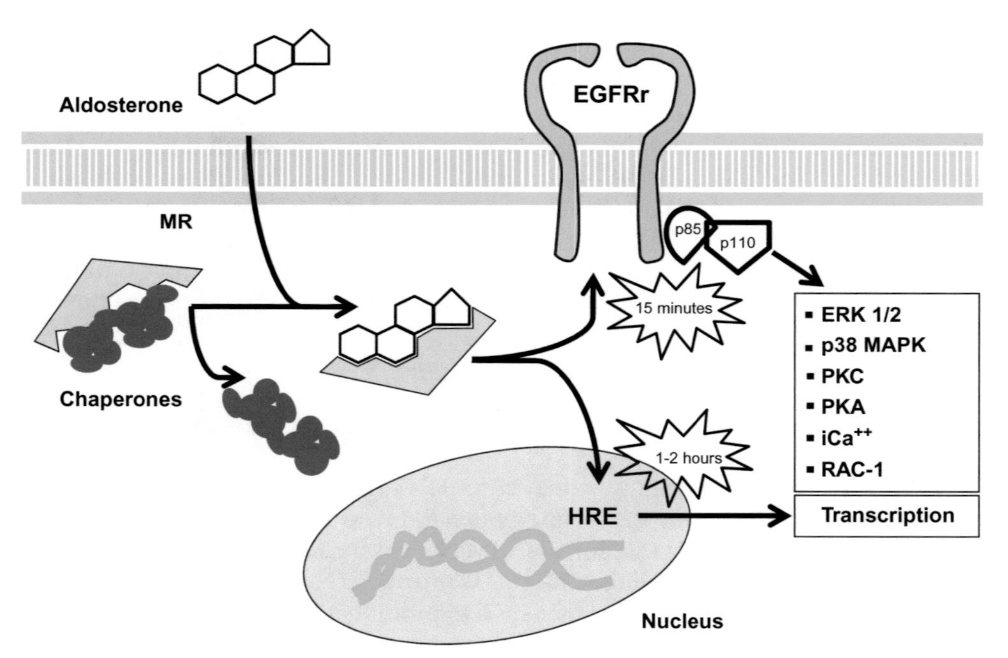

FIGURE 10.1 Genomic and nongenomic effects of MR. The activation of mineralcorticoid receptor (MR) triggers the translocation of this receptor within the nucleus and promotes the transcription of target genes through binding specific hormone response elements (HRE). Alongside this genomic pathway, also for mineralcorticoid hormones, have been recognized nongenomic effects. Compared to the prolonged time needed for the genomic response (hours to days), the nongenomic effects are quicker (minutes) and provide synergic responses involving the downstream pathways of endothelial growth factor receptor (EGFR) such as kinases (ERK 1/2, p38, PKC, PKA), intracellular Ca^{++}, and RAC-1, small G-protein of Rho family.

10.2.2 Inflammation

The pivotal role of vascular inflammation as a precursor to the development of cardiac fibrosis is now well established and reported in several experimental models of hyperaldosteronism. First, an immunostimulatory state occurs very early after mineralcorticoid administration in high-salt diet uninephrectomized rats [24]. In addition, MR activation enhances the expression of adhesion molecules on endothelial surface, thus promoting leukocyte adhesion and recruitment [25]. Thus, aldosterone induces vascular inflammation, characterized by generation of reactive oxygen species (ROS) and expression of several inflammatory markers and profibrotic factors. Consistent with these findings, macrophage-restricted deletion of MR gene prevents cardiac fibrosis in mice [26,27]. MR activation enhances oxidative stress by promoting synthesis and coupling of nicotinamide adenine dinucleotide phosphate (NADPH) oxidase subunits ($gp91^{phox}$, $gp22^{phox}$, and $gp47^{phox}$) [28] in macrophages [29], endothelial cells [30] as well as within the myocardial tissue [31]. Accordingly, mice knockout for $gp47^{phox}$ or NOX2 display insensitive ROS production triggered by aldosterone [32]. Moreover, by suppressing the endothelial expression of glucose-6-phosphate dehydrogenase [33], as well as the dephosphorylation of phosphatase 2A [34], MR activation impairs the activity of nitric oxide synthase, further promoting endothelial dysfunction. Furthermore, MR activation in dendritic cells was shown to induce the activation of $CD8^+$ T-cell, promote a shift of $CD4^+$ T-cell toward a proinflammatory phenotype [35], and suppress the regulatory T cells [36].

These proinflammatory effects of MR activation might also promote myocardial inflammation by enhancing the recruitment of activated inflammatory cells within the heart (Figure 10.2). Lother and coworkers showed that a selective deletion of MR in cardiomyocytes (but not in fibroblasts) prevented cardiac remodeling in a mouse model of chronic pressure overload achieved by aortic constriction [37]. Fraccarollo and colleagues suggested that the oxidative stress response might be the leading pathway of detrimental MR activity within the myocardial tissue. The authors showed that myocardial MR ablation prevented the increase of superoxide ($O_2^{\bullet-}$) and the upregulation of the NADPH oxidases (NOX) 2 and 4 in a mouse model of myocardial infarction [38]. Furthermore, additional detrimental effects induced by MR activation in cardiomyocytes were recognized by comparing the response to deoxycorticosterone/salt in MR-null as compared to wild type mice. In this model, MR signaling suppression was associated with a reduced expression of oxidative stress markers ($p22^{phox}$), inflammatory markers (CCR5, CD14, and CD81),

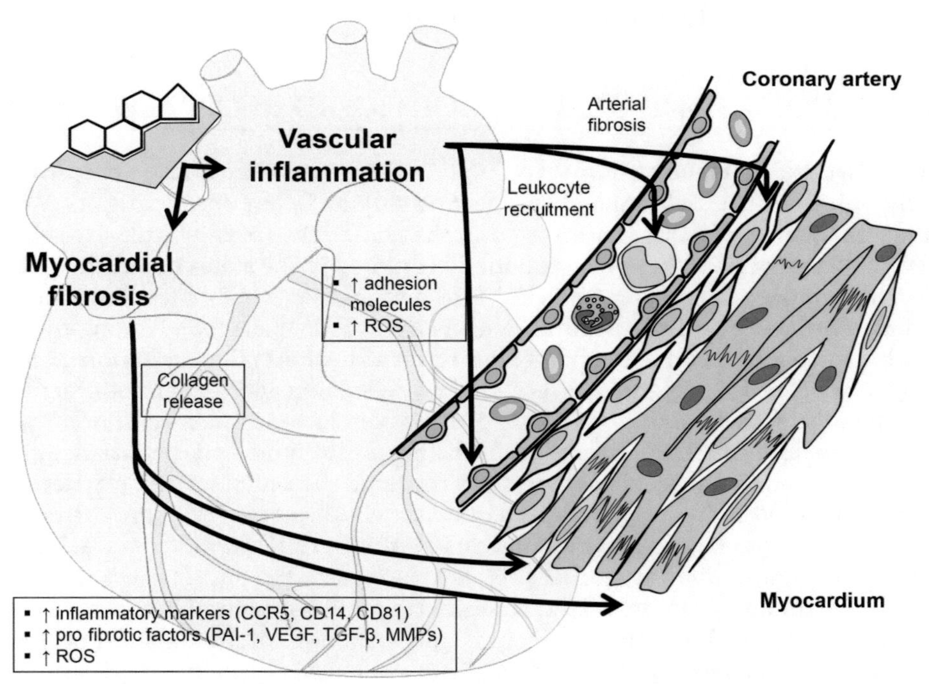

FIGURE 10.2 The main detrimental effects induced by mineralcorticoid receptor activation (MRA) on cardiovascular systems include the vascular inflammation and the cardiac fibrosis. Vascular inflammation is promoted by the proinflammatory activation of endothelial cells leading to increased production of reactive oxygen species (ROS) and leukocyte recruitment. Ultimately, the endothelial dysfunction sustains the shift of vascular smooth-muscle cells toward a profibrotic phenotype. On the other hand, MRA promotes myocardial fibrosis mainly by a direct effect of cardiomyocytes that include the increased expression of chemokine receptor (CCR5), profibrotic factor (plasminogen activator inhibitor (PAI)-1, vascular endothelial growth factor (VEGF), transforming growth factor (TGF)-β), matrix metalloproteinases (MMPs), and ROS.

and profibrotic factors (plasminogen activation inhibitor [PAI]-1, vascular endothelial growth factor, p22phox, of transforming growth factor [TGF]-β, integrin β1, and matrix metalloproteinases-2 and -9) [39]. Finally, Nagase and coworkers recently confirmed in cultured cardiomyocytes the pivotal role of Rac-1 in MR signaling [40]. Rac-1 is known to be a critical mediator of cell stress responses, including cytoskeleton remodeling, adhesion, migration, cell cycle progression, and gene expression. Nagase and colleagues suggested that oxidative stress may enhance MR activity through Rac-1 signaling, thus leading to the formation of a feed-forward loop. These interesting findings from basic research merit additional confirmation in clinical studies.

10.2.3 Cardiac Fibrosis

Cardiac remodeling is characterized by the accumulation of collagen fibers (typically collagen types I and III) that alter structure, shape, and contractility of the heart, thus markedly worsening ventricular contractility, valvular functioning, and electrical conduction [41]. Although MR activation may directly stimulate collagen gene transcription in fibroblast [42,43], a growing body of evidences suggests a paracrine activation of fibrogenesis [44]. Several upstream pathways modulate cardiac remodeling, and MR is now a well-established regulator of this signaling network [39]. Among these, TGF-β1—probably the best-known fibrogenic factor—was found highly expressed in experimental models of cardiac fibrosis [45] as well as in fibrotic human hearts [46,47]. Accordingly, experimental MR blockade slightly attenuates cardiac fibrosis in association with TGF-β1 and collagen-III suppression [48].

Synergistically with TGF-β1, upregulation in cultured cardiomyocyte [49,50] and VSMCs [51,52], aldosterone also increases the expression of PAI-1, the major inhibitor of extracellular matrix degradation. This occurs in fibroblast [53] as well as in endothelial cells [54], VSMCs [54], cardiomyocyte [55], and monocyte [56]. Furthermore, MR activation may promote cardiac fibrosis also by enhancing myocardial oxidative stress [57], and this may explain the protective effect of NADPH oxidase inhibitor apocynin [58] and other antioxidants [59] on cardiac remodeling. On the other hand, the suppression of cardiac T-cell recruitment through inhibition of stromal-derived factor-1/CXCR4 axis prevents cardiac fibrosis in experimental model of mineralcorticoid excess [60]. Other positive markers linking MR signaling and cardiac fibrosis are endothelin-1 [61,62], osteopontin [63–65], and galactin-3 [66]. Overall, as recently reported by Azibani and coworkers, hyperaldosteronism favors the macrophage infiltration in the heart and enhances the transcription of profibrotic molecules, further suppressing the expression of anti-fibrotic factor such as bone morphogenetic protein 4 and B-type natriuretic peptide [67].

10.3 PHARMACOLOGY OF MINERALCORTICOID RECEPTOR ANTAGONIST

At present, the only two MR antagonists approved on the market for clinical use are spironolactone and eplerenone. However, although the term "MR antagonists" is currently accepted, these drugs do not act as aldosterone antagonists, but rather as inverse agonists. Synthesized in the late 1950s, spironolactone was introduced in clinical practice in 1960, and for 40 years its use has been essentially confined to the states of aldosterone excess.

Spironolactone is characterized by a short half-life (1.3-1.4 h), but its metabolites (7α-thiomethylspironolactone and canrenone) prolong natriuretic and antikaliuretic effects up to 13.8-16.5 h [68]. Among these, the canrenone (as well as its water-soluble form, potassium canrenoate) has fewer side effects than spironolactone, but it is marketed only in some countries of Europe [69]. In addition, both liver dysfunction and renal impairment (common in HF patients) may further extend the activity of this drug to 24-50 h. Spironolactone is a progesterone derivative compound and, thus, it may induce progesterone-like side effects, including mastodynia and disturbance of the menstrual cycle in woman. Furthermore, also antiandrogenic activity was recognized in men, causing gynecomastia, erectile dysfunction, and possibly decreased libido [70]. This lack of specificity has been fully understood through molecular analysis of steroid receptors, recognizing a close homology in terms of full-length sequence, ligand-binding domain, and the helices forming the ligand-binding pocket [71]. In this regard, Hu and Funder also suggested a potential evolutionary drift among steroidal receptors [71,72]. Considering these structural homologies, as well as the failure of the following generation of both physiological ligands and synthetic compounds to improve the specificity for MR, Kolkhof and coworkers have recently updated evidence on some compounds under investigation, including mespirenone, spirorenone, and drospirenone (currently prescribed as contraceptive) [68].

In 2002, the Food and Drug Administration (FDA) approved for clinical use the eplerenone, characterized by improved selectivity, but also lower affinity for MR and less efficiency in lowering blood pressure [73]. Eplerenone has no active metabolites so its half-life is shorter than spironolactone (by about 3 h) [74]. However, an increased Na^+/K^+ ratio has been recognized in healthy patients also up to 12 h after eplerenone administration [75], suggesting a delayed destabilizing effect on MR [76]. Eplerenone is extensively metabolized by the cytochrome P450 (isoenzyme CYP3A4) so that its concentrations may increase in the presence of other drugs inhibiting CYP3A4 [77].

Overall, these long-term effects explain the efficacy of a daily administration of MR antagonist in HF patients, but they should also encourage caution about the risk of hyperkalemia, especially in concomitant liver and/or kidney failure as well as combined therapy with RAAS inhibitors.

10.4 CLINICAL EVIDENCES

10.4.1 Systolic HF

The first report demonstrating beneficial effects of the treatment with a MR antagonist in HF dates back from 1964 [78], but only in 1995 did the researchers at Michigan University design the first large, prospective, randomized, placebo-controlled trial to validate this approach [79]. In the RALES (Randomized Aldactone Evaluation Study), 1663 patients with severe HF (left ventricular ejection fraction (LVEF) <35% and New York Heart Association (NYHA) class III or IV) were randomized to receive daily placebo or 25 mg up to 50 mg of spironolactone [80]. After a mean follow-up of 24 months, the trial was stopped by the safety monitoring board, because the prespecified efficacy boundary for mortality was crossed [80]. In the treated group, spironolactone reduced overall mortality risk by 30% (relative risk (RR) 0.70 [95% CI 0.60-0.82]; $p<0.001$) as well as sudden death (RR 0.71 [95% CI 0.54-0.95]; $p=0.02$) and death related to HF progression (RR 0.64 [95% CI 0.51-0.80]; $p<0.001$). In addition, the treatment group had fewer rehospitalization for any cause (RR 0.77 [95% CI 0.68-0.86]; $p<0.001$) or cardiac cause (RR 0.68 [95% CI 0.60-0.77]; $p<0.001$) [80] as compared to the placebo group. The same research group investigated the cardioprotective effects of aldosterone in ischemic HF. In the EPHESUS (Eplerenone Post-Acute Myocardial Infarction Heart Failure and Survival Study), 6632 patients with recent acute myocardial infarction (3-14 days after) and LVEF <40% were randomly assigned to receive eplerenone (25 mg/day titrated up to 50 mg daily) or placebo in addition to the optimal medical therapy [81]. Also in this cohort, treatment with MR antagonist achieved the primary endpoints, reducing mortality for any cause (RR 0.85 [95% CI 0.75-0.96]; $p=0.008$) and hospitalization for CV causes (RR 0.87 [95% CI 0.79-0.95]; $p=0.002$) [81]. More recently, eplerenone was investigated in a large cohort of mild systolic HF patients. In EMPHASIS-HF (Eplerenone in Mild Patients Hospitalization and Survival Study in Heart Failure), 2737 patients having LVEF <35% and NYHA class II symptoms, in addition to recent CV hospitalization or rise of brain natriuretic peptide, were randomized to placebo or eplerenone dosed accordingly to renal function [82]. Eplerenone achieved the composite endpoint of reducing CV death and HF hospitalization (hazard ratio [HR] 0.63 [95% CI 0.54-0.74]; $p<0.001$), but it also significantly decreased the risk of overall mortality (HR 0.76 [95% CI 0.62-0.93]; $p=0.008$) [82]. Although a direct comparison between these three trials is not appropriate (due to the fact that different drugs were evaluated in different patient populations), these results had a huge scientific impact. First, they provided the basis for further studies, substantially confirming the protective effect of MR antagonist in patients with systolic HF [83–91]. Furthermore, the large sample sizes of these trials have enabled subgroup analysis (such as for African Americans) that showed less clinical benefit from MR blockade [92]. Also, the concerns regarding the safety of these treatments were resolved by these large trials. The incidence of hyperkalemia ranged from the 1% in EMPHASIS-HF to the 3% in EPHESUS, without significant difference compared to placebo groups. The safety of treatment with MR antagonists was also recently confirmed in a subgroup analysis of high-risk patients from EMPHASIS-HF [93] as well as in population-based studies [94,95]. However, it should be noted that the rate of hyperkalemia and related mortality may be higher in clinical practice, as suggested by Juurlink and coworkers [96]. Similarly, a post-hoc analysis of EPHESUS recognized an increased incidence of renal impairment [97], but several data from non-HF patients rather suggest a protection from aldosterone-induced kidney injury [98,99]. Ultimately, treatments with MR antagonist significantly improves outcome of systolic HF patients (even in those already receiving high doses of standard background therapies) [100], and this appears to be a highly cost-effective strategy for the management of HF [101,102]. New ongoing multicenter clinical studies, such as the REMINDER (clinicalTrials.gov, Identifier NCT 01176968) and the ALBATROS (clinicalTrials.gov, Identifier NCT 01059136), trials will be able to further support the use of MR antagonist in systolic HF patients.

10.4.2 Diastolic HF

On the other hand, currently under investigation is the effectiveness of MR blockade in patients with heart failure with preserved ejection fraction (HFpEF). About a decade ago, a pilot study recognized that spironolactone improved left ventricular diastolic dysfunction and chamber stiffness [103,104]. Furthermore, these echocardiographic findings were found to be associated with biochemical markers of fibrosis regression (such as serum ratio between carboxyl-terminal propeptide (PIP) and carboxyl-terminal telopeptide (CITP) of collagen type I) [104]. However, only recently randomized clinical trials were designed with this specific aim. In the first study, called RAAM-PEF (Randomized Aldosterone Antagonism in Heart Failure with Preserved Ejection Fraction), 44 patients were randomized to receive placebo or eplerenone. The study confirmed the improvement of diastolic function whether by biochemical or ultrasound assessment [105]. Similar results were provided by the Aldo-DHF (aldosterone in diastolic HF) trial over a prolonged follow-up (12 months) [106]. Unfortunately, these trials failed to prove a functional improvement investigated by 6 min walking test [105], patient symptoms, or quality of life [106]. Likewise, a subgroup analysis of 8013 patients from the OPTIMIZE-HF (Organized Program to Initiate Lifesaving Treatment in Hospitalized Patients with Heart Failure) trial failed to prove an association between treatment with MR antagonist and CV outcome defined as all-cause mortality (HR 1.03 [CI 95% 0.89-1.20]; $p=0.69$) or hospitalization for HF (HR 0.88 [95% CI, 0.73-1.07]; $p=0.188$) [107]. The results of the ongoing randomized controlled trial TOPCAT (Treatment of Preserved Cardiac Function Heart Failure with an Aldosterone Antagonist; clinicalTrials.gov, Identifier NCT 00094302) will provide further insights into the effect of the MR antagonist in HF-PEF [108,109].

10.4.3 Arrhythmias

Several clinical and experimental studies found a close association between aldosterone and arrhythmias. For instance, high incidence of atrial fibrillation (AF) was found in patients with primary hyperaldosteronism [110], whereas increased circulating aldosterone concentrations [111] were detected in patients with AF. Although electrolyte and autonomic imbalance might contribute to impair the electrical activity of the myocardium, the myocardial fibrosis is the main mechanism by which aldosterone enhances arrhythmic risk [112]. Fibrotic myocardium is a low-voltage tissue that potentially affects atrial activation thus promoting the occurrence of AF. In addition, the reduction of myocardial strain and the shortening of the effective refractory period support re-entry circuits sustaining arrhythmias [113]. Accordingly, clinical studies provided evidences that MR antagonists prevent sudden cardiac death [114] as shown in both RALES (RR 0.71 [95% CI 0.54-0.95]; $p=0.02$) [79] and EPHESUS (RR 0.79 [95% CI 0.64-0.97]; $p=0.03$) [81] trials. Furthermore, a reduced incidence of new onset AF or atrial flutter was recently reported in a subanalysis from EMPHASIS study (HR 0.58 [95% CI 0.35-0.96]; $p=0.03$) [115]. However, other therapies reducing left atrial size or stretch (e.g., RAAS inhibitors or beta-blockers) also are associated with a reduced incidence of arrhythmia, so that these finding might not be specific to MR antagonists [116].

10.5 CONCLUSION AND FUTURE PERSPECTIVES

Hyperkalemia and nonselective endocrine properties are probably the main concern limiting the clinical use of MR antagonists. Thus, in addition to improved adherence to international guidelines, several new compounds are under investigation to improve the pharmacologic profile of MR antagonists. In early 2000, several pharmaceutical companies discovered that nonsteroidal compounds also may act as MR antagonists. Among pyrazolines (the first identified class of nonsteroidal MR antagonists), PF-3882845 is currently under investigation [117], but already some concerns about safety have emerged (clinicalTrials.gov, Identifier NCT 00845258 and NCT 01314898). More recently, the dihydropyridinic compounds (already known as calcium channel blockers from which the BR-4628 has been developed) have been investigated. This compound had a potent and selective activity toward MR, acting by "bulky antagonism" [118]. On the other hand, aldosterone synthase (CYP11B2) inhibitors also are emerging as new treatment options in addition to MR blockade. FAD 286A was the first CYP11B2 inhibitor available [119]. The synthesis of the oral form of this drug (called LCI699) allowed the clinical assessment [120]. To date, two controlled phase-2 trials showed the feasibility and safety of this approach [121,122], but further large randomized clinical trials are ongoing (clinicalTrials.gov, Identifier NCT 00817414 and NCT 01331239).

References

[1] Sun Y. Myocardial repair/remodelling following infarction: roles of local factors. Cardiovasc Res 2009;81(3):482–90.
[2] Borghi C, Boschi S, Ambrosioni E, Melandri G, Branzi A, Magnani B. Evidence of a partial escape of renin-angiotensin-aldosterone blockade in patients with acute myocardial infarction treated with ACE inhibitors. J Clin Pharmacol 1993;33(1):40–5.
[3] Conn JW. Primary aldosteronism. J Lab Clin Med 1955;45(4):661–4.

[4] Brown NJ. Contribution of aldosterone to cardiovascular and renal inflammation and fibrosis. Nat Rev Nephrol 2013;9(8):459–69.
[5] Zannad F, Gattis Stough W, Rossignol P, et al. Mineralocorticoid receptor antagonists for heart failure with reduced ejection fraction: integrating evidence into clinical practice. Eur Heart J 2012;33(22):2782–95.
[6] Brilla CG, Weber KT. Mineralocorticoid excess, dietary sodium, and myocardial fibrosis. J Lab Clin Med 1992;120(6):893–901.
[7] Zannad F, Alla F, Dousset B, Perez A, Pitt B. Limitation of excessive extracellular matrix turnover may contribute to survival benefit of spironolactone therapy in patients with congestive heart failure: insights from the randomized aldactone evaluation study (RALES). Rales Investigators. Circulation 2000;102(22):2700–6.
[8] Young MJ, Rickard AJ. Mechanisms of mineralocorticoid salt-induced hypertension and cardiac fibrosis. Mol Cell Endocrinol 2012;350(2):248–55.
[9] Bassett MH, White PC, Rainey WE. The regulation of aldosterone synthase expression. Mol Cell Endocrinol 2004;217(1–2):67–74.
[10] Holland OB, Carr B. Modulation of aldosterone synthase messenger ribonucleic acid levels by dietary sodium and potassium and by adrenocorticotropin. Endocrinology 1993;132(6):2666–73.
[11] Ronconi V, Turchi F, Bujalska IJ, Giacchetti G, Boscaro M. Adipose cell-adrenal interactions: current knowledge and future perspectives. Trends Endocrinol Metab 2008;19(3):100–3.
[12] Ganguly A. Atrial natriuretic peptide-induced inhibition of aldosterone secretion: a quest for mediator(s). Am J Physiol 1992;263(2 Pt 1):E181–94.
[13] Viengchareun S, Le Menuet D, Martinerie L, Munier M, Pascual-Le Tallec L, Lombes M. The mineralocorticoid receptor: insights into its molecular and (patho)physiological biology. Nucl Recept Signal 2007;5:e012.
[14] Huyet J, Pinon GM, Fay MR, Rafestin-Oblin ME, Fagart J. Structural determinants of ligand binding to the mineralocorticoid receptor. Mol Cell Endocrinol 2012;350(2):187–95.
[15] So AY, Chaivorapol C, Bolton EC, Li H, Yamamoto KR. Determinants of cell- and gene-specific transcriptional regulation by the glucocorticoid receptor. PLoS Genet 2007;3(6):e94.
[16] Le Menuet D, Viengchareun S, Penfornis P, Walker F, Zennaro MC, Lombes M. Targeted oncogenesis reveals a distinct tissue-specific utilization of alternative promoters of the human mineralocorticoid receptor gene in transgenic mice. J Biol Chem 2000;275(11):7878–86.
[17] Wehling M. Specific, nongenomic actions of steroid hormones. Annu Rev Physiol 1997;59:365–93.
[18] Dooley R, Harvey BJ, Thomas W. Non-genomic actions of aldosterone: from receptors and signals to membrane targets. Mol Cell Endocrinol 2012;350(2):223–34.
[19] Acconcia F, Ascenzi P, Bocedi A, et al. Palmitoylation-dependent estrogen receptor alpha membrane localization: regulation by 17beta-estradiol. Mol Biol Cell 2005;16(1):231–7.
[20] Grossmann C, Husse B, Mildenberger S, Schreier B, Schuman K, Gekle M. Colocalization of mineralocorticoid and EGF receptor at the plasma membrane. Biochim Biophys Acta 2010;1803(5):584–90.
[21] Manegold JC, Falkenstein E, Wehling M, Christ M. Rapid aldosterone effects on tyrosine phosphorylation in vascular smooth muscle cells. Cell Mol Biol (Noisy-le-grand) 1999;45(6):805–13.
[22] Stockand JD, Meszaros JG. Aldosterone stimulates proliferation of cardiac fibroblasts by activating Ki-RasA and MAPK1/2 signaling. Am J Physiol Heart Circ Physiol 2003;284(1):H176–84.
[23] Okoshi MP, Yan X, Okoshi K, et al. Aldosterone directly stimulates cardiac myocyte hypertrophy. J Card Fail 2004;10(6):511–8.
[24] Gerling IC, Sun Y, Ahokas RA, et al. Aldosteronism: an immunostimulatory state precedes proinflammatory/fibrogenic cardiac phenotype. Am J Physiol Heart Circ Physiol 2003;285(2):H813–21.
[25] Caprio M, Newfell BG, la Sala A, et al. Functional mineralocorticoid receptors in human vascular endothelial cells regulate intercellular adhesion molecule-1 expression and promote leukocyte adhesion. Circ Res 2008;102(11):1359–67.
[26] Usher MG, Duan SZ, Ivaschenko CY, et al. Myeloid mineralocorticoid receptor controls macrophage polarization and cardiovascular hypertrophy and remodeling in mice. J Clin Invest 2010;120(9):3350–64.
[27] Bienvenu LA, Morgan J, Rickard AJ, et al. Macrophage mineralocorticoid receptor signaling plays a key role in aldosterone-independent cardiac fibrosis. Endocrinology 2012;153(7):3416–25.
[28] Park YM, Lim BH, Touyz RM, Park JB. Expression of NAD(P)H oxidase subunits and their contribution to cardiovascular damage in aldosterone/salt-induced hypertensive rat. J Korean Med Sci 2008;23(6):1039–45.
[29] Keidar S, Kaplan M, Pavlotzky E, et al. Aldosterone administration to mice stimulates macrophage NADPH oxidase and increases atherosclerosis development: a possible role for angiotensin-converting enzyme and the receptors for angiotensin II and aldosterone. Circulation 2004;109(18):2213–20.
[30] Iwashima F, Yoshimoto T, Minami I, Sakurada M, Hirono Y, Hirata Y. Aldosterone induces superoxide generation via Rac1 activation in endothelial cells. Endocrinology 2008;149(3):1009–14.
[31] Sun Y, Zhang J, Lu L, Chen SS, Quinn MT, Weber KT. Aldosterone-induced inflammation in the rat heart: role of oxidative stress. Am J Pathol 2002;161(5):1773–81.
[32] Johar S, Cave AC, Narayanapanicker A, Grieve DJ, Shah AM. Aldosterone mediates angiotensin II-induced interstitial cardiac fibrosis via a Nox2-containing NADPH oxidase. Faseb J 2006;20(9):1546–8.
[33] Leopold JA, Dam A, Maron BA, et al. Aldosterone impairs vascular reactivity by decreasing glucose-6-phosphate dehydrogenase activity. Nat Med 2007;13(2):189–97.
[34] Nagata D, Takahashi M, Sawai K, et al. Molecular mechanism of the inhibitory effect of aldosterone on endothelial NO synthase activity. Hypertension 2006;48(1):165–71.
[35] Herrada AA, Contreras FJ, Marini NP, et al. Aldosterone promotes autoimmune damage by enhancing Th17-mediated immunity. J Immunol 2010;184(1):191–202.
[36] Kasal DA, Barhoumi T, Li MW, et al. T regulatory lymphocytes prevent aldosterone-induced vascular injury. Hypertension 2012;59(2):324–30.
[37] Lother A, Berger S, Gilsbach R, et al. Ablation of mineralocorticoid receptors in myocytes but not in fibroblasts preserves cardiac function. Hypertension 2011;57(4):746–54.
[38] Fraccarollo D, Berger S, Galuppo P, et al. Deletion of cardiomyocyte mineralocorticoid receptor ameliorates adverse remodeling after myocardial infarction. Circulation 2011;123(4):400–8.

[39] Rickard AJ, Morgan J, Bienvenu LA, et al. Cardiomyocyte mineralocorticoid receptors are essential for deoxycorticosterone/salt-mediated inflammation and cardiac fibrosis. Hypertension 2012;60(6):1443–50.
[40] Nagase M, Ayuzawa N, Kawarazaki W, et al. Oxidative stress causes mineralocorticoid receptor activation in rat cardiomyocytes: role of small GTPase Rac1. Hypertension 2012;59(2):500–6.
[41] Kong P, Christia P, Frangogiannis NG. The pathogenesis of cardiac fibrosis. Cell Mol Life Sci 2013;71(4):549–74.
[42] Yamahara H, Kishimoto N, Nakata M, et al. Direct aldosterone action as a profibrotic factor via ROS-mediated SGK1 in peritoneal fibroblasts. Kidney Blood Press Res 2009;32(3):185–93.
[43] Chen D, Chen Z, Park C, et al. Aldosterone stimulates fibronectin synthesis in renal fibroblasts through mineralocorticoid receptor-dependent and independent mechanisms. Gene 2013;531(1):23–30.
[44] Fujisawa G, Dilley R, Fullerton MJ, Funder JW. Experimental cardiac fibrosis: differential time course of responses to mineralocorticoid-salt administration. Endocrinology 2001;142(8):3625–31.
[45] Deten A, Holzl A, Leicht M, Barth W, Zimmer HG. Changes in extracellular matrix and in transforming growth factor beta isoforms after coronary artery ligation in rats. J Mol Cell Cardiol 2001;33(6):1191–207.
[46] Li G, Li RK, Mickle DA, et al. Elevated insulin-like growth factor-I and transforming growth factor-beta 1 and their receptors in patients with idiopathic hypertrophic obstructive cardiomyopathy. A possible mechanism. Circulation 1998;98(19 Suppl):II144–9, discussion II9-50.
[47] Sanderson JE, Lai KB, Shum IO, Wei S, Chow LT. Transforming growth factor-beta(1) expression in dilated cardiomyopathy. Heart 2001;86(6):701–8.
[48] Wahed MI, Watanabe K, Ma M, et al. Effects of eplerenone, a selective aldosterone blocker, on the progression of left ventricular dysfunction and remodeling in rats with dilated cardiomyopathy. Pharmacology 2005;73(2):81–8.
[49] Chun TY, Bloem LJ, Pratt JH. Aldosterone inhibits inducible nitric oxide synthase in neonatal rat cardiomyocytes. Endocrinology 2003;144(5):1712–7.
[50] Sohn HJ, Yoo KH, Jang GY, et al. Aldosterone modulates cell proliferation and apoptosis in the neonatal rat heart. J Korean Med Sci 2010;25(9):1296–304.
[51] Krug AW, Allenhofer L, Monticone R, et al. Elevated mineralocorticoid receptor activity in aged rat vascular smooth muscle cells promotes a proinflammatory phenotype via extracellular signal-regulated kinase 1/2 mitogen-activated protein kinase and epidermal growth factor receptor-dependent pathways. Hypertension 2010;55(6):1476–83.
[52] Zhu CJ, Wang QQ, Zhou JL, et al. The mineralocorticoid receptor-p38MAPK-NFkappaB or ERK-Sp1 signal pathways mediate aldosterone-stimulated inflammatory and profibrotic responses in rat vascular smooth muscle cells. Acta Pharmacol Sin 2012;33(7):873–8.
[53] Huang W, Xu C, Kahng KW, Noble NA, Border WA, Huang Y. Aldosterone and TGF-beta1 synergistically increase PAI-1 and decrease matrix degradation in rat renal mesangial and fibroblast cells. Am J Physiol Renal Physiol 2008;294(6):F1287–95.
[54] Brown NJ, Kim KS, Chen YQ, et al. Synergistic effect of adrenal steroids and angiotensin II on plasminogen activator inhibitor-1 production. J Clin Endocrinol Metab 2000;85(1):336–44.
[55] Chun TY, Pratt JH. Aldosterone increases plasminogen activator inhibitor-1 synthesis in rat cardiomyocytes. Mol Cell Endocrinol 2005;239(1–2):55–61.
[56] Calo LA, Zaghetto F, Pagnin E, et al. Effect of aldosterone and glycyrrhetinic acid on the protein expression of PAI-1 and p22(phox) in human mononuclear leukocytes. J Clin Endocrinol Metab 2004;89(4):1973–6.
[57] He BJ, Joiner ML, Singh MV, et al. Oxidation of CaMKII determines the cardiotoxic effects of aldosterone. Nat Med 2011;17(12):1610–8.
[58] Li YQ, Li XB, Guo SJ, et al. Apocynin attenuates oxidative stress and cardiac fibrosis in angiotensin II-induced cardiac diastolic dysfunction in mice. Acta Pharmacol Sin 2013;34(3):352–9.
[59] Yoshida K, Kim-Mitsuyama S, Wake R, et al. Excess aldosterone under normal salt diet induces cardiac hypertrophy and infiltration via oxidative stress. Hypertens Res 2005;28(5):447–55.
[60] Chu PY, Zatta A, Kiriazis H, et al. CXCR4 antagonism attenuates the cardiorenal consequences of mineralocorticoid excess. Circ Heart Fail 2011;4(5):651–8.
[61] Park JB, Schiffrin EL. Cardiac and vascular fibrosis and hypertrophy in aldosterone-infused rats: role of endothelin-1. Am J Hypertens 2002;15(2 Pt 1):164–9.
[62] Kozakova M, Buralli S, Palombo C, et al. Myocardial ultrasonic backscatter in hypertension: relation to aldosterone and endothelin. Hypertension 2003;41(2):230–6.
[63] Sam F, Xie Z, Ooi H, et al. Mice lacking osteopontin exhibit increased left ventricular dilation and reduced fibrosis after aldosterone infusion. Am J Hypertens 2004;17(2):188–93.
[64] Matsui Y, Jia N, Okamoto H, et al. Role of osteopontin in cardiac fibrosis and remodeling in angiotensin II-induced cardiac hypertrophy. Hypertension 2004;43(6):1195–201.
[65] Zhang YL, Zhou SX, Lei J, Yuan GY, Wang JF. Blockades of angiotensin and aldosterone reduce osteopontin expression and interstitial fibrosis infiltration in rats with myocardial infarction. Chin Med J (Engl) 2008;121(21):2192–6.
[66] Calvier L, Miana M, Reboul P, et al. Galectin-3 mediates aldosterone-induced vascular fibrosis. Arterioscler Thromb Vasc Biol 2013;33(1):67–75.
[67] Azibani F, Benard L, Schlossarek S, et al. Aldosterone inhibits antifibrotic factors in mouse hypertensive heart. Hypertension 2012;59(6):1179–87.
[68] Kolkhof P, Borden SA. Molecular pharmacology of the mineralocorticoid receptor: prospects for novel therapeutics. Mol Cell Endocrinol 2012;350(2):310–7.
[69] Sadee W, Dagcioglu M, Schroder R. Pharmacokinetics of spironolactone, canrenone and canrenoate-K in humans. J Pharmacol Exp Ther 1973;185(3):686–95.
[70] Garthwaite SM, McMahon EG. The evolution of aldosterone antagonists. Mol Cell Endocrinol 2004;217(1–2):27–31.
[71] Hu X, Funder JW. The evolution of mineralocorticoid receptors. Mol Endocrinol 2006;20(7):1471–8.
[72] Funder JW, Mihailidou AS. Aldosterone and mineralocorticoid receptors: clinical studies and basic biology. Mol Cell Endocrinol 2009;301(1–2):2–6.

[73] Muldowney 3rd JA, Schoenhard JA, Benge CD. The clinical pharmacology of eplerenone. Expert Opin Drug Metab Toxicol 2009;5(4):425–32.
[74] Cook CS, Berry LM, Bible RH, Hribar JD, Hajdu E, Liu NW. Pharmacokinetics and metabolism of [14C]eplerenone after oral administration to humans. Drug Metab Dispos 2003;31(11):1448–55.
[75] Thosar SS, Gokhale RD, Tolbert DS. Immediate release eplerenone compositions. US 6, 558, 707(B1), GD Searle & Co, USA; 2003.
[76] Couette B, Lombes M, Baulieu EE, Rafestin-Oblin ME. Aldosterone antagonists destabilize the mineralocorticosteroid receptor. Biochem J 1992;282(Pt 3):697–702.
[77] Cook CS, Berry LM, Kim DH, Burton EG, Hribar JD, Zhang L. Involvement of CYP3A in the metabolism of eplerenone in humans and dogs: differential metabolism by CYP3A4 and CYP3A5. Drug Metab Dispos 2002;30(12):1344–51.
[78] Sanders LL, Melby JC. Aldosterone and the edema of congestive heart failure. Arch Intern Med 1964;113:331–41.
[79] Pitt D. ACE inhibitor co-therapy in patients with heart failure: rationale for the Randomized Aldactone Evaluation Study (RALES). Eur Heart J 1995;16 Suppl N:107–10.
[80] Pitt B, Zannad F, Remme WJ, et al. The effect of spironolactone on morbidity and mortality in patients with severe heart failure. Randomized Aldactone Evaluation Study Investigators. N Engl J Med 1999;341(10):709–17.
[81] Pitt B, Remme W, Zannad F, et al. Eplerenone, a selective aldosterone blocker, in patients with left ventricular dysfunction after myocardial infarction. N Engl J Med 2003;348(14):1309–21.
[82] Zannad F, McMurray JJ, Krum H, et al. Eplerenone in patients with systolic heart failure and mild symptoms. N Engl J Med 2011;364(1):11–21.
[83] Mariotti R, Borelli G, Coceani M, et al. Aldosterone receptor antagonism and heart failure: insights from an outpatient clinic. J Clin Pharm Ther 2008;33(4):349–56.
[84] Birocchi S, Cernuschi GC. Eplerenone, an aldosterone antagonist, reduces hospitalization and death in heart failure patients with NYHA class II and an ejection fraction of less than 30%. Intern Emerg Med 2011;6(5):453–4.
[85] Boccanelli A, Mureddu GF, Cacciatore G, et al. Anti-remodelling effect of canrenone in patients with mild chronic heart failure (AREA IN-CHF study): final results. Eur J Heart Fail 2009;11(1):68–76.
[86] Udelson JE, Feldman AM, Greenberg B, et al. Randomized, double-blind, multicenter, placebo-controlled study evaluating the effect of aldosterone antagonism with eplerenone on ventricular remodeling in patients with mild-to-moderate heart failure and left ventricular systolic dysfunction. Circ Heart Fail 2010;3(3):347–53.
[87] de Simone G, Chinali M, Mureddu GF, et al. Effect of canrenone on left ventricular mechanics in patients with mild systolic heart failure and metabolic syndrome: the AREA-in-CHF study. Nutr Metab Cardiovasc Dis 2011;21(10):783–91.
[88] Hamaguchi S, Kinugawa S, Tsuchihashi-Makaya M, et al. Spironolactone use at discharge was associated with improved survival in hospitalized patients with systolic heart failure. Am Heart J 2010;160(6):1156–62.
[89] Hernandez AF, Mi X, Hammill BG, et al. Associations between aldosterone antagonist therapy and risks of mortality and readmission among patients with heart failure and reduced ejection fraction. Jama 2012;308(20):2097–107.
[90] Frankenstein L, Katus HA, Grundtvig M, et al. Association between spironolactone added to beta-blockers and ACE inhibition and survival in heart failure patients with reduced ejection fraction: a propensity score-matched cohort study. Eur J Clin Pharmacol 2013;69(10):1747–55.
[91] Ferreira JP, Santos M, Almeida S, Marques I, Bettencourt P, Carvalho H. Mineralocorticoid receptor antagonism in acutely decompensated chronic heart failure. Eur J Intern Med 2013;25(1):67–72.
[92] Vardeny O, Cavallari LH, Claggett B, et al. Race influences the safety and efficacy of spironolactone in severe heart failure. Circ Heart Fail 2013;6(5):970–6.
[93] Eschalier R, McMurray JJ, Swedberg K, et al. Safety and efficacy of eplerenone in patients at high risk for hyperkalemia and/or worsening renal function: analyses of the EMPHASIS-HF study subgroups (Eplerenone in Mild Patients Hospitalization and SurvIval Study in Heart Failure). J Am Coll Cardiol 2013;62(17):1585–93.
[94] Wei L, Struthers AD, Fahey T, Watson AD, Macdonald TM. Spironolactone use and renal toxicity: population based longitudinal analysis. BMJ 2010;340:c1768.
[95] Rossi R, Crupi N, Coppi F, Monopoli D, Sgura F. Importance of the time of initiation of mineralocorticoid receptor antagonists on risk of mortality in patients with heart failure. J Renin Angiotensin Aldosterone Syst, in press; http://dx.doi.org/10.1177/1470320313482603.
[96] Juurlink DN, Mamdani MM, Lee DS, et al. Rates of hyperkalemia after publication of the Randomized Aldactone Evaluation Study. N Engl J Med 2004;351(6):543–51.
[97] Rossignol P, Cleland JG, Bhandari S, et al. Determinants and consequences of renal function variations with aldosterone blocker therapy in heart failure patients after myocardial infarction: insights from the Eplerenone Post-Acute Myocardial Infarction Heart Failure Efficacy and Survival Study. Circulation 2012;125(2):271–9.
[98] Sato A, Hayashi K, Saruta T. Antiproteinuric effects of mineralocorticoid receptor blockade in patients with chronic renal disease. Am J Hypertens 2005;18(1):44–9.
[99] Navaneethan SD, Nigwekar SU, Sehgal AR, Strippoli GF. Aldosterone antagonists for preventing the progression of chronic kidney disease: a systematic review and meta-analysis. Clin J Am Soc Nephrol 2009;4(3):542–51.
[100] Krum H, Shi H, Pitt B, et al. Clinical benefit of eplerenone in patients with mild symptoms of systolic heart failure already receiving optimal best practice background drug therapy: analysis of the EMPHASIS-HF study. Circ Heart Fail 2013;6(4):711–8.
[101] de Pouvourville G, Solesse A, Beillat M. Cost-effectiveness analysis of aldosterone blockade with eplerenone in patients with heart failure after acute myocardial infarction in the French context: the EPHESUS study. Arch Cardiovasc Dis 2008;101(9):515–21.
[102] McKenna C, Burch J, Suekarran S, et al. A systematic review and economic evaluation of the clinical effectiveness and cost-effectiveness of aldosterone antagonists for postmyocardial infarction heart failure. Health Technol Assess 2010;14(24):1–162.
[103] Orea-Tejeda A, Colin-Ramirez E, Castillo-Martinez L, et al. Aldosterone receptor antagonists induce favorable cardiac remodeling in diastolic heart failure patients. Rev Invest Clin 2007;59(2):103–7.
[104] Izawa H, Murohara T, Nagata K, et al. Mineralocorticoid receptor antagonism ameliorates left ventricular diastolic dysfunction and myocardial fibrosis in mildly symptomatic patients with idiopathic dilated cardiomyopathy: a pilot study. Circulation 2005;112(19):2940–5.

[105] Deswal A, Richardson P, Bozkurt B, Mann DL. Results of the randomized aldosterone antagonism in heart failure with preserved ejection fraction trial (RAAM-PEF). J Card Fail 2011;17(8):634–42.
[106] Edelmann F, Wachter R, Schmidt AG, et al. Effect of spironolactone on diastolic function and exercise capacity in patients with heart failure with preserved ejection fraction: the Aldo-DHF randomized controlled trial. Jama 2013;309(8):781–91.
[107] Patel K, Fonarow GC, Kitzman DW, et al. Aldosterone antagonists and outcomes in real-world older patients with heart failure and preserved ejection fraction. JACC Heart Fail 2013;1(1):40–7.
[108] Desai AS, Lewis EF, Li R, et al. Rationale and design of the treatment of preserved cardiac function heart failure with an aldosterone antagonist trial: a randomized, controlled study of spironolactone in patients with symptomatic heart failure and preserved ejection fraction. Am Heart J 2011;162(6): 966-972.e10.
[109] Shah SJ, Heitner JF, Sweitzer NK, et al. Baseline characteristics of patients in the treatment of preserved cardiac function heart failure with an aldosterone antagonist trial. Circ Heart Fail 2013;6(2):184–92.
[110] Milliez P, Girerd X, Plouin PF, Blacher J, Safar ME, Mourad JJ. Evidence for an increased rate of cardiovascular events in patients with primary aldosteronism. J Am Coll Cardiol 2005;45(8):1243–8.
[111] Goette A, Hoffmanns P, Enayati W, Meltendorf U, Geller JC, Klein HU. Effect of successful electrical cardioversion on serum aldosterone in patients with persistent atrial fibrillation. Am J Cardiol 2001;88(8):906–9, A8.
[112] de Jong S, van Veen TA, van Rijen HV, de Bakker JM. Fibrosis and cardiac arrhythmias. J Cardiovasc Pharmacol 2011;57(6):630–8.
[113] Tanaka K, Zlochiver S, Vikstrom KL, et al. Spatial distribution of fibrosis governs fibrillation wave dynamics in the posterior left atrium during heart failure. Circ Res 2007;101(8):839–47.
[114] Pitt B, Pitt GS. Added benefit of mineralocorticoid receptor blockade in the primary prevention of sudden cardiac death. Circulation 2007;115(23):2976–82, discussion 82.
[115] Swedberg K, Zannad F, McMurray JJ, et al. Eplerenone and atrial fibrillation in mild systolic heart failure: results from the EMPHASIS-HF (Eplerenone in Mild Patients Hospitalization And SurvIval Study in Heart Failure) study. J Am Coll Cardiol 2012;59(18):1598–603.
[116] Dabrowski R, Szwed H. Antiarrhythmic potential of aldosterone antagonists in atrial fibrillation. Cardiol J 2012;19(3):223–9.
[117] Meyers MJ, Arhancet GB, Hockerman SL, et al. Discovery of (3S,3aR)-2-(3-chloro-4-cyanophenyl)-3-cyclopentyl-3,3a,4,5-tetrahydro-2H-benzo[g] indazole-7-carboxylic acid (PF-3882845), an orally efficacious mineralocorticoid receptor (MR) antagonist for hypertension and nephropathy. J Med Chem 2010;53(16):5979–6002.
[118] Fagart J, Hillisch A, Huyet J, et al. A new mode of mineralocorticoid receptor antagonism by a potent and selective nonsteroidal molecule. J Biol Chem 2010;285(39):29932–40.
[119] Menard J, Pascoe L. Can the dextroenantiomer of the aromatase inhibitor fadrozole be useful for clinical investigation of aldosterone-synthase inhibition? J Hypertens 2006;24(6):993–7.
[120] Amar L, Azizi M, Menard J, Peyrard S, Watson C, Plouin PF. Aldosterone synthase inhibition with LCI699: a proof-of-concept study in patients with primary aldosteronism. Hypertension 2010;56(5):831–8.
[121] Calhoun DA, White WB, Krum H, et al. Effects of a novel aldosterone synthase inhibitor for treatment of primary hypertension: results of a randomized, double-blind, placebo- and active-controlled phase 2 trial. Circulation 2011;124(18):1945–55.
[122] Andersen K, Hartman D, Peppard T, et al. The effects of aldosterone synthase inhibition on aldosterone and cortisol in patients with hypertension: a phase II, randomized, double-blind, placebo-controlled, multicenter study. J Clin Hypertens (Greenwich) 2012;14(9):580–7.

C H A P T E R

11

PPARs as Modulators of Cardiac Metabolism and Inflammation

Anna Planavila[1], *Marc van Bilsen*[2]

[1]Departament de Bioquímica i Biologia Molecular, Institut de Biomedicina de la Universitat de Barcelona (IBUB), Universitat de Barcelona and CIBER Fisiopatología de la Obesidad y Nutrición (CIBEROBN), Barcelona, Spain

[2]Department of Physiology, Cardiovascular Research Institute Maastricht (CARIM), Maastricht University, Maastricht, The Netherlands

11.1 INTRODUCTION

Heart failure is a complex disorder in which the heart is unable to meet the metabolic demands of the peripheral tissues and represents a major cause of morbidity and mortality worldwide [1]. Over the past decade, chronic low-grade inflammation has been considered an important mediator during not only the final stages of heart failure but also during early stages such as the development of cardiac hypertrophy [2–4]. Moreover, growing evidence has linked this chronic state of inflammation with metabolic dysfunction since cardiac disorders are commonly associated with activation of inflammation and changes in metabolic pathways that are in fact linked and are interdependent [5].

Hypertrophy is one of the main ways in which cardiomyocytes respond to mechanical and neurohormonal stimuli. Multiple interacting signaling pathways have been shown to connect these stress signals with cardiac gene expression through a set of signal-dependent transcription factors [6–10]. The activation of these upstream events results in changes in transcription, translation, and sarcomeric organization. One of the signal transduction pathways that has been shown to play a crucial role in the hypertrophic growth of the myocardium is the nuclear factor-κB (NF-κB). This transcription factor regulates the expression of many genes involved in inflammation and their activation has been involved as a causal event in the cardiac hypertrophic response [4,11–13]. Moreover, cardiac hypertrophy and the subsequent development of heart failure are characterized by a shift in the source of energy from fatty acids to glucose [14]. The genes involved in the transport and metabolism of fatty acids and glucose are under the transcriptional control of nuclear hormone receptors, specifically the peroxisome proliferator-activated receptors (PPARs). Moreover, the PPARs have the capacity to attenuate the induction of the proinflammatory pathways, connecting in this way, metabolism and inflammation [15–18].

The mechanisms by which PPAR activity is regulated during the development of heart failure are not entirely clear. Here, we will review recent advances in our understanding of the role of PPARs as integrators of metabolic and inflammatory signaling pathways in cardiac disorders. In addition, we will briefly discuss the role of other factors related to PPARs that have recently been involved in the control of cardiac energy metabolism and inflammation during the development of cardiac diseases such as the deacetylase Sirtuin-1 (Sirt1) and the secreted protein fibroblast growth factor-21 (FGF21).

http://dx.doi.org/10.1016/B978-0-12-800039-7.00011-6

© 2015 Elsevier Inc. All rights reserved.

11.2 PEROXISOME PROLIFERATOR-ACTIVATED RECEPTORS

PPARs are members of the nuclear hormone receptor superfamily of ligand-activated transcription factors. PPARs act as metabolic sensors, enabling the organism to respond to environmental changes by inducing the expression of the appropriate metabolic genes [19]. They share a close structural homology with other members of this family like, for instance, the thyroid hormone and retinoic acid receptor. The amino-terminal domain is less conserved and includes a ligand-independent transactivation function (AF-1), containing putative phosphorylation sites. The DNA binding domain is highly conserved and includes two zinc finger motifs. The carboxyl-terminal ligand-binding domain encompasses a ligand-dependent transactivation function (AF-2). In addition to ligand binding, this region is required for nuclear localization, receptor dimerization, and the interaction with proteins acting as coactivators or corepressors such as PGC1α or NCOR, respectively.

Three PPAR isoforms, PPARα (NR1C1 according to the unified nomenclature system for the nuclear receptor superfamily), PPARβ/δ (NR1C2), and PPARγ (NR1C3) [20], have been identified in vertebrates with differing tissue distribution. PPARα is expressed in tissues with a high level of fatty acid catabolism including the liver, heart, kidney, skeletal muscle, and brown adipose tissue. PPARβ/δ is ubiquitously expressed and appears to have overlapping activity with that of both PPARα and PPARγ when coexpressed with those genes. PPARγ is expressed predominantly in adipose tissue and is largely involved in adipocyte differentiation and adipogenesis.

PPARs are activated by a wide range of naturally occurring or metabolized lipids derived from the diet or from intracellular signaling pathways, which include saturated and unsaturated fatty acids and fatty acids derivatives such as prostaglandins and leukotriens [21,22]. However, how specific endogenous lipid ligands activate the PPARs and whether ligand preference varies in a tissue-specific manner is not entirely clear. It is known that PPARα shows high affinity for unsaturated fatty acids while PPARγ is preferentially activated by eicosanoids, prostaglandins, and leukotriens. Recently, it has been shown that hydrolysis of cellular triglycerides by adipose triglyceride lipase (ATGL) in cardiac muscle is essential for the biological activation of PPARα but not for PPARβ/δ [23], revealing different PPAR activation ligands and pathways in the heart.

PPARs are also very important therapeutic targets, and specific synthetic ligands exist for the different isotypes. In fact, two classes of PPAR ligands are currently being applied in the treatment of metabolic disorders. The fibrates—the classical peroxisome proliferators—act as ligands for the PPARα isoform and are being prescribed for the treatment of hyperlipidemia. The second group includes the thiazolidinediones (TZDs), which are specific synthetic activators for PPARγ. TZDs have insulin sensitizing properties and are successfully being used in the treatment of type 2 diabetes. Several specific synthetic ligands for PPARβ/δ have also been developed, including L-165041, GW501516, and GW0742 [24,25]. However, clinical studies are required to determine the efficacy and safety of these compounds.

PPARs regulate gene expression through two different mechanisms—transactivation and transrepression (Figure 11.1). In order to be transcriptionally active, PPARs need to heterodimerize with the 9-*cis* retinoic acid receptor (RXR) [26–30]. PPAR/RXR heterodimers bind to specific response elements located in the 5′ end region of their target genes. Peroxisome proliferator-response elements (PPREs) are composed of the direct repetition of two half-sites of consensus sequence AGGTCA, spaced by one nucleotide. PPAR and RXR bind to the 5′ and 3′ half-sites of this element, respectively, and the 5′ flanking region of the PPRE contributes to the selectivity of binding of the different PPAR isotypes. In the absence of ligand, PPAR/RXR heterodimers recruit corepressors and associated histone deacetylases and chromatin-modifying enzymes thereby silencing gene expression by so-called active repression. Ligand binding induces a conformational change in PPAR/RXR complexes, releasing corepressors in exchange for coactivators. Ligand-activated complexes recruit the basal transcriptional machinery, resulting in enhanced gene expression [16].

Additionally, the regulation of gene transcription by PPARs extends beyond their ability to transactivate specific target genes in an agonist-dependent manner [31]. PPARs are also able to regulate gene expression independent of binding to PPREs. Most of the anti-inflammatory properties that have been ascribed to these transcription factors result from the ability of activated PPARs to repress the transcriptional activation of proinflammatory genes by other transcription factors, such as activator protein 1 (AP1), NF-κB, signal transducers and activators of transcription (STATs), and nuclear factor of activated T-cells (NFAT) [32,33], a process referred to as transrepression. It has been experimentally shown that all three PPAR isoforms are able to participate in the regulation of inflammatory responses. Depending on the affected tissue and on which PPAR isoforms are involved, these transcription factors can modulate the intensity, duration, and consequences of inflammatory events.

There are three main ways by which ligand-activated PPAR-RXR complexes can negatively regulate the activities of other transcription factors. The first mechanism involves the ability of the PPARs to compete for limiting amounts

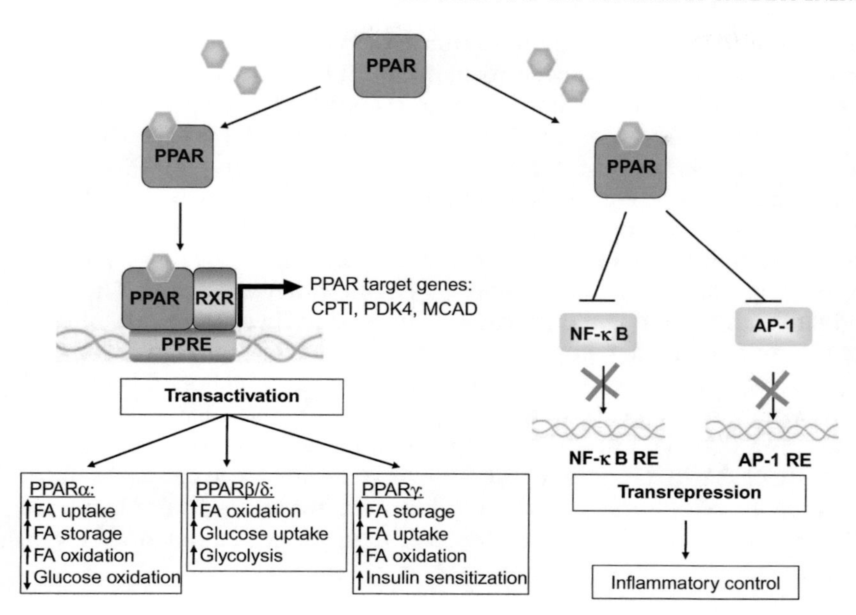

FIGURE 11.1 PPARs regulate gene expression through two mechanisms. The first of these mechanisms is DNA-dependent and is called transactivation. Initially PPARs need to heterodimerize with the 9-*cis* retinoic acid receptor (RXR) and then these heterodimers bind to DNA-specific sequences called peroxisome proliferator-response elements (PPREs), which are located in the promoter regions of genes involved in glucose and fatty acid metabolism. PPARs may also regulate gene expression independent of binding to DNA by blocking the activity of other transcription factors involved in inflammation, a mechanism called transrepression.

of coactivator proteins in a cell, making these coactivators unavailable to other transcription factors [34,35]. In the second mechanism, activated PPAR-RXR heterodimers are believed to act through physical interaction with other transcription factors. This association prevents the transcription factor from binding to its response element and thereby inhibits its ability to induce gene transcription [36]. The last transrepression mechanism relies on the ability of activated PPAR-RXR heterodimers to inhibit activation of certain members of the mitogen-activated protein kinase (MAPK) cascade. This prevents the MAPK from phosphorylating and activating downstream transcription factors [37].

Therefore, the biological and therapeutic activities of PPARs are the result of the combination of both transactivating and transrepressing properties of these receptors. In addition, post-translational modifications are important regulatory controls for all three PPAR isoforms. SUMOylation, ubiquitination (for a detailed review see Ref. [38]), and acetylation regulate transrepressive and transactivating activities of some nuclear receptors [39] and phosphorylation may inhibit transrepression by PPARα [40]. The ability of PPARs to regulate transcription is also a function of promoter architecture, ligand structure, cell type, and physiological and pathological conditions.

11.3 PPARs AND THE CONTROL OF CARDIAC ENERGY METABOLISM

PPARs regulate cardiac metabolism at the transcriptional level and play an important role during the development of cardiac disease [41,42]. The healthy adult myocardium uses fatty acid and glucose oxidation as its main energy sources, each covering approximately 65% and 30% of the energy needs, respectively [14]. However, in contrast to other tissues, the heart is a "promiscuous" substrate consumer. That is, it adapts its metabolism depending on the types of substrate that are available in order to maintain the constant pump function. For example, the fetal heart, which functions in a relatively hypoxic environment, derives energy from the catabolism of glucose and lactate. Immediately after birth, when the mammalian diet is composed almost entirely of lipids, the heart uses mainly fatty acids for myocardial energy production. Cardiac hypertrophy and the progressive development of cardiac failure are usually associated with suppression of fatty acid oxidation (FAO) and metabolic reversion to increase glucose utilization. This shift in the source of energy from fatty acids to glucose is accompanied by a dramatic fall in the cardiac expression of genes involved in fatty acid metabolism [43]. These genes involved in FAO are regulated primarily by the PPARs (Figure 11.1).

11.3.1 Role of PPARα

PPARα is highly expressed in tissues, such as the heart, with an elevated capacity to oxidize fatty acids. PPARα induces numerous genes critical for fatty acid handling including: fatty acid translocase/CD36, involved in fatty acid import into the cell; carnitine palmitoyltransferase I (CPT1), implicated in the import of fatty acids into mitochondria; medium-chain acyl-CoA dehydrogenase (MCAD), the rate-limiting enzyme in medium chain fatty acid β-oxidation; and pyruvate dehydrogenase kinase 4 (PDK4), the reciprocal inhibitor of pyruvate entry into mitochondria. Definitive evidence for PPARα as a key regulator of cardiac energy metabolism has been provided by the PPARα "knockout" mouse studies [44,45]. Hearts of these mice exhibit a decreased FAO and lower constitutive expression levels of FAO enzymes in the heart and a concomitant increase in glucose oxidation [46,47]. Despite these metabolic derangements, cardiac function is maintained in unstressed adult animals. However, a fasting stress causes pronounced hypoglycemia and hepatic and cardiac triglyceride accumulation in PPARα$^{-/-}$ mice [48]. Conversely, mice with heart-restricted overexpression of PPARα (MHC-PPARα mice) exhibit cardiac hypertrophy and ventricular dysfunction, a phenotype mimicking diabetic cardiomyopathy [49,50]. Moreover, metabolic studies showed that FAO is increased, whereas glucose utilization is reciprocally decreased in these mice. Despite an increase in myocardial FAO rates, hearts from these animals develop accumulation of neutral lipids in the fasted state and when placed on a high-fat diet, suggesting a mismatch between the uptake of fatty acids and the FAO capacity. Consistent with this, crossing of mice overexpressing PPARα with mice lacking cardiac CD36 rescued the cardiac toxicity of PPARα overexpression indicating that the intracellular lipid accumulation is toxic for cardiac function [51]. Collectively, these results indicate that PPARα regulates expression of genes involved in cardiac fatty acid and glucose metabolism in response to diverse physiological stimuli in order to maintain a tight lipid balance and that both infra- and supra-physiological activation of PPARα is maladaptive, suggesting that balanced levels of PPARα are needed to maintain cardiac function.

11.3.2 Role of PPARβ/δ

This PPAR subtype has also been relatively well studied in the heart. Gilde *et al.* [52] demonstrated that both PPARα and PPARβ/δ were expressed in comparable levels in the heart, whereas PPARγ was hardly detectable. It has been shown that PPARβ/δ selective ligands induce expression of mitochondrial FAO enzymes and increase palmitate oxidation rates in neonatal and adult cardiac myocytes [52,53]. Recently, it has been reported that in contrast to PPARα, mice with cardiac specific PPARβ/δ overexpression showed increased myocardial glucose utilization with no lipid accumulation and displayed normal cardiac function [44]. Indeed, it has been shown that constitutive cardiac overexpression of PPARβ/δ increases the myocardial oxidative mechanism, while myocardial glycogen content and the activity of AMPK are markedly reduced, therefore improving cardiac function [54]. Conversely, hearts from transgenic mice with heart-restricted PPARβ/δ deletion exhibit a considerable reduction of FAO capacity, increased cardiac lipid accumulation, and they develop lipotoxic cardiomyopathy [55]. Collectively, these results seem to indicate that PPARβ/δ is crucial for normal cardiac function and point out that PPARβ/δ, in contrast to PPARα, may be a therapeutic target for the treatment of diabetic cardiac dysfunction.

11.3.3 Role of PPARγ

The involvement of PPARγ in the heart remains controversial. It was generally accepted that PPARγ modulated fatty acid utilization through its effects on extra-cardiac tissues. Direct regulation of cardiac metabolism by PPARγ is still a matter of debate, although recent studies have attributed an important direct role of this subtype in cardiac metabolism. In contrast to PPARα and PPARβ/δ, PPARγ is only expressed in the heart in limited amounts, and PPARγ ligands do not affect metabolic gene expression or FAO rates in cultured cardiac myocytes [52]. Nevertheless, cardiomyocyte-specific deletion of PPARγ does result in cardiac hypertrophy, indicating an important role of this subtype of PPAR for the cardiac cells [56]. Furthermore, cardiac overexpression of PPARγ led to an increase in lipid uptake and oxidation in the heart, increased expression of glucose metabolism genes, and systolic dysfunction [57,58]. Recently, Luo *et al.* have shown that in cardiomyocyte-restricted PPARγ knockout mice, the levels of proteins involved in fatty acid uptake and oxidation are reduced in the heart, resulting in decreased fatty acid utilization [59]. Moreover, the hearts of these mice developed cardiac hypertrophy and heart failure. Collectively, all these data suggest that despite the low levels of PPARγ in the cardiac tissue, this PPAR subtype is also required for basal myocardial fatty acid utilization in the adult heart. By contrast, a recent study in humans associates the treatment with the PPARγ activator TZDs to the intramyocardial lipid accumulation [60] in accordance with previous studies showing prohypertrophic effects of the TZDs [61,62]. Although fluid retention may be the major contributor to the negative impact of TZDs on the failing heart, further studies are needed to clarify the role of PPARγ in this organ.

Taken together, all these data suggest that incorrect cardiac substrate utilization may be involved in the development of cardiac hypertrophy and failure [63] and that each of the three PPAR isoforms appear to be involved in cardiac disease.

11.4 PPARs AND CARDIAC INFLAMMATION

Despite their metabolic actions, PPARs are also able to modulate inflammatory responses in the heart that have been associated with the progressive pathological development of heart failure. Heart failure patients present elevated plasma levels of proinflammatory cytokines such as TNFα, MCP-1, and IL-6 [64] suggesting that chronic inflammation could play an important role during the development of cardiac hypertrophy. Proinflammatory cytokine expression is under control of the NF-κB transcription factor, which in fact is activated by hypertrophic and inflammatory stimuli, thus connecting both pathways.

NF-κB is a pleiotropic transcription factor involved in the regulation of diverse biological phenomena, including apoptosis, cell survival, cell growth, cell division, cellular differentiation, and the cellular responses to stress. Traditionally, the NF-κB pathway has been implicated as a pivotal intracellular mediator of the inflammatory response. Several studies have also implicated NF-κB activation as a causal event in the cardiac hypertrophic response, as modeled in cultured cardiac myocytes. NF-κB inhibition is able to block or attenuate the hypertrophic response either *in vitro* [4,11–13] or *in vivo* [65]. The NF-κB family consists of the members' p50, p52, p65 (RelA), c-Rel, and RelB, which form various homo- and heterodimers. Each possesses a ≈300-residue N-terminal Rel-homology-domain, responsible for dimerization, nuclear translocation, and DNA binding. p65, RelB, and c-Rel also contain a C-terminal transactivation domain. The most common active form is the p50-p65 dimer. Binding of most NF-κB complexes to gene promoter sequences supports transcription [66]. In quiescent cells, NF-κB dimers reside in the cytoplasm in an inactive form bound to inhibitory proteins known as IκB. At least six IκB proteins are involved in controlling the activity of the NF-κB dimer. These inhibitory proteins sequester NF-κB in the cytoplasm preventing its nuclear translocation. However, in addition to retaining NF-κB in the cytoplasm, IκBα also seems to be involved in the removal of NF-κB proteins from the nucleus [67]. Thus, IκBα has both cytoplasmatic and nuclear roles in regulating the NF-κB pathway. Stimulation of cells with a variety of inducers such as lipopolysaccharide (LPS), the inflammatory cytokines tumor necrosis factor-α (TNFα), or interleukin-1 (IL-1), leads to the phosphorylation of IκB by the IκB kinase (IKK) complex [68]. IKK activity resides in a large protein complex comprising two catalytic subunits, IKKα and IKKβ, and a scaffolding subunit, IKKγ/NEMO. The phosphorylation of IκB proteins is followed by the binding of the ubiquitin ligase complex which polyubiquitinylates IκB and targets it for degradation by the 26S proteasome [69]. This event releases the NF-κB heterodimer which then translocates to the nucleus and regulates the expression of genes involved in inflammatory and immune processes (e.g., TNFα, IκBα, MCP-1, or IL-6) [70].

There is a growing body of evidence demonstrating the effectiveness of PPAR agonists as anti-inflammatory agents. For instance, it has been reported that both PPARα and PPARγ ligands inhibit cardiac expression of the inflammatory cytokine TNFα and attenuate the activation of NF-κB induced by LPS or mechanical load [42,71]. Moreover it has been shown that PPARβ/δ may exert an anti-inflammatory effect through repressing TNFα transcription in response to inflammatory stimuli in cardiomyocytes [72]. In the next section, we will review recent advances on the anti-inflammatory properties of PPARs during the cardiac hypertrophy development, independently of their well-characterized metabolic functions.

11.4.1 PPARα

It has been shown that the PPARα activator fenofibrate suppresses NF-κB activity in the heart of rats with cardiac hypertrophy [73,74]. The anti-inflammatory effect of PPARα was further supported by studies using PPARα-null mice. Hearts from PPARα-null mice subjected to TAC-induced cardiac hypertrophy, differentially expressed gene clusters related to inflammatory signaling pathways [75]. Moreover, PPARα-null mice exhibited increased mRNA levels of inflammatory markers after hypertrophy induction, which was associated with an enhanced hypertrophic response [76]. Furthermore, in the context of cardiac hypertrophy, the mRNA levels of NF-κB target genes were decreased in cardiomyocytes after activation with a PPARα synthetic ligand or with PPARα-adenoviral overexpression [77]. These data point out a pivotal role for PPARα in limiting the inflammatory response by transrepression of NF-κB thus preventing the development of cardiac hypertrophy and heart failure.

11.4.2 PPARβ/δ

The anti-inflammatory properties of PPARβ/δ have been studied in recent years. Studies *in vitro* using neonatal cardiomyocytes have shown inhibition of LPS-induced TNFα expression after adenoviral-mediated overexpression of PPARβ/δ [72]. Moreover, it has been shown that PPARβ/δ is able to mitigate cardiomyocyte hypertrophy *in vitro* by inhibiting NF-κB activation [77]. In accordance, pharmacological activation of PPARβ/δ prevented the inflammatory responses in the heart after high-fat diet conditions, thus confirming the anti-inflammatory properties of PPARβ/δ [78]. Studies using PPARβ/δ-null mice have definitively identified PPARβ/δ as a critical factor in the control of proinflammatory pathways in the heart [72,78]. Lack of PPARβ/δ further exaggerated LPS and high-fat diet-induced proinflammatory cytokine production in the heart. A recent study has demonstrated protective effects of PPARβ/δ agonist against myocardial infarction associated with suppression of proinflammatory cytokines and neutrophil accumulation [79]. Collectively, these studies indicate that PPARβ/δ activators may represent a promising avenue for the treatment of cardiac disorders involving an inflammatory response.

11.4.3 PPARγ

Besides the metabolic effects of PPARγ, activation of this PPAR subtype has been associated with potent anti-inflammatory responses in the heart. In agreement with this, PPARγ activation negatively regulated both inflammatory NF-κB and AP1 transcription factors in angiotensin II-induced cardiomyocyte hypertrophy *in vitro*, and in spontaneously hypertensive rats [41,74]. In addition, PPARγ agonists inhibit cardiomyocyte inflammatory responses in culture [72,80]. Yamamoto *et al.* nicely demonstrated that activation of PPARγ by its ligands TZDs or the putative natural ligand 15d-PGJ_2 attenuated hypertrophy, most likely via the NF-κB pathway [42]. The pivotal role of PPARγ as an anti-inflammatory agent was further demonstrated using cardiac-specific $PPAR\gamma^{-/-}$ mice that spontaneously developed cardiac hypertrophy accompanied with an increased activation of the NF-κB pathway [56].

Collectively, these findings suggest that PPARs may serve as a therapeutic target to attenuate the inflammation that is involved in either cardiac pathological progression or hypertrophy.

11.5 CROSS TALK BETWEEN CARDIAC METABOLISM AND INFLAMMATION

An increasing body of evidence suggests a potential link between chronic inflammation and metabolic dysregulation during the development of heart failure. For instance, a clear relationship between NF-κB activation and the fall in FAO during the development of cardiac hypertrophy has been demonstrated [81]. However, the underlying mechanisms linking inflammation and heart failure are complex, since they are coupled to metabolic abnormalities and changes in cardiac phenotype.

11.5.1 Modulation of Inflammation/Metabolism by PPARs

It has been shown that either PPARα or PPARβ/δ is reduced in a rat model of cardiac hypertrophy [82]. Therefore, during the development of cardiac hypertrophy, the fall in the expression of both PPAR subtypes may be necessary to downregulate the expression of genes involved in fatty acid metabolism. Interestingly, the hypertrophy-associated changes in the expression of genes involved in fatty acid metabolism were abrogated when NF-κB was inhibited [81]. All these data pointed out a link between inflammation and metabolism during the development of cardiac hypertrophy. Moreover, in the same study these authors showed that the NF-κB signaling pathway, which plays a pivotal role in the hypertrophic growth of the myocardium, downregulates PPARβ/δ activity *in vitro* and *in vivo* through a mechanism that involves enhanced protein-protein interaction of the p65 subunit of NF-κB with this PPAR subtype. This association prevents PPARβ/δ from binding to its response element and thereby inhibits its ability to induce gene transcription, leading to a reduction in the expression of PDK-4. These findings are in agreement with the results reported by Westergaard *et al.* [83], who showed that PPARβ/δ physically interacts with p65 in psoriatic lesions. Transfection experiments using neonatal cardiomyocytes in cell culture showed that p65 inhibited PPARα and PPARβ/δ activity and vice versa, therefore confirming the interaction between both pathways [77]. Moreover, it has been reported that treatment with the PPARβ/δ agonist L-165041 prevented the reduction in the transcript level of genes involved in lipid metabolism, including CPT-1 and PDK-4 [84]. The PPARβ/δ ligand also inhibited PE-induced expression of the NF-κB-target gene MCP-1 (Monocyte Chemoattractant Protein 1), suggesting that the anti-hypertrophic effect of this compound involved downregulation of the NF-κB signaling pathway. Enhanced myocardial MCP-1 has been described in the hypertrophied and failing heart [85] and may lead to the infiltration and activation of inflammatory cells, such as monocytes/macrophages and lymphocytes. In addition, it has been reported that activation of MCP-1 expression contributes to left ventricular remodeling and failure after myocardial infarction [86]. Further, it was shown that L-165041 inhibited LPS-induced NF-κB activation through enhanced physical interaction of PPARβ/δ with the p65 subunit of NF-κB in H9c2 cells [84]. These data suggest that the beneficial effects of PPARβ/δ activation on fatty acid metabolism during the development of cardiac hypertrophy could be indirect through downregulation of the NF-κB signaling pathway. The involvement of PPARβ/δ linking inflammation and fatty acid metabolism was further confirmed in adult cardiomyocytes [87,88]. In these studies it was demonstrated that prolonged exposure of cardiomyocytes to angiotensin-II reduced FAO genes that can be prevented by PPARβ/δ activation or NF-κB inhibition. Finally, it has been demonstrated that the PPARα ligand fenofibrate increases the association between PPARα and NFATc4 [89]. This association decreases the binding NFATc4-NF-κB involved in the development of inflammation and cardiac hypertrophy, linking metabolic and inflammatory pathways through NFATc4.

Collectively, these findings indicate that pharmacological PPAR activation might be a therapeutic approach for cardiac disorders like heart failure, given their ability to promote myocardial FAO capacity and to reduce proinflammatory signaling.

11.5.2 Sirt1 Couples Inflammation and Metabolism in the Heart

The deacetylase sirtuin-1 (Sirt1) has emerged as a key protein linking metabolism and inflammation [90]. Besides its ability to inhibit the NF-κB signaling pathway [91–93], Sirt1 may associate with and deacetylate PGC-1α, leading to enhancement of its transcriptional activity [94,95]. Several studies have attributed an important role of Sirt1 in the heart [96]. The complete loss of Sirt1 through targeted ablation of the Sirt1 gene in mice leads to a substantial increase in the rate of perinatal mortality in association with cardiac malformations, whereas heterozygous Sirt1$^{+/-}$ mice survive to adulthood [97]. Experimental Sirt1 overexpression in the heart has revealed complex effects as both protective and damaging roles of Sirt1 in relation to cardiac hypertrophy and fibrosis have been found depending on the extent of Sirt1 overexpression [96]. Moreover, Sirt1 protects against hypertrophy and ischemia-reperfusion injury in the heart [96,98]. Furthermore, it has been shown that Sirt1 prevents hypertrophy and the associated fatty acid dysregulation in cardiomyocytes in culture in response to the hypertrophic stimulus phenylephrine [99]. Accordingly, in this study, Sirt1 overexpression led to enhanced PPARα binding to the p65 subunit of NF-κB and subsequent p65-deacetylation, thus blocking NF-κB activity. Consistent with this, isoproterenol-induced cardiac hypertrophy, metabolic dysregulation, and inflammation were prevented by the Sirt1 activator resveratrol in wild-type but not in PPARα-null mice. By contrast, Sirt1 deficiency prevents the development of cardiac hypertrophy after TAC-induced cardiac hypertrophy in mice [100,101] and leads to dilated cardiomyopathy without cardiac cell growth in 6-month old mice [102]. All these data point out a crucial role for Sirt1 in linking metabolism and inflammation with the development of cardiac disorders.

Recent findings suggest that the complex effects of Sirt1 on the heart can be explained by the transcription factor estrogen-related receptor (ERR)-α [100] as Sirt1 was found to regulate the ERR transcriptional pathway during the progression of heart failure. During pressure overload-induced cardiac hypertrophy and heart failure, PPARα was found to bind and recruit Sirt1 to the ERR response element, thereby suppressing the transcription of ERR target genes involved in mitochondrial function. Lack of Sirt1 or PPARα avoided the repression of the ERR genes and in this way prevented the development of heart failure. These data suggest that the extent of Sirt1 expression is crucial for maintaining appropriate heart function (see Figure 11.2). However, further research is needed to clearly elucidate

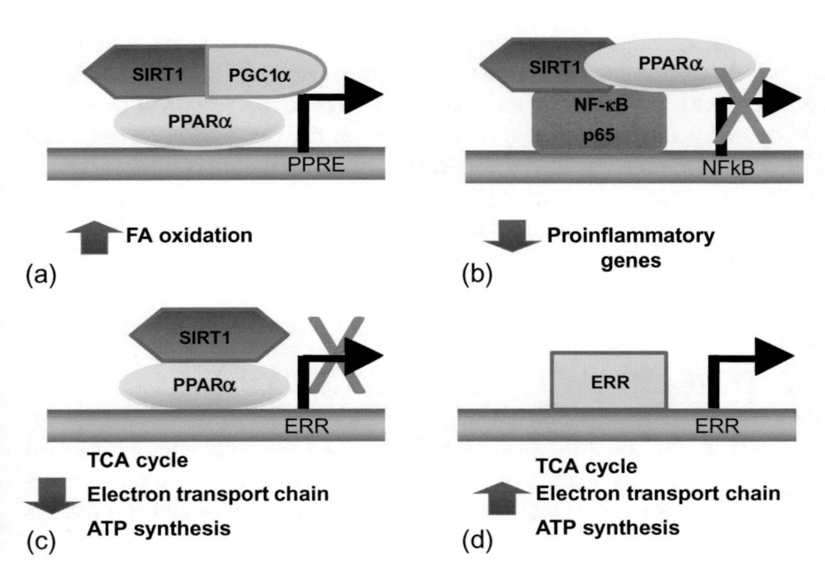

FIGURE 11.2 The amount of Sirt1 determines the response of cardiac cells to hypertrophy stimuli. Moderate levels of Sirt1 protect against cardiac hypertrophy by potentiating fatty acid oxidation gene expression (a) and repress proinflammatory pathways in the heart (b). At high levels of Sirt1, the formation of a complex between Sirt1 and PPARα leads to cardiac hypertrophy and failure due to the repressive effect exerted on ERR target genes mainly involved in energy production (c). By contrast, low levels of Sirt1 attenuate the development of cardiac hypertrophy probably as the expression of ERR-dependent genes, many of which are associated with mitochondrial function, is not repressed (d).

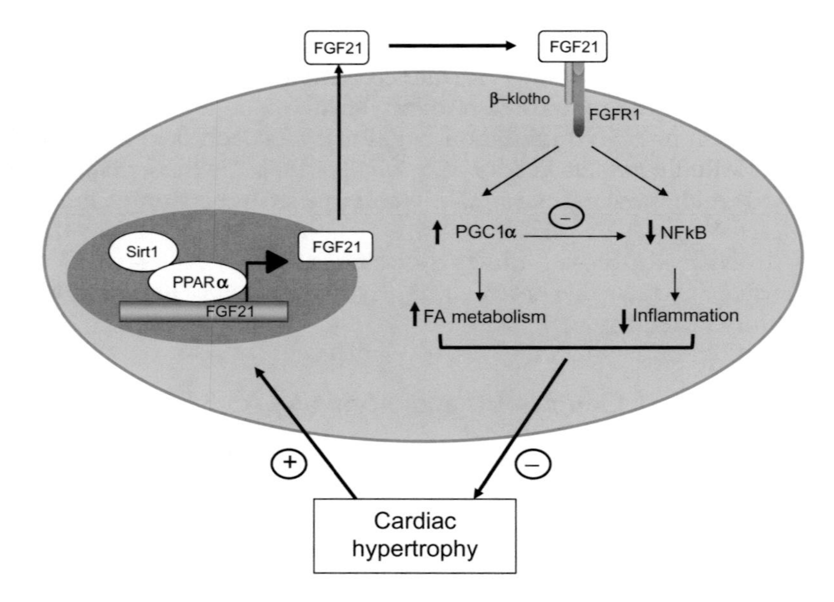

FIGURE 11.3 Schematic representation of the transcriptional regulation and autocrine function of FGF21 in the heart. In response to hypertrophy stimuli and under control of the Sirt1-PPARα pathway, the heart produces FGF21 which in turn acts on cardiac cells preventing metabolic dysregulation and inflammation. Therefore, the heart is a target of systemic and locally generated FGF21 as well as a potential source of FGF21.

the interactions between Sirt1, PPARα, and ERR-α, on the one hand, and their effect on cardiac metabolism and inflammation during the development of heart failure, on the other.

11.5.3 The FGF21 Autocrine Pathway

Several lines of evidence indicate that communication among cardiac cells via secreted factors may contribute to myocardial hypertrophic remodeling [103,104]. Recently, the term cardiomyokine has emerged to describe proteins secreted from the heart that have autocrine, paracrine, and/or endocrine functions crucial for the maintenance of cardiac function [105]. The number of cardiomyokines has been estimated at between 30 and 60 and includes growth factors, endocrine hormones, and cytokines [106,107].

FGF21 is a secreted protein that acts as a metabolic regulator and is involved in the control of glucose homeostasis, insulin sensitivity, and ketogenesis [108,109]. FGF21 expression is under the control of PPARα, and the liver is considered the main site of production and release of FGF21 into the blood [108,110]. Extra-hepatic tissues, such as white and brown adipose tissues and skeletal muscle, also express FGF21 [111–113]. Endocrine actions of FGF21 involve the promotion of glucose uptake by white adipocytes through induction of the glucose transporter Glut1 [114] and activation of brown fat thermogenic activity [115]. FGF21 also has autocrine/paracrine effects, such as induction of hepatic ketogenesis [108]. The action of FGF21 on target cells requires FGF receptors (mainly FGFR1 and FGFR4 in adipose tissue and liver, respectively) and β-Klotho, a single-pass transmembrane protein that functions as an obligate cofactor for FGF21 signaling [116,117]. In a recent study we demonstrated that FGF21 has an important role in cardiac remodeling [118]. We showed that in response to isoproterenol infusion, cardiac hypertrophy was more enhanced in FGF21 knockout mice. Furthermore, FGF21 treatment reversed cardiac hypertrophy development, enhanced FAO, and prevented the induction of proinflammatory pathways in the heart. Moreover, it was shown that FGF21 is secreted by the cardiac cells in response to cardiac stress and activated through the Sirt1-PPARα pathway and that FGF21 secretion was able to inhibit cardiac damage. This study demonstrated that via the Sirt1-PPARα pathway the heart locally generates FGF21 which acts in an autocrine manner preventing hypertrophy, metabolic dysregulation, and the activation of proinflammatory pathways in cardiac tissue (Figure 11.3). Collectively, this study describes a new mechanism through which the Sirt1-PPARα pathway controls inflammation and metabolism by locally producing FGF21 in the heart.

11.6 PPAR AGONISTS AND HEART FAILURE TREATMENT

Despite improvements in treating cardiovascular pathologic conditions, such as hypertension, hyperlipidemia and acute coronary syndromes, the prevalence of heart failure continues to grow as the population ages. Mortality rates remain high and approach 65% at 5 years after myocardial infarction [119,120]. PPAR agonists, because of their ability to increase fatty acid metabolism and their anti-inflammatory properties, show promise of targeting maladaptive pathways currently not addressed by the standard therapies to treat heart failure at early stages. As previously described in this chapter, animal studies have shown potential benefits of PPAR agonists to treat cardiac disorders. However, to date, clinical use and development of PPAR ligands has been limited to treatment of metabolic diseases and not primary cardiac disorders, and their success in treating even cardiovascular disease has been dismal.

PPARα ligands have not been studied specifically in clinical trials for human heart failure. The two commercially available PPARα ligands, the fibrates gemfibrozil and fenofibrate, are not contraindicated for heart failure patients and both have demonstrated excellent cardiac safety profiles. Two large studies showed in the 1990s a decrease in death from coronary heart disease in patients treated with gemfibrozil [121,122]. In addition, fibrates display many pleiotropic properties that are likely beneficial in heart failure patients [123]. However, further studies to specifically validate the efficacy of fibrates in heart failure patients have not been performed so far.

Unlike fibrates, the PPARγ ligands such as rosiglitazone or pioglitazone show adverse side effects as they have been found to increase fluid retention. For this reason they are contraindicated for patients with heart failure symptoms [124,125]. However, a retrospective study of diabetic patients with heart failure showed no increase hospitalizations for heart failure or death for those patients taking either rosiglitazone or pioglitazone [126]. Nevertheless, the clinical experience with TZDs and their propensity for fluid retention have tempered enthusiasm for further development of PPARγ ligands specifically for heart failure treatment.

11.7 CONCLUSIONS AND PERSPECTIVES

Although a considerable effort has been devoted to improve therapy, heart failure remains a critical health problem; thus, the identification of underlying molecular targets is crucially important for improving the efficacy of therapeutic strategies. Experimental data suggest that PPARs are involved in the pathophysiology of cardiac hypertrophy and failure by modulation of crucial cellular processes. Each of the three PPAR isoforms has the capacity to modulate both energy metabolism and inflammation. A growing body of evidence indicates that these two seemingly unrelated processes are closely interconnected. The mechanism of action of PPARs provides at least one clue as to how inflammation and metabolism are intertwined. Therefore, the study of the mechanisms controlling PPAR function to fully understand how PPARs work and control inflammation and metabolism in the heart, represents a major challenge for the future in terms of therapeutic applications. Moreover, better understanding of tissue distribution of PPARs and adverse effects of current PPAR agonists should allow for the design of safer compounds to treat cardiac disease in the near future. Although the past experiences with PPAR agonists have not been successful, the future for new categories of PPAR ligands for the treatment of heart failure is far from futile. In summary, preclinical research has clearly demonstrated that PPARs play an important role in the development of cardiac hypertrophy and heart failure. However, further studies are needed to clarify their exact mechanism of action and whether the PPARs prove to be useful pharmacological targets for the treatment of heart failure in the clinical setting.

References

[1] Linseman JV, Bristow MR. Drug therapy and heart failure prevention. Circulation 2003;107:1234–6.
[2] Yndestad A, Damas JK, Oie E, Ueland T, Gullestad L, Aukrust P. Systemic inflammation in heart failure—the whys and wherefores. Heart Fail Rev 2006;11:83–92.
[3] Thaik CM, Calderone A, Takahashi N, Colucci WS. Interleukin-1 beta modulates the growth and phenotype of neonatal rat cardiac myocytes. J Clin Invest 1995;96:1093–9.
[4] Purcell NH, Tang G, Yu C, Mercurio F, DiDonato JA, Lin A. Activation of NF-kappa B is required for hypertrophic growth of primary rat neonatal ventricular cardiomyocytes. Proc Natl Acad Sci U S A 2001;98:6668–73.
[5] Plutzky J. Expansion and contraction: the mighty, mighty fatty acid. Nat Med 2009;15:618–9.
[6] Frey N, Katus HA, Olson EN, Hill JA. Hypertrophy of the heart: a new therapeutic target? Circulation 2004;109:1580–9.
[7] Devereux RB, Roman MJ. Left ventricular hypertrophy in hypertension: stimuli, patterns, and consequences. Hypertens Res 1999;22:1–9.
[8] Molkentin JD, Dorn II GW. Cytoplasmic signaling pathways that regulate cardiac hypertrophy. Annu Rev Physiol 2001;63:391–426.
[9] Dahlof B, Hansson L. Regression of left ventricular hypertrophy in previously untreated essential hypertension: different effects of enalapril and hydrochlorothiazide. J Hypertens 1992;10:1513–24.
[10] Olson EN, Schneider MD. Sizing up the heart: development redux in disease. Genes Dev 2003;17:1937–56.
[11] Hirotani S, Otsu K, Nishida K, Higuchi Y, Morita T, Nakayama H, et al. Involvement of nuclear factor-kappaB and apoptosis signal-regulating kinase 1 in G-protein-coupled receptor agonist-induced cardiomyocyte hypertrophy. Circulation 2002;105:509–15.
[12] Higuchi Y, Otsu K, Nishida K, Hirotani S, Nakayama H, Yamaguchi O, et al. Involvement of reactive oxygen species-mediated NF-kappa B activation in TNF-alpha-induced cardiomyocyte hypertrophy. J Mol Cell Cardiol 2002;34:233–40.
[13] Gupta S, Purcell NH, Lin A, Sen S. Activation of nuclear factor-kappaB is necessary for myotrophin-induced cardiac hypertrophy. J Cell Biol 2002;159:1019–28.

[14] van Bilsen M, van der Vusse GJ, Reneman RS. Transcriptional regulation of metabolic processes: implications for cardiac metabolism. Pflugers Arch 1998;437:2–14.
[15] Kersten S, Desvergne B, Wahli W. Roles of PPARs in health and disease. Nature 2000;405:421–4.
[16] Chinetti G, Fruchart JC, Staels B. Peroxisome proliferator-activated receptors (PPARs): nuclear receptors at the crossroads between lipid metabolism and inflammation. Inflamm Res 2000;49:497–505.
[17] Palomer X, Salvado L, Barroso E, Vazquez-Carrera M. An overview of the crosstalk between inflammatory processes and metabolic dysregulation during diabetic cardiomyopathy. Int J Cardiol 2013;168:3160–72.
[18] Lockyer P, Schisler JC, Patterson C, Willis MS. Minireview: won't get fooled again: the nonmetabolic roles of peroxisome proliferator-activated receptors (PPARs) in the heart. Mol Endocrinol 2010;24:1111–9.
[19] Feige JN, Gelman L, Michalik L, Desvergne B, Wahli W. From molecular action to physiological outputs: peroxisome proliferator-activated receptors are nuclear receptors at the crossroads of key cellular functions. Prog Lipid Res 2006;45:120–59.
[20] Braissant O, Foufelle F, Scotto C, Dauca M, Wahli W. Differential expression of peroxisome proliferator-activated receptors (PPARs): tissue distribution of PPAR-alpha, -beta, and -gamma in the adult rat. Endocrinology 1996;137:354–66.
[21] Krey G, Braissant O, L'Horset F, Kalkhoven E, Perroud M, Parker MG, et al. Fatty acids, eicosanoids, and hypolipidemic agents identified as ligands of peroxisome proliferator-activated receptors by coactivator-dependent receptor ligand assay. Mol Endocrinol 1997;11:779–91.
[22] Berger J, Moller DE. The mechanisms of action of PPARs. Annu Rev Med 2002;53:409–35.
[23] Haemmerle G, Moustafa T, Woelkart G, Buttner S, Schmidt A, van de Weijer T, et al. ATGL-mediated fat catabolism regulates cardiac mitochondrial function via PPAR-alpha and PGC-1. Nat Med 2011;17:1076–85.
[24] Sznaidman ML, Haffner CD, Maloney PR, Fivush A, Chao E, Goreham D, et al. Novel selective small molecule agonists for peroxisome proliferator-activated receptor delta (PPARdelta)—synthesis and biological activity. Bioorg Med Chem Lett 2003;13:1517–21.
[25] Oliver Jr. WR, Shenk JL, Snaith MR, Russell CS, Plunket KD, Bodkin NL, et al. A selective peroxisome proliferator-activated receptor delta agonist promotes reverse cholesterol transport. Proc Natl Acad Sci U S A 2001;98:5306–11.
[26] Zamir I, Zhang J, Lazar MA. Stoichiometric and steric principles governing repression by nuclear hormone receptors. Genes Dev 1997;11:835–46.
[27] Dowell P, Ishmael JE, Avram D, Peterson VJ, Nevrivy DJ, Leid M. Identification of nuclear receptor corepressor as a peroxisome proliferator-activated receptor alpha interacting protein. J Biol Chem 1999;274:15901–7.
[28] DiRenzo J, Soderstrom M, Kurokawa R, Ogliastro MH, Ricote M, Ingrey S, et al. Peroxisome proliferator-activated receptors and retinoic acid receptors differentially control the interactions of retinoid X receptor heterodimers with ligands, coactivators, and corepressors. Mol Cell Biol 1997;17:2166–76.
[29] Zhu Y, Qi C, Jain S, Rao MS, Reddy JK. Isolation and characterization of PBP, a protein that interacts with peroxisome proliferator-activated receptor. J Biol Chem 1997;272:25500–6.
[30] Yuan CX, Ito M, Fondell JD, Fu ZY, Roeder RG. The TRAP220 component of a thyroid hormone receptor- associated protein (TRAP) coactivator complex interacts directly with nuclear receptors in a ligand-dependent fashion. Proc Natl Acad Sci U S A 1998;95:7939–44.
[31] Daynes RA, Jones DC. Emerging roles of PPARs in inflammation and immunity. Nat Rev Immunol 2002;2:748–59.
[32] Li M, Pascual G, Glass CK. Peroxisome proliferator-activated receptor gamma-dependent repression of the inducible nitric oxide synthase gene. Mol Cell Biol 2000;20:4699–707.
[33] Kamei Y, Xu L, Heinzel T, Torchia J, Kurokawa R, Gloss B, et al. A CBP integrator complex mediates transcriptional activation and AP-1 inhibition by nuclear receptors. Cell 1996;85:403–14.
[34] Delerive P, De BK, Besnard S, Vanden BW, Peters JM, Gonzalez FJ, et al. Peroxisome proliferator-activated receptor alpha negatively regulates the vascular inflammatory gene response by negative cross-talk with transcription factors NF-kappaB and AP-1. J Biol Chem 1999;274:32048–54.
[35] Delerive P, Martin-Nizard F, Chinetti G, Trottein F, Fruchart JC, Najib J, et al. Peroxisome proliferator-activated receptor activators inhibit thrombin-induced endothelin-1 production in human vascular endothelial cells by inhibiting the activator protein-1 signaling pathway. Circ Res 1999;85:394–402.
[36] Desreumaux P, Dubuquoy L, Nutten S, Peuchmaur M, Englaro W, Schoonjans K, et al. Attenuation of colon inflammation through activators of the retinoid X receptor (RXR)/peroxisome proliferator-activated receptor gamma (PPARgamma) heterodimer. A basis for new therapeutic strategies. J Exp Med 2001;193:827–38.
[37] Johnson TE, Holloway MK, Vogel R, Rutledge SJ, Perkins JJ, Rodan GA, et al. Structural requirements and cell-type specificity for ligand activation of peroxisome proliferator-activated receptors. J Steroid Biochem Mol Biol 1997;63:1–8.
[38] Wadosky KM, Willis MS. The story so far: post-translational regulation of peroxisome proliferator-activated receptors by ubiquitination and SUMOylation. Am J Physiol Heart Circ Physiol 2012;302:H515–26.
[39] Pascual G, Fong AL, Ogawa S, Gamliel A, Li AC, Perissi V, et al. A SUMOylation-dependent pathway mediates transrepression of inflammatory response genes by PPAR-gamma. Nature 2005;437:759–63.
[40] Blanquart C, Mansouri R, Paumelle R, Fruchart JC, Staels B, Glineur C. The protein kinase C signaling pathway regulates a molecular switch between transactivation and transrepression activity of the peroxisome proliferator-activated receptor alpha. Mol Endocrinol 2004;18:1906–18.
[41] Asakawa M, Takano H, Nagai T, Uozumi H, Hasegawa H, Kubota N, et al. Peroxisome proliferator-activated receptor gamma plays a critical role in inhibition of cardiac hypertrophy in vitro and in vivo. Circulation 2002;105:1240–6.
[42] Yamamoto K, Ohki R, Lee RT, Ikeda U, Shimada K. Peroxisome proliferator-activated receptor gamma activators inhibit cardiac hypertrophy in cardiac myocytes. Circulation 2001;104:1670–5.
[43] Barger PM, Brandt JM, Leone TC, Weinheimer CJ, Kelly DP. Deactivation of peroxisome proliferator-activated receptor-alpha during cardiac hypertrophic growth. J Clin Invest 2000;105:1723–30.
[44] Burkart EM, Sambandam N, Han X, Gross RW, Courtois M, Gierasch CM, et al. Nuclear receptors PPARbeta/delta and PPARalpha direct distinct metabolic regulatory programs in the mouse heart. J Clin Invest 2007;117:3930–9.

[45] Duncan JG, Bharadwaj KG, Fong JL, Mitra R, Sambandam N, Courtois MR, et al. Rescue of cardiomyopathy in peroxisome proliferator-activated receptor-alpha transgenic mice by deletion of lipoprotein lipase identifies sources of cardiac lipids and peroxisome proliferator-activated receptor-alpha activators. Circulation 2010;121:426–35.
[46] Campbell FM, Kozak R, Wagner A, Altarejos JY, Dyck JR, Belke DD, et al. A role for peroxisome proliferator-activated receptor alpha (PPARalpha) in the control of cardiac malonyl-CoA levels: reduced fatty acid oxidation rates and increased glucose oxidation rates in the hearts of mice lacking PPARalpha are associated with higher concentrations of malonyl-CoA and reduced expression of malonyl-CoA decarboxylase. J Biol Chem 2002;277:4098–103.
[47] Watanabe K, Fujii H, Takahashi T, Kodama M, Aizawa Y, Ohta Y, et al. Constitutive regulation of cardiac fatty acid metabolism through peroxisome proliferator-activated receptor alpha associated with age-dependent cardiac toxicity. J Biol Chem 2000;275:22293–9.
[48] Leone TC, Weinheimer CJ, Kelly DP. A critical role for the peroxisome proliferator-activated receptor alpha (PPARalpha) in the cellular fasting response: the PPARalpha-null mouse as a model of fatty acid oxidation disorders. Proc Natl Acad Sci U S A 1999;96:7473–8.
[49] Finck BN, Lehman JJ, Leone TC, Welch MJ, Bennett MJ, Kovacs A, et al. The cardiac phenotype induced by PPARalpha overexpression mimics that caused by diabetes mellitus. J Clin Invest 2002;109:121–30.
[50] Finck BN, Han X, Courtois M, Aimond F, Nerbonne JM, Kovacs A, et al. A critical role for PPARalpha-mediated lipotoxicity in the pathogenesis of diabetic cardiomyopathy: modulation by dietary fat content. Proc Natl Acad Sci U S A 2003;100:1226–31.
[51] Yang J, Sambandam N, Han X, Gross RW, Courtois M, Kovacs A, et al. CD36 deficiency rescues lipotoxic cardiomyopathy. Circ Res 2007;100:1208–17.
[52] Gilde AJ, van der Lee KA, Willemsen PH, Chinetti G, van der Leij FR, van der Vusse GJ, et al. Peroxisome proliferator-activated receptor (PPAR) alpha and PPARbeta/delta, but not PPARgamma, modulate the expression of genes involved in cardiac lipid metabolism. Circ Res 2003;92:518–24.
[53] Cheng L, Ding G, Qin Q, Xiao Y, Woods D, Chen YE, et al. Peroxisome proliferator-activated receptor delta activates fatty acid oxidation in cultured neonatal and adult cardiomyocytes. Biochem Biophys Res Commun 2004;313:277–86.
[54] Liu J, Wang P, Luo J, Huang Y, He L, Yang H, et al. Peroxisome proliferator-activated receptor beta/delta activation in adult hearts facilitates mitochondrial function and cardiac performance under pressure-overload condition. Hypertension 2011;57:223–30.
[55] Cheng L, Ding G, Qin Q, Huang Y, Lewis W, He N, et al. Cardiomyocyte-restricted peroxisome proliferator-activated receptor-delta deletion perturbs myocardial fatty acid oxidation and leads to cardiomyopathy. Nat Med 2004;10:1245–50.
[56] Duan SZ, Ivashchenko CY, Russell MW, Milstone DS, Mortensen RM. Cardiomyocyte-specific knockout and agonist of peroxisome proliferator-activated receptor-gamma both induce cardiac hypertrophy in mice. Circ Res 2005;97:372–9.
[57] Son NH, Park TS, Yamashita H, Yokoyama M, Huggins LA, Okajima K, et al. Cardiomyocyte expression of PPARgamma leads to cardiac dysfunction in mice. J Clin Invest 2007;117:2791–801.
[58] Son NH, Yu S, Tuinei J, Arai K, Hamai H, Homma S, et al. PPARgamma-induced cardiolipotoxicity in mice is ameliorated by PPARalpha deficiency despite increases in fatty acid oxidation. J Clin Invest 2010;120:3443–54.
[59] Luo J, Wu S, Liu J, Li Y, Yang H, Kim T, et al. Conditional PPARgamma knockout from cardiomyocytes of adult mice impairs myocardial fatty acid utilization and cardiac function. Am J Transl Res 2010;3:61–72.
[60] Marfella R, Portoghese M, Ferraraccio F, Siniscalchi M, Babieri M, Di FC, et al. Thiazolidinediones may contribute to the intramyocardial lipid accumulation in diabetic myocardium: effects on cardiac function. Heart 2009;95:1020–2.
[61] Edgley AJ, Thalen PG, Dahllof B, Lanne B, Ljung B, Oakes ND. PPARgamma agonist induced cardiac enlargement is associated with reduced fatty acid and increased glucose utilization in myocardium of Wistar rats. Eur J Pharmacol 2006;538:195–206.
[62] Zhang H, Zhang A, Kohan DE, Nelson RD, Gonzalez FJ, Yang T. Collecting duct-specific deletion of peroxisome proliferator-activated receptor gamma blocks thiazolidinedione-induced fluid retention. Proc Natl Acad Sci U S A 2005;102:9406–11.
[63] Frey N, Olson EN. Modulating cardiac hypertrophy by manipulating myocardial lipid metabolism? Circulation 2002;105:1152–4.
[64] Ancey C, Menet E, Corbi P, Fredj S, Garcia M, Rucker-Martin C, et al. Human cardiomyocyte hypertrophy induced in vitro by gp130 stimulation. Cardiovasc Res 2003;59:78–85.
[65] Li Y, Ha T, Gao X, Kelley J, Williams DL, Browder IW, et al. NF-kappaB activation is required for the development of cardiac hypertrophy in vivo. Am J Physiol Heart Circ Physiol 2004;287:H1712–20.
[66] Moynagh PN. The NF-kappaB pathway. J Cell Sci 2005;118:4589–92.
[67] Tam WF, Lee LH, Davis L, Sen R. Cytoplasmic sequestration of rel proteins by IkappaBalpha requires CRM1-dependent nuclear export. Mol Cell Biol 2000;20:2269–84.
[68] Yamamoto Y, Gaynor RB. IkappaB kinases: key regulators of the NF-kappaB pathway. Trends Biochem Sci 2004;29:72–9.
[69] Karin M, Ben-Neriah Y. Phosphorylation meets ubiquitination: the control of NF-[kappa]B activity. Annu Rev Immunol 2000;18:621–63.
[70] Karin M. The beginning of the end: IkappaB kinase (IKK) and NF-kappaB activation. J Biol Chem 1999;274:27339–42.
[71] Takano H, Nagai T, Asakawa M, Toyozaki T, Oka T, Komuro I, et al. Peroxisome proliferator-activated receptor activators inhibit lipopolysaccharide-induced tumor necrosis factor-alpha expression in neonatal rat cardiac myocytes. Circ Res 2000;87:596–602.
[72] Ding G, Cheng L, Qin Q, Frontin S, Yang Q. PPARdelta modulates lipopolysaccharide-induced TNFalpha inflammation signaling in cultured cardiomyocytes. J Mol Cell Cardiol 2006;40:821–8.
[73] Maruyama S, Kato K, Kodama M, Hirono S, Fuse K, Nakagawa O, et al. Fenofibrate, a peroxisome proliferator-activated receptor alpha activator, suppresses experimental autoimmune myocarditis by stimulating the interleukin-10 pathway in rats. J Atheroscler Thromb 2002;9:87–92.
[74] Diep QN, Benkirane K, Amiri F, Cohn JS, Endemann D, Schiffrin EL. PPAR alpha activator fenofibrate inhibits myocardial inflammation and fibrosis in angiotensin II-infused rats. J Mol Cell Cardiol 2004;36:295–304.
[75] Smeets PJ, de Vogel-van den Bosch HM, Willemsen PH, Stassen AP, Ayoubi T, van der Vusse GJ, et al. Transcriptomic analysis of PPARalpha-dependent alterations during cardiac hypertrophy. Physiol Genomics 2008;36:15–23.
[76] Smeets PJ, Teunissen BE, Willemsen PH, van Nieuwenhoven FA, Brouns AE, Janssen BJ, et al. Cardiac hypertrophy is enhanced in PPAR alpha-/- mice in response to chronic pressure overload. Cardiovasc Res 2008;78:79–89.

[77] Smeets PJ, Teunissen BE, Planavila A, de Vogel-van den Bosch H, Willemsen PH, van der Vusse GJ, et al. Inflammatory pathways are activated during cardiomyocyte hypertrophy and attenuated by peroxisome proliferator-activated receptors PPARalpha and PPARdelta. J Biol Chem 2008;283:29109–18.
[78] Alvarez-Guardia D, Palomer X, Coll T, Serrano L, Rodriguez-Calvo R, Davidson MM, et al. PPARbeta/delta activation blocks lipid-induced inflammatory pathways in mouse heart and human cardiac cells. Biochim Biophys Acta 2011;1811:59–67.
[79] Collino M, Benetti E, Miglio G, Castiglia S, Rosa AC, Aragno M, et al. Peroxisome proliferator-activated receptor beta/delta agonism protects the kidney against ischemia/reperfusion injury in diabetic rats. Free Radic Biol Med 2011;50:345–53.
[80] Takeishi Y, Kubota I. Role of toll-like receptor mediated signaling pathway in ischemic heart. Front Biosci (Landmark Ed) 2009;14:2553–8.
[81] Planavila A, Laguna JC, Vazquez-Carrera M. Nuclear factor-kappaB activation leads to down-regulation of fatty acid oxidation during cardiac hypertrophy. J Biol Chem 2005;280:17464–71.
[82] Planavila A, Laguna JC, Vazquez-Carrera M. Atorvastatin improves peroxisome proliferator-activated receptor signaling in cardiac hypertrophy by preventing nuclear factor-kappa B activation. Biochim Biophys Acta 2005;1687:76–83.
[83] Westergaard M, Henningsen J, Johansen C, Rasmussen S, Svendsen ML, Jensen UB, et al. Expression and localization of peroxisome proliferator-activated receptors and nuclear factor kappaB in normal and lesional psoriatic skin. J Invest Dermatol 2003;121:1104–17.
[84] Planavila A, Rodriguez-Calvo R, Jove M, Michalik L, Wahli W, Laguna JC, et al. Peroxisome proliferator-activated receptor beta/delta activation inhibits hypertrophy in neonatal rat cardiomyocytes. Cardiovasc Res 2005;65:832–41.
[85] Shioi T, Matsumori A, Kihara Y, Inoko M, Ono K, Iwanaga Y, et al. Increased expression of interleukin-1 beta and monocyte chemotactic and activating factor/monocyte chemoattractant protein-1 in the hypertrophied and failing heart with pressure overload. Circ Res 1997;81:664–71.
[86] Hayashidani S, Tsutsui H, Shiomi T, Ikeuchi M, Matsusaka H, Suematsu N, et al. Anti-monocyte chemoattractant protein-1 gene therapy attenuates left ventricular remodeling and failure after experimental myocardial infarction. Circulation 2003;108:2134–40.
[87] Pellieux C, Montessuit C, Papageorgiou I, Lerch R. Inactivation of peroxisome proliferator-activated receptor isoforms alpha, beta/delta, and gamma mediate distinct facets of hypertrophic transformation of adult cardiac myocytes. Pflugers Arch 2007;455:443–54.
[88] Pellieux C, Montessuit C, Papageorgiou I, Lerch R. Angiotensin II downregulates the fatty acid oxidation pathway in adult rat cardiomyocytes via release of tumour necrosis factor-alpha. Cardiovasc Res 2009;82:341–50.
[89] Zou J, Le K, Xu S, Chen J, Liu Z, Chao X, et al. Fenofibrate ameliorates cardiac hypertrophy by activation of peroxisome proliferator-activated receptor-alpha partly via preventing p65-NFkappaB binding to NFATc4. Mol Cell Endocrinol 2013;370:103–12.
[90] Pfluger PT, Herranz D, Velasco-Miguel S, Serrano M, Tschop MH. Sirt1 protects against high-fat diet-induced metabolic damage. Proc Natl Acad Sci U S A 2008;105:9793–8.
[91] Yeung F, Hoberg JE, Ramsey CS, Keller MD, Jones DR, Frye RA, et al. Modulation of NF-kappaB-dependent transcription and cell survival by the SIRT1 deacetylase. EMBO J 2004;23:2369–80.
[92] Yoshizaki T, Milne JC, Imamura T, Schenk S, Sonoda N, Babendure JL, et al. SIRT1 exerts anti-inflammatory effects and improves insulin sensitivity in adipocytes. Mol Cell Biol 2009;29:1363–74.
[93] Lee JH, Song MY, Song EK, Kim EK, Moon WS, Han MK, et al. Overexpression of SIRT1 protects pancreatic beta-cells against cytokine toxicity by suppressing the nuclear factor-kappaB signaling pathway. Diabetes 2009;58:344–51.
[94] Gerhart-Hines Z, Rodgers JT, Bare O, Lerin C, Kim SH, Mostoslavsky R, et al. Metabolic control of muscle mitochondrial function and fatty acid oxidation through SIRT1/PGC-1alpha. EMBO J 2007;26:1913–23.
[95] Rodgers JT, Lerin C, Haas W, Gygi SP, Spiegelman BM, Puigserver P. Nutrient control of glucose homeostasis through a complex of PGC-1alpha and SIRT1. Nature 2005;434:113–8.
[96] Vinciguerra M, Fulco M, Ladurner A, Sartorelli V, Rosenthal N. SirT1 in muscle physiology and disease: lessons from mouse models. Dis Model Mech 2010;3:298–303.
[97] Cheng HL, Mostoslavsky R, Saito S, Manis JP, Gu Y, Patel P, et al. Developmental defects and p53 hyperacetylation in Sir2 homolog (SIRT1)-deficient mice. Proc Natl Acad Sci U S A 2003;100:10794–9.
[98] Hsu CP, Zhai P, Yamamoto T, Maejima Y, Matsushima S, Hariharan N, et al. Silent information regulator 1 protects the heart from ischemia/reperfusion. Circulation 2010;122:2170–82.
[99] Planavila A, Iglesias R, Giralt M, Villarroya F. Sirt1 acts in association with PPAR{alpha} to protect the heart from hypertrophy, metabolic dysregulation, and inflammation. Cardiovasc Res 2011;90:276–84.
[100] Oka S, Alcendor R, Zhai P, Park JY, Shao D, Cho J, et al. PPARalpha-Sirt1 complex mediates cardiac hypertrophy and failure through suppression of the ERR transcriptional pathway. Cell Metab 2011;14:598–611.
[101] Sundaresan NR, Pillai VB, Wolfgeher D, Samant S, Vasudevan P, Parekh V, et al. The deacetylase SIRT1 promotes membrane localization and activation of Akt and PDK1 during tumorigenesis and cardiac hypertrophy. Sci Signal 2011;4:ra46.
[102] Planavila A, Dominguez E, Navarro M, Vinciguerra M, Iglesias R, Giralt M, et al. Dilated cardiomyopathy and mitochondrial dysfunction in Sirt1-deficient mice: a role for Sirt1-Mef2 in adult heart. J Mol Cell Cardiol 2012;53:521–31.
[103] Fredj S, Bescond J, Louault C, Potreau D. Interactions between cardiac cells enhance cardiomyocyte hypertrophy and increase fibroblast proliferation. J Cell Physiol 2005;202:891–9.
[104] Gnecchi M, He H, Liang OD, Melo LG, Morello F, Mu H, et al. Paracrine action accounts for marked protection of ischemic heart by Akt-modified mesenchymal stem cells. Nat Med 2005;11:367–8.
[105] Doroudgar S, Glembotski CC. The cardiokine story unfolds: ischemic stress-induced protein secretion in the heart. Trends Mol Med 2011;17:207–14.
[106] Frost RJ, Engelhardt S. A secretion trap screen in yeast identifies protease inhibitor 16 as a novel antihypertrophic protein secreted from the heart. Circulation 2007;116:1768–75.
[107] Stastna M, Chimenti I, Marban E, Van Eyk JE. Identification and functionality of proteomes secreted by rat cardiac stem cells and neonatal cardiomyocytes. Proteomics 2010;10:245–53.
[108] Badman MK, Pissios P, Kennedy AR, Koukos G, Flier JS, Maratos-Flier E. Hepatic fibroblast growth factor 21 is regulated by PPARalpha and is a key mediator of hepatic lipid metabolism in ketotic states. Cell Metab 2007;5:426–37.

[109] Galman C, Lundasen T, Kharitonenkov A, Bina HA, Eriksson M, Hafstrom I, et al. The circulating metabolic regulator FGF21 is induced by prolonged fasting and PPARalpha activation in man. Cell Metab 2008;8:169–74.
[110] Inagaki T, Dutchak P, Zhao G, Ding X, Gautron L, Parameswara V, et al. Endocrine regulation of the fasting response by PPARalpha-mediated induction of fibroblast growth factor 21. Cell Metab 2007;5:415–25.
[111] Muise ES, Azzolina B, Kuo DW, El-Sherbeini M, Tan Y, Yuan X, et al. Adipose fibroblast growth factor 21 is up-regulated by peroxisome proliferator-activated receptor gamma and altered metabolic states. Mol Pharmacol 2008;74:403–12.
[112] Izumiya Y, Bina HA, Ouchi N, Akasaki Y, Kharitonenkov A, Walsh K. FGF21 is an Akt-regulated myokine. FEBS Lett 2008;582:3805–10.
[113] Hondares E, Iglesias R, Giralt A, Gonzalez FJ, Giralt M, Mampel T, et al. Thermogenic activation induces FGF21 expression and release in brown adipose tissue. J Biol Chem 2011;286:12983–90.
[114] Kharitonenkov A, Shiyanova TL, Koester A, Ford AM, Micanovic R, Galbreath EJ, et al. FGF-21 as a novel metabolic regulator. J Clin Invest 2005;115:1627–35.
[115] Hondares E, Rosell M, Gonzalez FJ, Giralt M, Iglesias R, Villarroya F. Hepatic FGF21 expression is induced at birth via PPARalpha in response to milk intake and contributes to thermogenic activation of neonatal brown fat. Cell Metab 2010;11:206–12.
[116] Kharitonenkov A, Dunbar JD, Bina HA, Bright S, Moyers JS, Zhang C, et al. FGF-21/FGF-21 receptor interaction and activation is determined by betaKlotho. J Cell Physiol 2008;215:1–7.
[117] Kurosu H, Choi M, Ogawa Y, Dickson AS, Goetz R, Eliseenkova AV, et al. Tissue-specific expression of betaKlotho and fibroblast growth factor (FGF) receptor isoforms determines metabolic activity of FGF19 and FGF21. J Biol Chem 2007;282:26687–95.
[118] Planavila A, Redondo I, Hondares E, Vinciguerra M, Munts C, Iglesias R, et al. Fibroblast growth factor 21 protects against cardiac hypertrophy in mice. Nat Commun 2013;4:2019.
[119] Curtis LH, Whellan DJ, Hammill BG, Hernandez AF, Anstrom KJ, Shea AM, et al. Incidence and prevalence of heart failure in elderly persons, 1994-2003. Arch Intern Med 2008;168:418–24.
[120] Curtis LH, Greiner MA, Hammill BG, Kramer JM, Whellan DJ, Schulman KA, et al. Early and long-term outcomes of heart failure in elderly persons, 2001-2005. Arch Intern Med 2008;168:2481–8.
[121] Frick MH, Elo O, Haapa K, Heinonen OP, Heinsalmi P, Helo P, et al. Helsinki Heart Study: primary-prevention trial with gemfibrozil in middle-aged men with dyslipidemia. Safety of treatment, changes in risk factors, and incidence of coronary heart disease. N Engl J Med 1987;317:1237–45.
[122] Rubins HB, Robins SJ, Collins D, Fye CL, Anderson JW, Elam MB, et al. Gemfibrozil for the secondary prevention of coronary heart disease in men with low levels of high-density lipoprotein cholesterol. Veterans Affairs High-Density Lipoprotein Cholesterol Intervention Trial Study Group. N Engl J Med 1999;341:410–8.
[123] Sarma S, Ardehali H, Gheorghiade M. Enhancing the metabolic substrate: PPAR-alpha agonists in heart failure. Heart Fail Rev 2012;17:35–43.
[124] Nissen SE, Wolski K. Rosiglitazone revisited: an updated meta-analysis of risk for myocardial infarction and cardiovascular mortality. Arch Intern Med 2010;170:1191–201.
[125] Nesto RW, Bell D, Bonow RO, Fonseca V, Grundy SM, Horton ES, et al. Thiazolidinedione use, fluid retention, and congestive heart failure: a consensus statement from the American Heart Association and American Diabetes Association. Circulation 2003;108:2941–8.
[126] Aguilar D, Bozkurt B, Pritchett A, Petersen NJ, Deswal A. The impact of thiazolidinedione use on outcomes in ambulatory patients with diabetes mellitus and heart failure. J Am Coll Cardiol 2007;50:32–6.

CHAPTER

12

Inflammatory Modulation by Statins and Heart Failure: From Pharmacological Data to Clinical Evidence

Nicoletta Ronda[1], *Elda Favari*[1], *Francesca Zimetti*[1], *Arrigo F.G. Cicero*[2]

[1]Pharmaceutical Sciences Department, University of Parma, Parma, Italy

[2]Medical and Surgical Sciences Department, University of Bologna, Bologna, Italy

Heart failure (HF) results from a variety of processes leading to deregulation and damage of cellular and interstitial components of the heart, such as endothelial cells, myocytes, fibroblasts, conduction system fibers, extracellular matrix, and structural interstitial proteins. Statins inhibit HMG-CoA reductase, the enzyme catalyzing the conversion of 3-hydroxy-3-methylglutaryl-coenzyme A to L-mevalonic acid, the rate-limiting step in cholesterol synthesis. The inhibition of cholesterol synthesis promotes LDL cholesterol plasma clearance through an increase in Low Density Lipoprotein (LDL) receptor expression in the liver [1]. However, mevalonate is a precursor of important isoprenoid intermediates, such as farnesylpyrophosphate and geranyl-geranyl pyrophosphate [2], that play a key role in post-translational modification of a range of proteins, including small GTPases (Ras, Rho, Rac). So, statins have the capacity to modulate many intracellular processes potentially involved in myocardial contractility and remodeling, in cardiac inflammatory processes, and in interstitial modifications, all contributing to HF development (Figure 12.1).

The reported effects of statins at the cellular level may be beneficial or, in some cases, potentially unfavorable in HF. In fact, on one side, the lipid-lowering action, anti-inflammatory effects, and immune-modulatory properties of statins are of widely proven efficacy in hindering atherosclerosis and coronary disease [3,4], which are among the major causes of HF, and in cardiac remodeling after injury [5]. On the other side, by inhibiting the mevalonate pathway, statins deplete cells of coenzyme Q10, an essential cofactor in mitochondrial oxidative phosphorylation and generation of adenosinetriphosphate [6]. Moreover, RhoA inactivation and Rho-kinase inhibition are involved in mitochondrial membrane depolarization and subsequent caspase-dependent cell death and myotoxicity/hepatotoxicity induced by statins [7,8]

In this chapter, we will summarize the evidence for the regulatory effects of statins on cellular pathways possibly implicated in HF, approaching, in particular, inflammation and the functions of immune cells, endothelial cells, fibroblasts, and cardiomyocytes. Then, we will briefly evaluate how this large amount of preclinical data relates to the clinical evidence of statin efficacy in HF patients.

12.1 INFLAMMATION AND IMMUNE CELLS

Inflammation and immune activation are important factors of HF pathogenesis, because they take part in the development of most, if not all, cardiac diseases leading to HF and because they may accelerate its progression (see also Chapter 2 of this book). For example, proinflammatory cytokines cause systolic dysfunction and myocardial hypertrophy, activate a fetal gene program in cardiomyocytes, contribute to extracellular matrix modifications, and induce cardiac cachexia [9,10]. Clinical data support the relationship between circulating cytokines and adverse prognosis

http://dx.doi.org/10.1016/B978-0-12-800039-7.00012-8
© 2015 Elsevier Inc. All rights reserved.

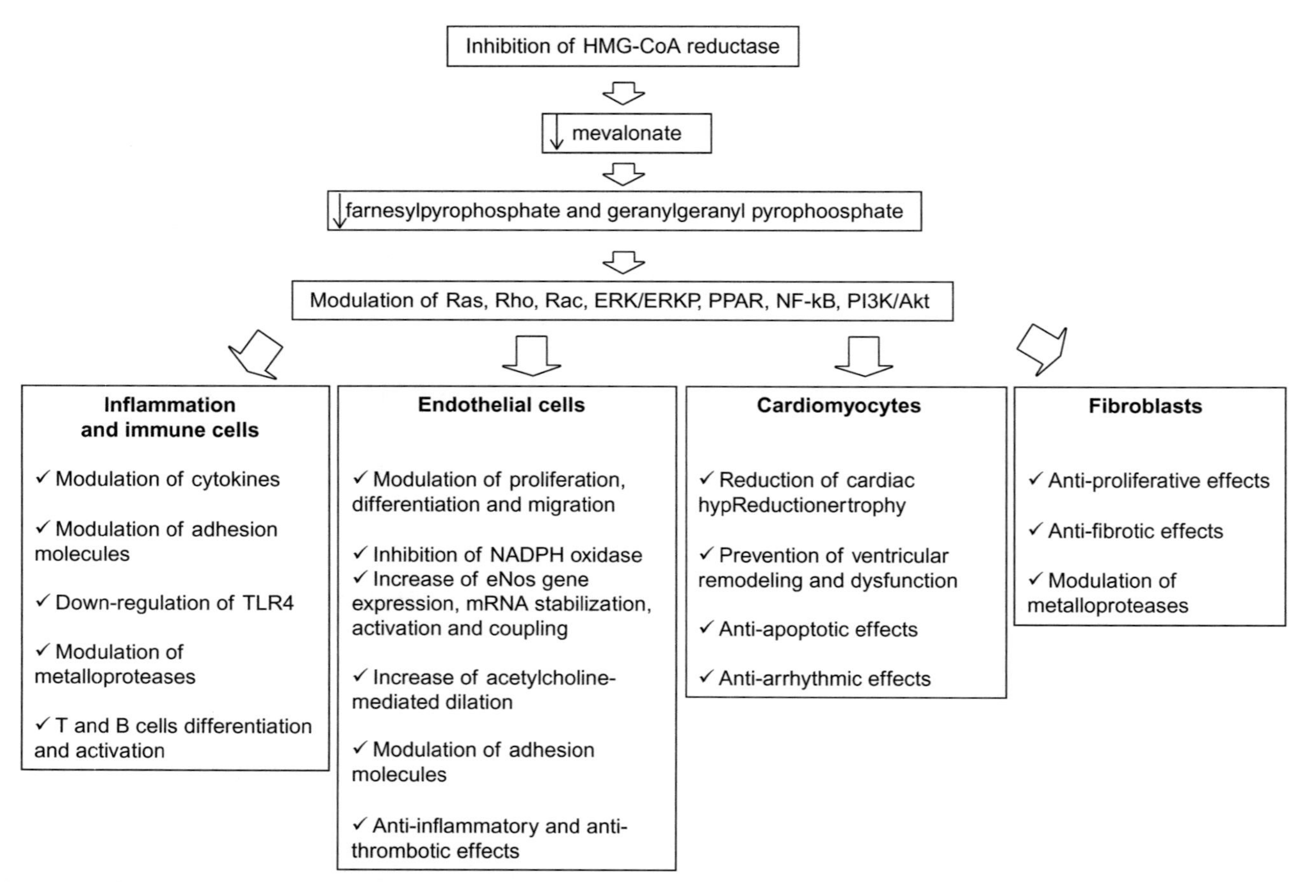

FIGURE 12.1 Potential effects of HMG-CoA reductase inhibition by statins on pathophysiology of heart failure.

of HF patients [11,12]. Immune system activation plays a major role in many cardiac diseases leading to HF and in the progression of cardiac remodeling in HF [13,14].

Statins have many anti-inflammatory and immune-regulating activities. They have been demonstrated to suppress cytokine secretion and adhesion molecules expression in many experimental settings. For example, they inhibit lipopolysaccharides-induced expression of tumor necrosis factor-α (TNF-α) and monocyte chemoattactant protein-1 (MCP-1) in macrophages through peroxisome proliferator-activated receptor (PPAR)α- and PPARγ-dependent pathway modulation [15,16]. The direct activation of PPARα and γ by statins has been demonstrated in inflammatory cells, platelets, cardiomyocytes, and vascular wall cells [17,18]. In addition, HMG-CoA reductase inhibition and subsequent isoprenoid depletion reduce TNF-α-induced upregulation of intercellular adhesion molecule-1 (ICAM-1) and vascular cell adhesion molecule-1 (VCAM-1) surface expression [19]. Another important anti-inflammatory mechanism of HMG-CoA reductase inhibitors is the reduction of mass and activity of lipoprotein-associated phospholipase A2 mass and activity, a proinflammatory enzyme whose levels predict incident cardiovascular disease [20,21]. Statins inhibitory activity on cytokine secretion and adhesion molecules expression has been confirmed in both animal models and human studies. Simvastatin reduced the levels of TNF-α, interleukin (IL)-1, IL-6 and in heart necrotic regions in rats after acute myocardial infarction [22], and reduced cardiopulmonary bypass-induced systemic and myocardial levels of proinflammatory cytokines through the stimulation of PPAR-γ receptors and the inhibition of the nuclear factor kB (NF-kB) expression in myocardial tissue [18]. In addition to reducing proinflammatory cytokines expression, statins have been shown to upregulate that of anti-inflammatory cytokine IL-10 and to ameliorate the balance between TNF-α/IL-10 after myocardial infarction in rats, resulting in improved remodeling [23]. Again, in a murine model of myocarditis, atorvastatin decreased myocardial TNF-α and interferon-γ (IFN-γ) and ameliorated survival [24]. With respect to HF, statin treatment is associated with a decrease in serum levels of C-reactive protein (CRP), IL-6, and TNF-α receptor II in patients with the systolic form [25]. Finally, in a randomized study on ischemic HF, atorvastatin treatment was associated with reduced ICAM-1 serum levels [26].

The anti-inflammatory action of statins is mediated also by the downregulation of toll-like receptor 4 (TLR4) expression in endothelial cells through the inactivation of ERK phosphorylation, which indirectly inhibits NF-κB activation [27]. Moreover, simvastatin and atorvastatin decrease LOX-1 expression, a lectin-like receptor on endothelial cells facilitating the uptake of oxidized-LDL [28]. HF is characterized by high production of reactive oxygen species (ROS), predominantly related to increased nicotinamide adenine dinucleotide phosphate (NADPH) oxidase and Rac1-GTPase activity, which is efficiently modified by statins [29].

Disturbances of metalloprotease (MMPs) composition and activity in the interstitial space have a significant role in cardiac remodeling after various injuries [30–33] and in modulating inflammation, collagen deposition, and left ventricular dilation after myocardial infarction [34,35]. (Refer to Chapter 4 for details.) MMPs activity is associated also with dilated cardiomyopathy [36,37]. Statins can modulate MMP activity in several cell types through the inhibition of small GTPases [38,39]. Atorvastatin reduced collagen synthesis and alpha(I)-procollagen mRNA as well as gene expression of the profibrotic peptide connective tissue growth factor 4 in rat and human cardiac fibroblasts (CFs) [40].

Statins modulate specific immune cell production and function. Besides reducing growth, differentiation, and activation of macrophages [41], they inhibit TLR4 expression in blood monocytes of patients with chronic HF [42]. Moreover, statins inhibit Fc receptor-mediated phagocytosis by macrophages [43]. Statins can influence the T-lymphocyte inhibiting cell activation and proliferation through disruption of T-cell receptor and blockade of the signaling cascade at the critical steps regulated by small Ras-likeGTPase [44,45]; they modulate Th1/Th2 differentiation and antagonize Th17-mediated response, directly suppressing the IL-17gene expression and protein secretion in $CD4^{+}$ cells [46]. Moreover, lovastatin inhibits IL-2, IL-4, and IFN-γ production from activated T cells, downregulating both activator protein-1 (AP-1) and NF-kB transcription factors [47]. Modulation of cytokine production from circulating T-helper-1 lymphocytes has been considered a possible explanation for the improvement of cardiac function after acute myocardial infarction with statin treatment [48]. Many animal models of myocarditis have contributed to demonstrate the ability of statins to improve cardiac function and the histopathological severity of the disease [49]. Statins also modulate B lymphocyte proliferation and functions. For example, lovastatin can inhibit the proliferation and differentiation of lipopolysaccharide-stimulated B cells [50].

12.2 ENDOTHELIAL CELLS

Endothelial dysfunction contributes to the pathogenesis of HF and can enhance adverse left ventricle (LV) remodeling and increase afterload in subjects with HF [51–55]. Vascular endothelial cells have a crucial role in the pathogenesis of inflammatory diseases and, similarly to immune cells, are a potential cellular target for statin action [56].

HMG-CoA reductase inhibitors increase endothelial progenitor cell mobilization, proliferation, migration, and differentiation in mice via the phosphatidylinositol 3-kinase (PI3K)/Akt pathway [57]. Statins may contribute to repair and regeneration of damaged endothelial cells, as demonstrated for simvastatin, which promotes endothelial healing in injured hamster arteries inducing vascular endothelial growth factor synthesis [58] and for rosuvastatin, which increases the number of circulating endothelial progenitor cells in patients with HF [59]. The intactness of the endothelial layer is also dependent on oxidative status. The ability of statins to inhibit the activity of NADPH oxidase, a group of several homologues (Nox 1-5 and Duox 1-2), is a class effect [60,61]. Statins downregulate Nox-1 mRNA expression and Rac1 translocation from the cytosol to cell membrane, which are required for the activation of NADPH oxidase [60]. The potential benefit of statin use in HF on EC function is supported by clinical studies. For example, in HF patients, rosuvastatin administration increases circulating endothelial progenitor cells [59] and improves endothelial function independently of lipid-lowering effects [62].

Endothelium-derived nitric oxide (NO) is an important factor in heart and vessel function and it is widely accepted that statins favorably affect important pathways regulating NO bioavailability by different mechanisms. Statins upregulate endothelial nitric oxide synthase (eNOS) gene expression in human endothelial cells (EC) in an L-mevalonate- and GGPP inhibitable way [63]. This effect is mediated by inhibition of Rho kinases geranylgeranyl-phosphorylation [63] and results in increased Kruppel-like factor 2 expression—a strong regulator of eNOS expression [64]. Statins induce eNOS activation also through the rapid triggering of the serine–threonine protein kinase Akt similarly to activation of the PI3K/Akt/eNOS pathway induced by the Rho-kinase inhibitor hydroxyfasudil [65]. Moreover, the beneficial effects of Akt activation are not limited to eNOS activation but also extend to the promotion of new blood vessels growth [66]. Both lipophilic and hydrophilic statins induce polyadenylation of eNOS mRNA in a Rho-dependent way by modulation of RNA polymerase II activity, a process that stabilizes eNOS mRNA [67]. Statins also downregulate caveolin-1 expression, a molecule that influences eNOS subcellular localization and inactivates eNOS [68]. These effects are reversed by mevalonate, highlighting the therapeutic potential of inhibiting cholesterol synthesis in peripheral cells to correct NO-dependent endothelial dysfunction associated with hypercholesterolemia and possibly other diseases [69]. In line with these findings, in isolated rat mesenteric resistance arteries, simvastatin can acutely modulate resistance arterial contractile function and enhance acetylcholine-mediated dilatations through mechanisms that involve the AMP-activated protein kinase/phospho-eNOS (Ser1177)/NO-dependent pathway [70]. Again, in a mouse model of hypertensive cardiomyopathy, pravastatin upregulated eNOS gene expression [71].

In addition to expression and activation of eNOS, statins favorably affect eNOS coupling. Asymmetrical dimethylarginine (ADMA) is believed to be a key mediator of inflammation-induced endothelial dysfunction [72], as it induces eNOS uncoupling in advanced atherosclerosis [73]. Statins interfere with ADMA metabolism upregulating the enzyme dimethylarginine dimethylaminohydrolase [74]. Moreover, statins increase the bioavailability of tetrahydrobiopterin (BH4), which is the critical eNOS cofactor that maintains the enzyme at its coupled form [75].

Statins may affect proinflammatory EC activation, which is one of the first steps in atherogenesis. Inhibition of Rho-kinase by statins is responsible for attenuation of adhesion molecules expression in endothelial cells, such as ICAM-1, independent of any effects on NO bioavailability [76]. It has also been demonstrated that statins inhibit many of the TNFα-intracellular effects, such as Rac-1 activation, ROS generation, NF-kB activation, and VCAM-1 and ICAM-1 expression, through ERK5 activation [77]. Other potential anti-thrombotic and anti-inflammatory effects of statins are related to the downregulation of von-Willebrand factor expression [78], the reduction of matrix metalloproteinase (MMP)-9 expression and activity [79], the suppression of angiopoietin-2release [80], the modification of EC redox state [81–83], and IL-6 activity [84]. Statins inhibit the pre-pro-endothelin-1 gene transcription in EC [85,86]; as endothelin-1 levels are correlated with disease severity and mortality [87,88] in HF, the beneficial effects of HMGCoA reductase inhibitors in HF may be further suggested.

12.3 CARDIOMYOCYTES

Independent of the causes of HF, cardiomyocytes in this condition are affected by trophism disturbances, inflammatory changes, and apoptosis [89]. Several studies on cultured cardiomyocytes reported a beneficial effect of statins on many of these processes. For example, the deleterious effect of TNF-α on cardiomyocytes is efficiently reversed by atorvastatin, with an inhibition of glucose oxidation [90], a phenomenon involved in the development of cardiac hypertrophy and dysfunction. Statins are also active on cardiac hypertrophy induced by various stimuli, such as angiotensin II and endothelin [91,92]. The mechanism of their action is again independent of their cholesterol lowering effect, but is rather related to the inhibition of prenilation of the GTPase belonging to Rho and Rac families that play a central role into hypertrophic mechanisms [93]. Alternatively, a postulated mechanism explaining the anti-hypertrophic actions of statins involves the ERK-mediated signaling pathway, as it has been observed in cultured cells and in experimental animal models [91,94,95] or the PI3K-AKT and JAK-STAT signaling pathways [96,97]. Statins are able to inhibit inflammation, as discussed above, and to attenuate the decrease of sarco/endoplasmic reticulum $Ca2^{+}$-ATPase in the peri-infarction zone, preventing left ventricular remodeling and dysfunction [98]. Finally, statins reduce apoptosis with a mechanism involving an inhibition of the small GTPase Rac1 [99,100]. Alternative pathways have been suggested to mediate statin cell-death inhibition, such as an interference with the PI3K-AKT signaling cascade or other kinases-mediated pathways, resulting in increased NO synthesis and decreased mitochondrial dysfunction [101]. More recent findings indicated that pravastatin may attenuate cardiac remodeling by inhibiting c-Jun N-terminal kinase (JNK)-dependent proapoptotic signaling [102].

Statins may have an anti-arrhythmic effect that is due in part to their anti-inflammatory properties but also to direct modulation of cell electrical activity. In fact, they have been shown to modulate Kv1.5 and Kv4.3 channel activity, thus modifying the myocardial action potential plateau [103], to attenuate reperfusion induced lethal ventricular arrhythmias by inhibition of calcium overload [104], and to regulate connexin 43 gene expression and phosphorylation, improving cardiomyocyte and intercellular junction integrity and significant increase of threshold for ventricular fibrillation [105]. Other possible anti-arrhythmic mechanisms are the suppression of sympathetic nervous activity through the upregulation of neuronal NO synthase expression in the central nervous system [106] and the downregulation of mRNA and protein expression of angiotensin II type 1 receptor [107]. Atorvastatin has been shown to inhibit Rac1-mediated activation of Nox2 NADPH-oxidase, to lower atrial superoxide generation, and to reduce the risk of atrial fibrillation after reperfusion [108]. Finally, in a rabbit animal model, statins showed direct effects on atrial ion currents inducing a favorable atrial electrical remodeling [109].

12.4 FIBROBLASTS

CFs are important for extracellular matrix homeostasis, for the structural integrity of interstitial space, and for the remodeling that occurs in response to pathological changes in HF [89]. Many of the functional modifications of CF in HF occur through their differentiation in myofibroblasts, cells that express contractile proteins and exhibit increased migratory, proliferative, and secretory properties [110]. Statins have consistently been proven to inhibit CF proliferation induced by various stimuli and leading to potential beneficial effects in the prevention of HF [111,112]. Recent studies reported that statins are able to inhibit myofibroblastic activity in an *in vitro* model of aortic stenosis [113]. Moreover, it has been demonstrated that atorvastatin inhibits aldosterone-induced proliferation and blocks cell-cycle progression in newborn rat CF through the suppression of ERK1/2 signaling pathway [114], similarly to what is described in other cell types [115,116]. CFs are involved in collagen deposition and fibrosis associated with HF. Several reports have highlighted an anti-fibrotic activity of statins, exerted through the reduction of collagen production and of a profibrotic peptide gene expression. This anti-fibrotic action may contribute to the anti-remodeling effect of statins. Moreover, the combination of pravastatin and pioglitazone was shown to dramatically reduce superoxide anion generation and the activation of both isoforms of mitogen-activated protein kinases (MAPKs) as well as NF-kB and AP-1 transcription factors. This was associated with dramatic reduction in procollagen-1 synthesis [117]. Transforming growth factor-beta 1 (TGF-β1) plays a causal role in promoting cardiac fibrosis. Atorvastatin attenuates the expression of endoglin, a membrane glycoprotein coreceptor induced by TGF- β1. This reduced expression led to an attenuation of cardiac fibrosis. The molecular mechanism underlying this effect is related to the PI3K, Akt, and Smad3 pathways [118]. In addition to the inhibitory activity on collagen production, TNF-α-induced MMP-9 appeared reduced in CF treated with statins [111].

12.5 A SUMMARY OF THE CLINICAL EVIDENCE

On the basis of the above reported extensive preclinical literature, we could expect a large clinical impact from the use of statins in HF patients. However, the clinical data are neither clear nor univocal. Evidence for statin therapy in HF mainly comes from nonrandomized studies that evaluated its effects on clinical outcomes in patients with HF and various cardiovascular conditions [119]. Subgroup and post-hoc analyses of statin trials in various cardiovascular conditions and HF trials that also evaluated other pharmacological agents provided substantial evidence for statin use in HF [120–123].

Several systematic reviews and meta-analyses have been conducted to synthesize evidence for statin therapy in reducing major adverse events in HF. The first systematic review [124], including mainly data from retrospective, nonrandomized trials and a few prospective randomized studies of statin treatment in HF, found that there is a paucity of prospective data required to determine the effect of statins on clinical outcomes in HF and concluded that available experimental, post-hoc data, observational data, and theoretical considerations are inconsistent. The authors reported that: (1) lower cholesterol levels are associated with poorer outcomes in HF patients and may be related to the function of cholesterol as a scavenger for harmful endotoxins; (2) statins in HF may adversely affect mitochondrial function through inhibition of ubiquinone; and (3) statins may decrease selenoproteins, which could result in decreased myocardial function. The researchers concluded that statin treatment may favor HF and recommended a large randomized clinical trial.

Another meta-analysis from 13 studies—11 retrospective and 2 prospective studies—reported that statin treatment favored HF with a significant (26%) decrease in relative risk of mortality.[125] Conversely, two recent meta-analyses performed on randomized clinical trials did not show improved survival with statins in HF.[126,127] Moreover, statin therapy seems to be safe and is not associated with increased risk of adverse events in HF patients compared to other patients. It appears that the majority of patient data included in these meta-analyses came from CORONA[128] and GISSI-HF[129] trials, which randomized older patients to low-dose rosuvastatin or matching placebo that may have skewed the summary statistic toward the results of these two large trials. From the various studies, low and moderate doses of statins seem to have better outcomes than high doses of statins in patients with HF. However, these claims were not confirmed when investigated with meta-regression models and subgroup analysis. Similarly, the age and sex of patients did not influence the outcomes of HF with statin therapy in any of the meta-analyses. These meta-analyses, always strongly influenced by the CORONA and GISSI-HF trials, also conclude that liposoluble statins could be more effective than hydrosoluble ones in improving HF prognosis. However, in a subgroup analysis of patients with ischemic systolic HF, patients with low levels of biomarkers such as N-terminal pro-brain natriuretic peptide and galectin-3, a marker of fibrosis, had a significant benefit with rosuvastatin (hazard ratio,

0.33; 95% confidence interval, 0.16-0.67). It could be hypothesized that the beneficial effect of this hydrophilic statin was, in part, related to improvement in atherosclerotic disease via a peripheral effect on carotid bodies decreasing hypoxic chemosensitivity or via possible anti-fibrotic effects. Although post-hoc analysis of these studies is only hypothesis-provoking, biomarkers may help identify a key subpopulation of HF patients with less advanced disease and less atherosclerotic lesion burden who may benefit from statin therapy. Patients with a high-fibrotic burden, as evidenced by a high level of galectin-3, may be at a degree of fibrosis too advanced to benefit.[130]

On the other hand, a recent meta-analysis limited to 11 trials carried out on 17,985 HF patients with preserved LVEF showed that statin use was associated with a 40% lower risk of mortality (RR 0.60, 95% CI 0.49-0.74, $p<0.001$), a subgroup of patients where no therapy was seen to improve prognosis.[131] A recent trial with pitavastatin showed that patients with LVEF between 30% and 45% experienced a significant improvement whereas patients with LVEF <30% tended to do worse. This suggests that the main predictor of outcome in HF treated patients is the baseline cardiac function.[132]

Some authors have also supposed that the effect of statins on a HF patient's mortality is the results of a balance between positive and negative effects, so that the improvement of prognosis appears only when this balance is in favor of positive effects. The paradoxical adverse effect of low cholesterol levels in patients with severe HF led Rauchhaus *et al.* [133] to propose the endotoxin-lipoprotein hypothesis: the increased mortality seen in HF patients with low cholesterol levels might occur because there is less serum lipoprotein to bind lipopolysaccharide (also known as endotoxin) absorbed from the gastrointestinal system. Lowering lipoprotein fractions may thus lead to increased proinflammatory cytokine and lipopolysaccharide concentrations, effects which are known to be detrimental to cardiac function in the HF setting.

Another hypothesis used to explain this paradoxical effect is known as the ubiquinone (coenzyme Q10) hypothesis.[134] As mevalonate concentrations fall with inhibition of HMG-CoA reductase, decreases in ubiquinone concentrations occur. Ubiquinone plays an important antioxidant and membrane-stabilizing role and is also involved in mitochondrial functioning, which is important for high-ATP-utilizing tissues such as the heart. Similarly, a decrease in mevalonate concentrations reduces selenoprotein concentrations, which are necessary for cell transcription and repair, processes that are important in cardiac cell functioning.[135]

Beyond the effect on mortality, clinical trials partially confirm the positive effect of statins on a relatively large number of pathophysiological mechanism involved in development and worsening of HF, and in particular on inflammation, heart and vascular remodeling, thrombotic phenomena, endothelial function, and autonomic function.[136]

Then, the meta-analysis by Lipinski *et al.* [127], which found that statin therapy does not decrease all-cause or cardiovascular mortality but significantly decreases hospitalization for worsening HF and increased left ventricular ejection function (LVEF) compared with placebo in patients with HF, suggests that the potential usefulness of statins in HF patients is not to be mainly evaluated on mortality but on other significant outcomes, such as for instance myocardial functionality and evolution of the disease.

Another promising field of investigation is the preventive effect of statins against the development of HF in patients taking cardiotoxic drugs, for instance, because of cancer. This is an extreme model, but preliminary data show that statins could have a significant protective effect in those frail subjects, again stressing the possible higher preventive efficacy of statins in the early HF phases.[137]

Given the key differences between the animal studies, human physiological studies, and randomized trial data, we suggest that it is too soon to give up on statin therapy for patients with HF, and more research is needed to determine whether statins should be part of the treatment for this increasingly prevalent and ultimately fatal form of heart disease.

References

[1] Ma PT, Gil G, Sudhof TC, Bilheimer DW, Goldstein JL, Brown MS. Mevinolin, an inhibitor of cholesterol synthesis, induces mRNA for low density lipoprotein receptor in livers of hamsters and rabbits. Proc Natl Acad Sci U S A 1986;83:8370–4.

[2] Liao Y, Zhao H, Ogai A, Kato H, Asakura M, Kim J, et al. Atorvastatin slows the progression of cardiac remodeling in mice with pressure overload and inhibits epidermal growth factor receptor activation. Hypertens Res 2008;31:335–44.

[3] Kjekshus J, Pedersen TR, Olsson AG, Faergeman O, Pyorala K. The effects of simvastatin on the incidence of heart failure in patients with coronary heart disease. J Card Fail 1997;3:249–54.

[4] Sever PS, Poulter NR, Dahlof B, Wedel H, Beevers G, Caulfield M, et al. The Anglo-Scandinavian Cardiac Outcomes Trial lipid lowering arm: extended observations 2 years after trial closure. Eur Heart J 2008;29:499–508.

[5] Tousoulis D, Oikonomou E, Siasos G, Stefanadis C. Statins in heart failure-With preserved and reduced ejection fraction. An update. Pharmacol Ther 2014;141:79–91.

[6] Vaklavas C, Chatzizisis YS, Ziakas A, Zamboulis C, Giannoglou GD. Molecular basis of statin-associated myopathy. Atherosclerosis 2009;202:18–28.

[7] Maeda A, Yano T, Itoh Y, Kakumori M, Kubota T, Egashira N, et al. Down-regulation of RhoA is involved in the cytotoxic action of lipophilic statins in HepG2 cells. Atherosclerosis 2010;208:112–8.
[8] Tanaka S, Sakamoto K, Yamamoto M, Mizuno A, Ono T, Waguri S, et al. Mechanism of statin-induced contractile dysfunction in rat cultured skeletal myofibers. J Pharmacol Sci 2010;114:454–63.
[9] Oikonomou E, Tousoulis D, Siasos G, Zaromitidou M, Papavassiliou AG, Stefanadis C. The role of inflammation in heart failure: new therapeutic approaches. Hellenic J Cardiol 2011;52:30–40.
[10] Tousoulis D, Siasos G, Maniatis K, Oikonomou E, Vlasis K, Papavassiliou AG, et al. Novel biomarkers assessing the calcium deposition in coronary artery disease. Curr Med Chem 2012;19:901–20.
[11] Deswal A, Petersen NJ, Feldman AM, Young JB, White BG, Mann DL. Cytokines and cytokine receptors in advanced heart failure: an analysis of the cytokine database from the Vesnarinone trial (VEST). Circulation 2001;103:2055–9.
[12] Tousoulis D, Oikonomou E, Siasos G, Chrysohoou C, Charakida M, Trikas A, et al. Predictive value of biomarkers in patients with heart failure. Curr Med Chem 2012;19:2534–47.
[13] Ismahil MA, Hamid T, Bansal SS, Patel B, Kingery JR, Prabhu SD. Remodeling of the mononuclear phagocyte network underlies chronic inflammation and disease progression in heart failure: critical importance of the cardiosplenic axis. Circ Res 2014;114:266–82.
[14] Koller L, Richter B, Goliasch G, Blum S, Korpak M, Zorn G, et al. CD4+ CD28(null) cells are an independent predictor of mortality in patients with heart failure. Atherosclerosis 2013;230:414–6.
[15] Yano M, Matsumura T, Senokuchi T, Ishii N, Murata Y, Taketa K, et al. Statins activate peroxisome proliferator-activated receptor gamma through extracellular signal-regulated kinase 1/2 and p38 mitogen-activated protein kinase-dependent cyclooxygenase-2 expression in macrophages. Circ Res 2007;100:1442–51.
[16] Paumelle R, Blanquart C, Briand O, Barbier O, Duhem C, Woerly G, et al. Acute antiinflammatory properties of statins involve peroxisome proliferator-activated receptor-alpha via inhibition of the protein kinase C signaling pathway. Circ Res 2006;98:361–9.
[17] Phipps RP, Blumberg N. Statin islands and PPAR ligands in platelets. Arterioscler Thromb Vasc Biol 2009;29:620–1.
[18] Shen Y, Wu H, Wang C, Shao H, Huang H, Jing H, et al. Simvastatin attenuates cardiopulmonary bypass-induced myocardial inflammatory injury in rats by activating peroxisome proliferator-activated receptor gamma. Eur J Pharmacol 2010;649:255–62.
[19] Landsberger M, Wolff B, Jantzen F, Rosenstengel C, Vogelgesang D, Staudt A, et al. Cerivastatin reduces cytokine-induced surface expression of ICAM-1 via increased shedding in human endothelial cells. Atherosclerosis 2007;190:43–52.
[20] Zalewski A, Macphee C. Role of lipoprotein-associated phospholipase A2 in atherosclerosis: biology, epidemiology, and possible therapeutic target. Arterioscler Thromb Vasc Biol 2005;25:923–31.
[21] White HD, Simes J, Stewart RA, Blankenberg S, Barnes EH, Marschner IC, et al. Changes in lipoprotein-associated phospholipase A2 activity predict coronary events and partly account for the treatment effect of pravastatin: results from the Long-Term Intervention with Pravastatin in Ischemic Disease study. J Am Heart Assoc 2013;2:e000360.
[22] Zhang J, Cheng X, Liao YH, Lu B, Yang Y, Li B, et al. Simvastatin regulates myocardial cytokine expression and improves ventricular remodeling in rats after acute myocardial infarction. Cardiovasc Drugs Ther 2005;19:13–21.
[23] Stumpf C, Petzi S, Seybold K, Wasmeier G, Arnold M, Raaz D, et al. Atorvastatin enhances interleukin-10 levels and improves cardiac function in rats after acute myocardial infarction. Clin Sci (Lond) 2009;116:45–52.
[24] Zhang A, Zhang H, Wu S. Immunomodulation by atorvastatin upregulates expression of gap junction proteins in coxsackievirus B3 (CVB3)-induced myocarditis. Inflamm Res 2010;59:255–62.
[25] Sola S, Mir MQ, Rajagopalan S, Helmy T, Tandon N, Khan BV. Statin therapy is associated with improved cardiovascular outcomes and levels of inflammatory markers in patients with heart failure. J Card Fail 2005;11:607–12.
[26] Tousoulis D, Oikonomou E, Siasos G, Chrysohoou C, Zaromitidou M, Kioufis S, et al. Dose-dependent effects of short term atorvastatin treatment on arterial wall properties and on indices of left ventricular remodeling in ischemic heart failure. Atherosclerosis 2013;227:367–72.
[27] Wang Y, Zhang MX, Meng X, Liu FQ, Yu GS, Zhang C, et al. Atorvastatin suppresses LPS-induced rapid upregulation of toll-like receptor 4 and its signaling pathway in endothelial cells. Am J Physiol Heart Circ Physiol 2011;300:H1743–52.
[28] Li DY, Chen HJ, Mehta JL. Statins inhibit oxidized-LDL-mediated LOX-1 expression, uptake of oxidized-LDL and reduction in PKB phosphorylation. Cardiovasc Res 2001;52:130–5.
[29] Maack C, Kartes T, Kilter H, Schafers HJ, Nickenig G, Bohm M, et al. Oxygen free radical release in human failing myocardium is associated with increased activity of rac1-GTPase and represents a target for statin treatment. Circulation 2003;108:1567–74.
[30] Sato S, Ashraf M, Millard RW, Fujiwara H, Schwartz A. Connective tissue changes in early ischemia of porcine myocardium: an ultrastructural study. J Mol Cell Cardiol 1983;15:261–75.
[31] Danielsen CC, Wiggers H, Andersen HR. Increased amounts of collagenase and gelatinase in porcine myocardium following ischemia and reperfusion. J Mol Cell Cardiol 1998;30:1431–42.
[32] Siasos G, Tousoulis D, Kioufis S, Oikonomou E, Siasou Z, Limperi M, et al. Inflammatory mechanisms in atherosclerosis: the impact of matrix metalloproteinases. Curr Top Med Chem 2012;12:1132–48.
[33] Tousoulis D, Kampoli AM, Papageorgiou N, Antoniades C, Siasos G, Latsios G, et al. Matrix metallopropteinases in heart failure. Curr Top Med Chem 2012;12:1181–91.
[34] Ducharme A, Frantz S, Aikawa M, Rabkin E, Lindsey M, Rohde LE, et al. Targeted deletion of matrix metalloproteinase-9 attenuates left ventricular enlargement and collagen accumulation after experimental myocardial infarction. J Clin Invest 2000;106:55–62.
[35] Creemers EE, Cleutjens JP, Smits JF, Daemen MJ. Matrix metalloproteinase inhibition after myocardial infarction: a new approach to prevent heart failure? Circ Res 2001;89:201–10.
[36] Li YY, Feldman AM, Sun Y, McTiernan CF. Differential expression of tissue inhibitors of metalloproteinases in the failing human heart. Circulation 1998;98:1728–34.
[37] Thomas CV, Coker ML, Zellner JL, Handy JR, Crumbley 3rd AJ, Spinale FG. Increased matrix metalloproteinase activity and selective upregulation in LV myocardium from patients with end-stage dilated cardiomyopathy. Circulation 1998;97:1708–15.
[38] Ikeda U, Shimpo M, Ohki R, Inaba H, Takahashi M, Yamamoto K, et al. Fluvastatin inhibits matrix metalloproteinase-1 expression in human vascular endothelial cells. Hypertension 2000;36:325–9.

[39] Bellosta S, Bernini F, Ferri N, Quarato P, Canavesi M, Arnaboldi L, et al. Direct vascular effects of HMG-CoA reductase inhibitors. Atherosclerosis 1998;137(Suppl):S101–9.
[40] Martin J, Denver R, Bailey M, Krum H. In vitro inhibitory effects of atorvastatin on cardiac fibroblasts: implications for ventricular remodelling. Clin Exp Pharmacol Physiol 2005;32:697–701.
[41] Aikawa M, Rabkin E, Sugiyama S, Voglic SJ, Fukumoto Y, Furukawa Y, et al. An HMG-CoA reductase inhibitor, cerivastatin, suppresses growth of macrophages expressing matrix metalloproteinases and tissue factor in vivo and in vitro. Circulation 2001;103:276–83.
[42] Foldes G, von Haehling S, Okonko DO, Jankowska EA, Poole-Wilson PA, Anker SD. Fluvastatin reduces increased blood monocyte toll-like receptor 4 expression in whole blood from patients with chronic heart failure. Int J Cardiol 2008;124:80–5.
[43] Loike JD, Shabtai DY, Neuhut R, Malitzky S, Lu E, Husemann J, et al. Statin inhibition of Fc receptor-mediated phagocytosis by macrophages is modulated by cell activation and cholesterol. Arterioscler Thromb Vasc Biol 2004;24:2051–6.
[44] Ghittoni R, Patrussi L, Pirozzi K, Pellegrini M, Lazzerini PE, Capecchi PL, et al. Simvastatin inhibits T-cell activation by selectively impairing the function of Ras superfamily GTPases. FASEB J 2005;19:605–7.
[45] Ghittoni R, Lazzerini PE, Pasini FL, Baldari CT. T lymphocytes as targets of statins: molecular mechanisms and therapeutic perspectives. Inflamm Allergy Drug Targets 2007;6:3–16.
[46] Zhang X, Markovic-Plese S. Statins' immunomodulatory potential against Th17 cell-mediated autoimmune response. Immunol Res 2008;41:165–74.
[47] Cheng SM, Lai JH, Yang SP, Tsao TP, Ho LJ, Liou JT, et al. Modulation of human T cells signaling transduction by lovastatin. Int J Cardiol 2010;140:24–33.
[48] Cheng X, Liao YH, Zhang J, Li B, Ge H, Yuan J, et al. Effects of Atorvastatin on Th polarization in patients with acute myocardial infarction. Eur J Heart Fail 2005;7:1099–104.
[49] Lazzerini PE, Capecchi PL, Laghi-Pasini F. Statins as a new therapeutic perspective in myocarditis and postmyocarditis dilated cardiomyopathy: editorial to "Pitavastatin regulates helper T-cell differentiation and ameliorates autoimmune myocarditis in mice" by K.Tajiri et al. Cardiovasc Drugs Ther 2013;27:365–9.
[50] Reedquist KA, Pope TK, Roess DA. Lovastatin inhibits proliferation and differentiation and causes apoptosis in lipopolysaccharide-stimulated murine B cells. Biochem Biophys Res Commun 1995;211:665–70.
[51] Drexler H. Endothelium as a therapeutic target in heart failure. Circulation 1998;98:2652–5.
[52] Drexler H. Nitric oxide synthases in the failing human heart: a doubled-edged sword? Circulation 1999;99:2972–5.
[53] Heitzer T, Baldus S, von Kodolitsch Y, Rudolph V, Meinertz T. Systemic endothelial dysfunction as an early predictor of adverse outcome in heart failure. Arterioscler Thromb Vasc Biol 2005;25:1174–9.
[54] Katz SD, Hryniewicz K, Hriljac I, Balidemaj K, Dimayuga C, Hudaihed A, et al. Vascular endothelial dysfunction and mortality risk in patients with chronic heart failure. Circulation 2005;111:310–4.
[55] Papageorgiou N, Tousoulis D, Androulakis E, Giotakis A, Siasos G, Latsios G, et al. Lifestyle factors and endothelial function. Curr Vasc Pharmacol 2012;10:94–106.
[56] Lin CP, Lin FY, Huang PH, Chen YL, Chen WC, Chen HY, et al. Endothelial progenitor cell dysfunction in cardiovascular diseases: role of reactive oxygen species and inflammation. Biomed Res Int 2013;2013:845037.
[57] Dimmeler S, Aicher A, Vasa M, Mildner-Rihm C, Adler K, Tiemann M, et al. HMG-CoA reductase inhibitors (statins) increase endothelial progenitor cells via the PI 3-kinase/Akt pathway. J Clin Invest 2001;108:391–7.
[58] Matsuno H, Takei M, Hayashi H, Nakajima K, Ishisaki A, Kozawa O. Simvastatin enhances the regeneration of endothelial cells via VEGF secretion in injured arteries. J Cardiovasc Pharmacol 2004;43:333–40.
[59] Tousoulis D, Andreou I, Tsiatas M, Miliou A, Tentolouris C, Siasos G, et al. Effects of rosuvastatin and allopurinol on circulating endothelial progenitor cells in patients with congestive heart failure: the impact of inflammatory process and oxidative stress. Atherosclerosis 2011;214:151–7.
[60] Wassmann S, Laufs U, Muller K, Konkol C, Ahlbory K, Baumer AT, et al. Cellular antioxidant effects of atorvastatin in vitro and in vivo. Arterioscler Thromb Vasc Biol 2002;22:300–5.
[61] Ishibashi Y, Matsui T, Takeuchi M, Yamagishi S. Rosuvastatin blocks advanced glycation end products-elicited reduction of macrophage cholesterol efflux by suppressing NADPH oxidase activity via inhibition of geranylgeranylation of Rac-1. Horm Metab Res 2011;43:619–24.
[62] Gounari P, Tousoulis D, Antoniades C, Kampoli AM, Stougiannos P, Papageorgiou N, et al. Rosuvastatin but not ezetimibe improves endothelial function in patients with heart failure, by mechanisms independent of lipid lowering. Int J Cardiol 2010;142:87–91.
[63] Laufs U, Liao JK. Post-transcriptional regulation of endothelial nitric oxide synthase mRNA stability by Rho GTPase. J Biol Chem 1998;273:24266–71.
[64] Sen-Banerjee S, Mir S, Lin Z, Hamik A, Atkins GB, Das H, et al. Kruppel-like factor 2 as a novel mediator of statin effects in endothelial cells. Circulation 2005;112:720–6.
[65] Wolfrum S, Dendorfer A, Rikitake Y, Stalker TJ, Gong Y, Scalia R, et al. Inhibition of Rho-kinase leads to rapid activation of phosphatidylinositol 3-kinase/protein kinase Akt and cardiovascular protection. Arterioscler Thromb Vasc Biol 2004;24:1842–7.
[66] Kureishi Y, Luo Z, Shiojima I, Bialik A, Fulton D, Lefer DJ, et al. The HMG-CoA reductase inhibitor simvastatin activates the protein kinase Akt and promotes angiogenesis in normocholesterolemic animals. Nat Med 2000;6:1004–10.
[67] Kosmidou I, Moore JP, Weber M, Searles CD. Statin treatment and 3' polyadenylation of eNOS mRNA. Arterioscler Thromb Vasc Biol 2007;27:2642–9.
[68] Plenz GA, Hofnagel O, Robenek H. Differential modulation of caveolin-1 expression in cells of the vasculature by statins. Circulation 2004;109:e7–8, author reply e7-8.
[69] Feron O, Dessy C, Desager JP, Balligand JL. Hydroxy-methylglutaryl-coenzyme A reductase inhibition promotes endothelial nitric oxide synthase activation through a decrease in caveolin abundance. Circulation 2001;103:113–8.
[70] Rossoni LV, Wareing M, Wenceslau CF, Al-Abri M, Cobb C, Austin C. Acute simvastatin increases endothelial nitric oxide synthase phosphorylation via AMP-activated protein kinase and reduces contractility of isolated rat mesenteric resistance arteries. Clin Sci (Lond) 2011;121:449–58.

[71] Xu Z, Okamoto H, Akino M, Onozuka H, Matsui Y, Tsutsui H. Pravastatin attenuates left ventricular remodeling and diastolic dysfunction in angiotensin II-induced hypertensive mice. J Cardiovasc Pharmacol 2008;51:62–70.
[72] Antoniades C, Demosthenous M, Tousoulis D, Antonopoulos AS, Vlachopoulos C, Toutouza M, et al. Role of asymmetrical dimethylarginine in inflammation-induced endothelial dysfunction in human atherosclerosis. Hypertension 2011;58:93–8.
[73] Antoniades C, Shirodaria C, Leeson P, Antonopoulos A, Warrick N, Van-Assche T, et al. Association of plasma asymmetrical dimethylarginine (ADMA) with elevated vascular superoxide production and endothelial nitric oxide synthase uncoupling: implications for endothelial function in human atherosclerosis. Eur Heart J 2009;30:1142–50.
[74] Ivashchenko CY, Bradley BT, Ao Z, Leiper J, Vallance P, Johns DG. Regulation of the ADMA-DDAH system in endothelial cells: a novel mechanism for the sterol response element binding proteins, SREBP1c and -2. Am J Physiol Heart Circ Physiol 2010;298:H251–8.
[75] Antoniades C, Shirodaria C, Crabtree M, Rinze R, Alp N, Cunnington C, et al. Altered plasma versus vascular biopterins in human atherosclerosis reveal relationships between endothelial nitric oxide synthase coupling, endothelial function, and inflammation. Circulation 2007;116:2851–9.
[76] Takeuchi S, Kawashima S, Rikitake Y, Ueyama T, Inoue N, Hirata K, et al. Cerivastatin suppresses lipopolysaccharide-induced ICAM-1 expression through inhibition of Rho GTPase in BAEC. Biochem Biophys Res Commun 2000;269:97–102.
[77] Wu K, Tian S, Zhou H, Wu Y. Statins protect human endothelial cells from TNF-induced inflammation via ERK5 activation. Biochem Pharmacol 2013;85:1753–60.
[78] Fish RJ, Yang H, Viglino C, Schorer R, Dunoyer-Geindre S, Kruithof EK. Fluvastatin inhibits regulated secretion of endothelial cell von Willebrand factor in response to diverse secretagogues. Biochem J 2007;405:597–604.
[79] Massaro M, Zampolli A, Scoditti E, Carluccio MA, Storelli C, Distante A, et al. Statins inhibit cyclooxygenase-2 and matrix metalloproteinase-9 in human endothelial cells: anti-angiogenic actions possibly contributing to plaque stability. Cardiovasc Res 2010;86:311–20.
[80] Hilbert T, Poth J, Frede S, Klaschik S, Hoeft A, Baumgarten G, et al. Anti-atherogenic effects of statins: impact on angiopoietin-2 release from endothelial cells. Biochem Pharmacol 2013;86:1452–60.
[81] Antoniades C, Bakogiannis C, Tousoulis D, Reilly S, Zhang MH, Paschalis A, et al. Preoperative atorvastatin treatment in CABG patients rapidly improves vein graft redox state by inhibition of Rac1 and NADPH-oxidase activity. Circulation 2010;122:S66–73.
[82] Antoniades C, Bakogiannis C, Leeson P, Guzik TJ, Zhang MH, Tousoulis D, et al. Rapid, direct effects of statin treatment on arterial redox state and nitric oxide bioavailability in human atherosclerosis via tetrahydrobiopterin-mediated endothelial nitric oxide synthase coupling. Circulation 2011;124:335–45.
[83] Ortego M, Bustos C, Hernandez-Presa MA, Tunon J, Diaz C, Hernandez G, et al. Atorvastatin reduces NF-kappaB activation and chemokine expression in vascular smooth muscle cells and mononuclear cells. Atherosclerosis 1999;147:253–61.
[84] Jougasaki M, Ichiki T, Takenoshita Y, Setoguchi M. Statins suppress interleukin-6-induced monocyte chemo-attractant protein-1 by inhibiting Janus kinase/signal transducers and activators of transcription pathways in human vascular endothelial cells. Br J Pharmacol 2010;159:1294–303.
[85] Hernandez-Perera O, Perez-Sala D, Soria E, Lamas S. Involvement of Rho GTPases in the transcriptional inhibition of preproendothelin-1 gene expression by simvastatin in vascular endothelial cells. Circ Res 2000;87:616–22.
[86] Hernandez-Perera O, Perez-Sala D, Navarro-Antolin J, Sanchez-Pascuala R, Hernandez G, Diaz C, et al. Effects of the 3-hydroxy-3-methylglutaryl-CoA reductase inhibitors, atorvastatin and simvastatin, on the expression of endothelin-1 and endothelial nitric oxide synthase in vascular endothelial cells. J Clin Invest 1998;101:2711–9.
[87] Wei CM, Lerman A, Rodeheffer RJ, McGregor CG, Brandt RR, Wright S, et al. Endothelin in human congestive heart failure. Circulation 1994;89:1580–6.
[88] Milo-Cotter O, Cotter-Davison B, Lombardi C, Sun H, Bettari L, Bugatti S, et al. Neurohormonal activation in acute heart failure: results from VERITAS. Cardiology 2011;119:96–105.
[89] Porter KE, Turner NA. Statins and myocardial remodelling: cell and molecular pathways. Expert Rev Mol Med 2011;13:e22.
[90] Gao F, Ni Y, Luo Z, Liang Y, Yan Z, Xu X, et al. Atorvastatin attenuates TNF-alpha-induced increase of glucose oxidation through PGC-1alpha upregulation in cardiomyocytes. J Cardiovasc Pharmacol 2012;59:500–6.
[91] Oi S, Haneda T, Osaki J, Kashiwagi Y, Nakamura Y, Kawabe J, et al. Lovastatin prevents angiotensin II-induced cardiac hypertrophy in cultured neonatal rat heart cells. Eur J Pharmacol 1999;376:139–48.
[92] Nishikimi T, Tadokoro K, Wang X, Mori Y, Asakawa H, Akimoto K, et al. Cerivastatin, a hydroxymethylglutaryl coenzyme A reductase inhibitor, inhibits cardiac myocyte hypertrophy induced by endothelin. Eur J Pharmacol 2002;453:175–81.
[93] Brown JH, Del Re DP, Sussman MA. The Rac and Rho hall of fame: a decade of hypertrophic signaling hits. Circ Res 2006;98:730–42.
[94] Xu X, Zhang L, Liang J. Rosuvastatin prevents pressure overload induced myocardial hypertrophy via inactivation of the Akt, ERK1/2 and GATA4 signaling pathways in rats. Mol Med Rep 2013;8:385–92.
[95] Zhang WB, Du QJ, Li H, Sun AJ, Qiu ZH, Wu CN, et al. The therapeutic effect of rosuvastatin on cardiac remodelling from hypertrophy to fibrosis during the end-stage hypertension in rats. J Cell Mol Med 2012;16:2227–37.
[96] Liu J, Shen Q, Wu Y. Simvastatin prevents cardiac hypertrophy in vitro and in vivo via JAK/STAT pathway. Life Sci 2008;82:991–6.
[97] Hauck L, Harms C, Grothe D, An J, Gertz K, Kronenberg G, et al. Critical role for FoxO3a-dependent regulation of p21CIP1/WAF1 in response to statin signaling in cardiac myocytes. Circ Res 2007;100:50–60.
[98] Yang Y, Mou Y, Hu SJ, Fu M. Beneficial effect of rosuvastatin on cardiac dysfunction is associated with alterations in calcium-regulatory proteins. Eur J Heart Fail 2009;11:6–13.
[99] Webster KA. Programmed death as a therapeutic target to reduce myocardial infarction. Trends Pharmacol Sci 2007;28:492–9.
[100] Ito M, Adachi T, Pimentel DR, Ido Y, Colucci WS. Statins inhibit beta-adrenergic receptor-stimulated apoptosis in adult rat ventricular myocytes via a Rac1-dependent mechanism. Circulation 2004;110:412–8.
[101] Verma S, Rao V, Weisel RD, Li SH, Fedak PW, Miriuka S, et al. Novel cardioprotective effects of pravastatin in human ventricular cardiomyocytes subjected to hypoxia and reoxygenation: beneficial effects of statins independent of endothelial cells. J Surg Res 2004;119:66–71.
[102] Cao S, Zeng Z, Wang X, Bin J, Xu D, Liao Y. Pravastatin slows the progression of heart failure by inhibiting the c-Jun N-terminal kinase-mediated intrinsic apoptotic signaling pathway. Mol Med Rep 2013;8:1163–8.
[103] Vaquero M, Caballero R, Gomez R, Nunez L, Tamargo J, Delpon E. Effects of atorvastatin and simvastatin on atrial plateau currents. J Mol Cell Cardiol 2007;42:931–45.

[104] Thuc LC, Teshima Y, Takahashi N, Nishio S, Fukui A, Kume O, et al. Cardioprotective effects of pravastatin against lethal ventricular arrhythmias induced by reperfusion in the rat heart. Circ J 2011;75:1601–8.
[105] Bacova B, Radosinska J, Knezl V, Kolenova L, Weismann P, Navarova J, et al. Omega-3 fatty acids and atorvastatin suppress ventricular fibrillation inducibility in hypertriglyceridemic rat hearts: implication of intracellular coupling protein, connexin-43. J Physiol Pharmacol 2010;61:717–23.
[106] Gao L, Wang W, Zucker IH. Simvastatin inhibits central sympathetic outflow in heart failure by a nitric-oxide synthase mechanism. J Pharmacol Exp Ther 2008;326:278–85.
[107] Gao L, Wang W, Li YL, Schultz HD, Liu D, Cornish KG, et al. Simvastatin therapy normalizes sympathetic neural control in experimental heart failure: roles of angiotensin II type 1 receptors and NAD(P)H oxidase. Circulation 2005;112:1763–70.
[108] Reilly SN, Jayaram R, Nahar K, Antoniades C, Verheule S, Channon KM, et al. Atrial sources of reactive oxygen species vary with the duration and substrate of atrial fibrillation: implications for the antiarrhythmic effect of statins. Circulation 2011;124:1107–17.
[109] Laszlo R, Menzel KA, Bentz K, Schreiner B, Kettering K, Eick C, et al. Atorvastatin treatment affects atrial ion currents and their tachycardia-induced remodeling in rabbits. Life Sci 2010;87:507–13.
[110] Porter KE, Turner NA. Cardiac fibroblasts: at the heart of myocardial remodeling. Pharmacol Ther 2009;123:255–78.
[111] Porter KE, Turner NA, O'Regan DJ, Balmforth AJ, Ball SG. Simvastatin reduces human atrial myofibroblast proliferation independently of cholesterol lowering via inhibition of RhoA. Cardiovasc Res 2004;61:745–55.
[112] Tian B, Liu J, Bitterman P, Bache RJ. Angiotensin II modulates nitric oxide-induced cardiac fibroblast apoptosis by activation of AKT/PKB. Am J Physiol Heart Circ Physiol 2003;285:H1105–12.
[113] Benton JA, Kern HB, Leinwand LA, Mariner PD, Anseth KS. Statins block calcific nodule formation of valvular interstitial cells by inhibiting alpha-smooth muscle actin expression. Arterioscler Thromb Vasc Biol 2009;29:1950–7.
[114] Wang Q, Cui W, Zhang HL, Hu HJ, Zhang YN, Liu DM, et al. Atorvastatin suppresses aldosterone-induced neonatal rat cardiac fibroblast proliferation by inhibiting ERK1/2 in the genomic pathway. J Cardiovasc Pharmacol 2013;61:520–7.
[115] Montecucco F, Mach F. Common inflammatory mediators orchestrate pathophysiological processes in rheumatoid arthritis and atherosclerosis. Rheumatology (Oxford) 2009;48:11–22.
[116] Taraseviciene-Stewart L, Scerbavicius R, Choe KH, Cool C, Wood K, Tuder RM, et al. Simvastatin causes endothelial cell apoptosis and attenuates severe pulmonary hypertension. Am J Physiol Lung Cell Mol Physiol 2006;291:L668–76.
[117] Chen J, Mehta JL. Angiotensin II-mediated oxidative stress and procollagen-1 expression in cardiac fibroblasts: blockade by pravastatin and pioglitazone. Am J Physiol Heart Circ Physiol 2006;291:H1738–45.
[118] Shyu KG, Wang BW, Chen WJ, Kuan P, Hung CR. Mechanism of the inhibitory effect of atorvastatin on endoglin expression induced by transforming growth factor-beta1 in cultured cardiac fibroblasts. Eur J Heart Fail 2010;12:219–26.
[119] Charach G, Rabinovich A, Ori A, Weksler D, Sheps D, Charach L, et al. Low levels of low-density lipoprotein cholesterol: a negative predictor of survival in elderly patients with advanced heart failure. Cardiology 2014;127:45–50.
[120] Bonsu KO, Kadirvelu A, Reidpath DD. Statins in heart failure: do we need another trial? Vasc Health Risk Manag 2013;9:303–19.
[121] Krum H, Latini R, Maggioni AP, Anand I, Masson S, Carretta E, et al. Statins and symptomatic chronic systolic heart failure: a post-hoc analysis of 5010 patients enrolled in Val-HeFT. Int J Cardiol 2007;119:48–53.
[122] Krum H, Bailey M, Meyer W, Verkenne P, Dargie H, Lechat P, et al. Impact of statin therapy on mortality in CHF patients according to beta-blocker use: results of CIBIS II. Cardiology 2007;108:28–34.
[123] Anker SD, Clark AL, Winkler R, Zugck C, Cicoira M, Ponikowski P, et al. Statin use and survival in patients with chronic heart failure: results from two observational studies with 5200 patients. Int J Cardiol 2006;112:234–42.
[124] van der Harst P, Voors AA, van Gilst WH, Böhm M, van Veldhuisen D. Statins in the treatment of chronic heart failure: a systematic review. PLoS Med 2006;3:e333.
[125] Ramasubbu K, Estep J, White DL, Deswal A, Mann DL. Experimental and clinical basis for the use of statins in patients with ischemic and nonischemic cardiomyopathy. J Am Coll Cardiol 2008;51:415–26.
[126] Zhang S, Zhang L, Sun A, Jiang H, Qian J, Ge J. Efficacy of statin therapy in chronic systolic cardiac insufficiency: a meta-analysis. Eur J Intern Med 2011;22:478–84.
[127] Lipinski MJ, Cauthen CA, Biondi-Zoccai GG, Abbate A, Vrtovec B, Khan BV, et al. Meta-analysis of randomized controlled trials of statins versus placebo in patients with heart failure. Am J Cardiol 2009;104:1708–16.
[128] Kjekshus J, Apetrei E, Barrios V, CORONA Group, et al. Rosuvastatin in older patients with systolic heart failure. N Engl J Med 2007;357(22):2248–61.
[129] Tavazzi L, Maggioni AP, Marchioli R, et al. Gissi-HF Investigators. Effect of rosuvastatin in patients with chronic heart failure (the GISSI-HF trial): a randomised, double-blind, placebo-controlled trial. Lancet 2008;372(9645):1231–9.
[130] Gullestad L, Ueland T, Kjekshus J, Nymo SH, Hulthe J, Muntendam P, et al. Galectin-3 predicts response to statin therapy in the Controlled Rosuvastatin Multinational Trial in Heart Failure (CORONA). Eur Heart J 2012;33:2290–6.
[131] Liu G, Zheng XX, Xu YL, Ru J, Hui RT, Huang XO. Meta-analysis of the effect of statins on mortality in patients with preserved ejection fraction. Am J Cardiol 2014;113(7):1198–204. http://dx.doi.org/10.1016/j.amjcard.2013.12.023.
[132] Takano H, Mizuma H, Kuwabara Y, Sato Y, Shindo S, Kotooka N, et al. Effects of pitavastatin on Japanese patients with chronic heart failure: the Pitavastatin Heart Failure Study (PEARL Study). Circ J 2013;77:917–25, on behalf of the PEARL Study Investigators.
[133] Rauchhaus M, Koloczek V, Volk H, Kemp M, Niebauer J, Francis DP, et al. Inflammatory cytokines and the possible immunological role for lipoproteins in chronic heart failure. Int J Cardiol 2000;76:125–33.
[134] Soja AM, Mortensen SA. Treatment of congestive heart failure with coenzyme Q10 illuminated by meta-analyses of clinical trials. Mol Asp Med 1997;18:S159–68.
[135] Rauchova H, Drahota Z, Lenaz G. Function of coenzyme Q in the cell: some biochemical and physiological properties. Physiol Res 1995;44:209–16.
[136] De Gennaro L, Brunetti ND, Correale M, Buquicchio F, Caldarola P, Di Biase M. Statin therapy in heart failure: for good, for bad, or indifferent? Curr Atheroscler Rep 2014;16:377–85.
[137] Seicean S, Seicean A, Plana JC, Thomas Budd GT, Marwick TH. Effect of statin therapy on the risk for incident heart failure in patients with breast cancer receiving anthracycline chemotherapy. An observational clinical cohort study. J Am Coll Cardiol 2012;60:2384–90.

CHAPTER

13

Small but Smart: microRNAs in the Center of Inflammatory Processes During Cardiovascular Diseases, the Metabolic Syndrome, and Aging*

Blanche Schroen, Stephane Heymans

Center for Heart Failure Research, Cardiovascular Research Institute Maastricht, Maastricht University, Maastricht, The Netherlands

13.1 INTRODUCTION

The incidence of cancer and cardiovascular diseases in Western societies is rising due to the aging of the population. While cancer is seen as a "gain-of-function" disease, in which cells adopt novel mechanisms to survive, cardiovascular diseases are of the "degenerative" type, in which cells become increasingly dysfunctional and as a consequence tissue functions decline [1]. Interestingly, all aging-associated diseases share a common denominator: inflammation (Figure 13.1) [31]. Here, we must distinguish acute from chronic inflammation, where the chronic type is associated with the aging process. While acute inflammation involves immune cell influx to address injury or infection, chronic inflammation does not only involve the presence of immune cells but more importantly it is a state in which cells such as fibroblasts, epithelial, and endothelial cells (ECs) enter a senescent state and produce inflammatory mediators, changing the tissue microenvironment [31,32]. Cellular senescence is at the basis of the chronic inflammatory state during aging and is characterized by the mitotic exit of dividing cells while they remain metabolically active. Cellular senescence is thought to be a major player in the cardiovascular aging process [33].

MicroRNAs have been implicated in the full range of processes of cellular senescence, inflammation, and cardiovascular diseases [2,23,34]. They are the most studied class of noncoding RNA molecules so far. This review focuses on their role in cardiovascular diseases with emphasis on their implication in the chronic inflammatory processes that accompany heart failure (HF), atherosclerosis, and coronary artery disease (CAD). Inflammation and cardiovascular morbidity converge into the process of aging, or vice versa, aging accelerates inflammation and cardiovascular diseases. The metabolic syndrome with diabetes, obesity, and hyperlipidemia as a trigger for inflammation, HF, and atherosclerosis will be addressed. The role of microRNAs in the inflammatory components of these pathologies will be the subject of the next paragraphs. The organization of the review is outlined in Figure 13.1.

* This article is part of the review focus on "The Role of MicroRNA in Cardiovascular Biology and Disease." Reprinted from Schroen, Blanche and Heymans, Stephane, Cardiovascular Research, Vol. 93(4), 605-613, Oxford Journals, 2012.

http://dx.doi.org/10.1016/B978-0-12-800039-7.00013-X

Published by Elsevier Inc.

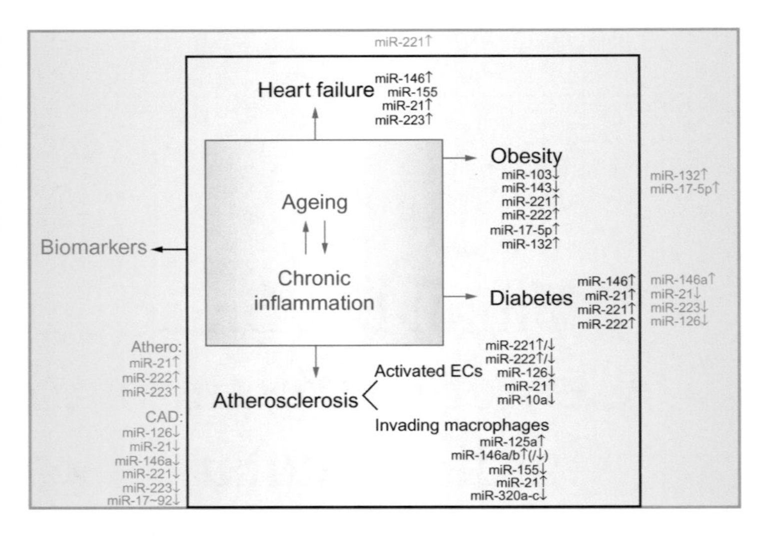

FIGURE 13.1 Guide to the build-up of the review. The processes of aging and chronic inflammation are indistinguishably intertwined. Both entities are also crucially associated with HF, atherosclerosis, and the metabolic syndrome, which all have a major inflammatory component and which incidence rises dramatically in the elderly. Inflammation-related microRNAs have a clear role in these aging- and inflammation-associated pathologies and also appear in the circulation. In addition, aging-associated as well as immune-associated microRNAs are often regulated in HF.

13.2 ROLE OF INFLAMMATION-RELATED microRNAs IN HF

Recent advances in microRNA research technologies have ensured major progress in our understanding of the role of myocytic and fibrotic microRNAs in the development of HF (reviewed elsewhere in this issue). In contrast, only one study describes a role for a microRNA in inflammation during myocardial infarction (MI)-induced HF. Using computational analyses, elegantly integrating microRNA with mRNA expression data, Zhu *et al.* [15] found a role for miR-98 in MI, during which it is a central hub in inflammation. In brief, they generated cardiac microRNA and mRNA expression data of MI in rats, built regulatory networks taking into account gene abundance, and searched these networks for functional enrichment. In this way, miR-98 was found to be a potential regulator of inflammatory pathways following MI. However, further *in vivo* proof is needed to determine whether miR-98 function indeed is central in the inflammatory response during MI. Interestingly, but not surprisingly, these authors found gene abundance to have an impact on the regulatory performance of a microRNA using this *in silico* approach, a modality that already has been recognized in the early days of prediction of microRNA-mRNA interactions [35]. For the integration of microRNA with mRNA expression data, this is a factor to be taken into account.

To conclude, a role for microRNAs in the inflammatory response during HF is still open to discoveries. On the other hand, there are a few studies implicating inflammatory microRNAs in the development of specific forms of HF (Figure 13.1), focusing on their role in modulating myocyte function.

MiR-146a is a well-known player in the immune system. Its expression in inflammatory cells is induced by lipopolysaccharide and is nuclear factor-kappaB (NF-kB)-dependent, and it acts as a negative feedback regulator of the innate immune response by targeting the proinflammatory adapter proteins, TNF receptor-associated factor 6 and interleukin-1 receptor-associated kinase 1 [30]. MiR-146a expression is increased in doxorubicin-induced HF. These authors show that it targets the v-erb-a erythroblastic leukemia viral oncogene homolog 4 (ErbB4) in the cardiac myocyte and may thereby cause cardiac myocyte dysfunction [36]. However, *in vivo* knockout or anti-miR studies in doxorubicin-induced HF or other forms of HF is still lacking.

Another example is miR-223, which was originally described as a myeloid lineage-specific microRNA [28]. MiR-223 may modulate diabetic HF by regulating glucose metabolism in cardiac myocytes [29]. The miR-223 target MADS box transcription enhancer factor 2c (Mef2c) is not involved in this cardiac process, while the glucose transporter Glut4 was found to be induced by cardiac myocyte miR-223, facilitating myocytic glucose uptake. MiR-223 was also found upregulated in atrial biopsies of patients with atrial fibrillation and rheumatic heart disease, a complication of the autoimmune disease rheumatic fever [37].

Another well-known central regulator of immune activation is miR-155 [5]. A polymorphism in its target gene angiotensin II type 1 receptor (AT1R) is associated with increased risk of adverse outcome in hypertensive subjects [38,39]. However, the implication of miR-155 in HF, possibly via its target gene AT1R, which has major roles in cardiovascular as well as immune cell functions, is unknown. Finally, miR-21 has a role in multiple pathologies including cardiovascular diseases and inflammation [40], and was recently extensively studied for its function in cardiac fibrosis and concomitant HF [41]. The field is awaiting first data on real genetic evidence for the role of immune cell-derived microRNAs in mediating the cross talk between inflammatory cells and the different cardiac cells during HF. No publications till date exist on the role of microRNAs in acute viral myocarditis, or in chronic inflammation affecting diabetic, ischemic, hypertensive, or autoimmune heart disease. An unmet medical need exists to develop novel RNA-based therapies specifically targeting the uncontrolled inflammatory reaction in viral or autoimmune myocarditis, the most aggressive form of inflammatory heart disease affecting young previously healthy individuals [42]. MicroRNA-based medicines may also help dampen the chronic inflammatory processes in the more aging-dependent cardiovascular disease processes caused by diabetes, long-term smoking, hyperlipidemia, and hypertension.

13.3 MicroRNAs AS REGULATORS OF THE INFLAMMATORY RESPONSE DURING ATHEROGENESIS

13.3.1 MicroRNAs Modulate Inflammatory Function of ECs

Inflammation is a hallmark of all stages of atherosclerosis, whereby the initial stage is characterized by leucocyte recruitment to activated ECs [43]. Interestingly, microRNA-deficient ECs, following Dicer knockdown, showed proliferative defects [44,45] and were hallmarked by decreased levels of inflammatory chemokines and cytokines, including interleukin (IL)-8, IL-1 β, and chemokine (C-X-C motif) ligands (CXCL) 1 and 3, suggesting that microRNAs play a central role in EC function [45]. Restoration of levels of microRNAs 221 and 222, which are among the highest expressed microRNAs in ECs [46], could not restore the proliferative capacity of Dicer-deficient ECs [45].

Monocyte adhesion upon EC activation is a critical step in inflammatory invasion of an atherosclerotic lesion [43]. Several microRNAs are differentially expressed in activated ECs (Figure 13.1) and modulate adhesion molecule expression, pinpointing their central role in atherogenesis; i.e., in human umbilical vein ECs (HUVECs), miR-126 was found the highest expressed by microarray analysis, and inhibited leucocyte adherence through the direct regulation of the vascular cell adhesion molecule-1 (VCAM-1) [46]. Also in HUVECs, miR-21 expression increases upon oscillatory shear stress (used to induce monocyte adhesion to ECs) and this response depends on c-jun/activator protein-1 (AP-1) [47]. MiR-21 promotes monocyte adhesion by enhancing adhesion molecule expression, including VCAM-1 and monocyte chemotactic protein-1 (MCP-1). MiR-21 also targets peroxisome proliferator-activated receptor alpha (PPARa), and becomes part of a positive feedback loop with lowered PPARa allowing increased AP-1 activity.

MicroRNA array analysis of athero-susceptible vs. athero-protected parts of swine aorta yielded 27 upregulated miRs (among which are miRs 221 and 21) and 7 downregulated miRs [48]. Here, EC miR-10a is the most downregulated and targets mitogen-activated protein kinase 7 and β-transducin repeat-containing gene, two factors involved in IkBa degradation. Therefore, decreased miR-10a in athero-susceptible regions allows NF-kB activation and the consequent increased expression of the inflammatory biomarkers MCP-1, IL-6 and -8, VCAM-1, and E-selectin in ECs.

MicroRNA array analysis of HUVECs undergoing OS vs. laminar shear stress resulted in 10 confirmed differentially expressed microRNAs, including miR-1275, -638 and -663 (up), and miR-320a, b, c, -151-3p, -195, -139-5p, and -27b (down) [49]. The OS-induced regulation of an endothelial microRNA, miR-663, was linked to the induction of proinflammatory gene expression and to increased adhesion of monocytes. There was no overlap between this *in vitro* microarray experiment and the microRNAs regulated in the *in vivo* model of athero-susceptibility; [48] more peculiarly, miRs-27b and -151-3p showed inverse regulations *in vivo* and *in vitro*, speculatively due to the mixed cell types *in vivo* or the artificial nature of culturing conditions *in vitro*. This reinforces the requirement of *in vivo* validation experiments to confirm a role for genes found by expression studies.

The inflammation-linked miRs 155 and 221/222 were found highly expressed in HUVECs and all target v-ets erythroblastosis virus E26 oncogene homolog 1 (Ets-1), an important endothelial transcription factor that robustly regulates endothelial inflammation, angiogenesis, and vascular remodeling [50]. Angiotensin II, used to activate proinflammatory signaling by ECs, increased Ets-1 expression, leading to upregulation of Ets-1 downstream genes, including VCAM-1, MCP-1, and fms-related tyrosine kinase 1 (FLT-1), the vascular endothelial growth factor/vascular permeability factor receptor. MiRs 155 and 221/222 inhibited monocyte adhesion by targeting Ets-1, while miR-155 inhibited HUVEC migration, possibly by targeting the AT1R. While miR-221 was found upregulated in athero-susceptible tissue [48] and both miR-221 and -222 were significantly higher in endothelial progenitor cells of patients with CAD [51], it is unclear what effect AngII had on miR-155 and -221/-222 expression in HUVECs. However, stimulation of HUVECs with other proinflammatory moderators, basic fibroblast growth factor (bFGF) and IL-3, reduced miRs 221 and 222 [52]. These authors show that miR-222 but not miR-221 inhibited proliferation and migration of stimulated HUVECs, possibly by targeting of signal transducer and activator of transcription 5A, the downstream signaler used by bFGF and IL-3 to trigger vascular EC morphogenesis. Furthermore, bFGF and IL-3 were both found to decrease miR-126 and -296, and miR-21 and -17-5p showed a trend to increase. All these studies have been done *in vitro*, excluding their interaction and cross talk with other cells. The relevance of these findings *in vivo* in disease models of ischemia, diabetes, or hypertension has not been addressed yet. In view of previous paradoxical findings in *in vitro* setups compared with *in vivo* animal studies, extrapolating *in vitro* findings to the disease process itself has to be done with extreme care.

13.3.2 MicroRNAs Modulate the Macrophage Response to Oxidized LDL

Activation of macrophages with oxidized low-density lipoproteins (oxLDLs) is used as an *in vitro* model for macrophages present in atherosclerotic plaques. A first study addressing the role of microRNAs in this system performed microarray analysis on oxLDL-activated human primary peripheral blood monocytes [53]. Both miR-125a-3p and -5p were significantly upregulated. MiR-125a-5p inhibition increased inflammatory cytokine secretion and lipid uptake, possibly via its target oxysterol binding protein-related protein 9 (ORP9), which is involved in lipid metabolism and membrane transport.

The inflammatory miRs 146a and 146b were also found upregulated both by microarray and quantitative PCR in this study [53]. Another study, however, showed that miR-146a expression decreased upon oxLDL stimulation of THP-1 macrophages [54]. Here, decreased miR-146a in oxLDL-activated macrophages was linked to an increase of its target toll-like receptor 4 (TLR4), involved in lipid uptake and inflammatory cytokine secretion, thereby allowing the accumulation of oxLDL on the one hand and an inflammatory response characterized by IL-6 and -8, chemokine (C-C motif) ligand 2, and matrix metalloproteinase-9 on the other hand. Thus, miR-146a would be anti-inflammatory in THP-1 macrophages. However, it is striking that the expression pattern of activated THP-1 macrophages is completely the opposite of often-observed changes in miR expression; i.e., miR-21 and -155 are often found upregulated and miRs-320a-c are found downregulated in activated cells in an atherosclerotic setting, that is, in activated ECs [47–49,52], and in oxLDL macrophages [53,55]. The divergence of these data may reflect the dissimilarity of cell types used. Also, peripheral blood mononuclear cells (PBMCs) of patients with acute coronary syndrome [56] and peripheral monocytes of patients with CAD [57] had enhanced miR-146a expression. In CAD, high levels of miR-146a and TLR4 in circulating mononuclear cells were independent predictors of cardiac events after a 12-month follow-up [57]. In acute coronary syndrome, miR-146a had a dual function in peripheral monocytes: (i) it increases the transcription factor T-bet thereby leading to increased Th1 differentiation and (ii) it induces proinflammatory cytokine (TNFα, MCP-1) and NF-kB p65 production [56]. In conclusion, even though the general consensus is that miR-146a upregulation counterbalances activated innate immunity [58–63], its role in diverse pathological conditions is not always consistent.

Monocytes and macrophages activated with oxidized LDL increased their miR-155 expression [53,55]. Inhibiting miR-155 in activated macrophages leads to enhanced uptake of oxLDL and increased expression of scavenger receptors, such as CD36 and CD68, and promoted IL-6, -8, and TNFα cytokine release, presumably via myeloid differentiation primary response gene (88) (MyD88) and NF-kB signaling [55]. Therefore, in these atherosclerotic conditions, miR-155 appears to be anti-inflammatory, opposite to the generally accepted association of miR-155 with a proinflammatory state [64]. However, its implication *in vivo* in atherosclerosis needs to be addressed, since the function of microRNAs may differ depending on the phenotype of the macrophage and the influence of these inflammatory cell-mediated microRNAs on neighboring endothelial or smooth muscle cells.

13.4 MicroRNAs IN THE METABOLIC SYNDROME

Chronic inflammation is central in the metabolic syndrome, including obesity, type II diabetes mellitus (DM), hypertension, and atherosclerosis [65]. The role of microRNAs in the metabolic syndrome and associated etiologies was reviewed before [66]. Here, we go into the link of obesity and diabetes with inflammation and associated microRNAs (Figure 13.1).

During obesity, the adipose tissue environment is characterized by a chronic inflammatory state, with increased production of the inflammatory cytokine TNFα by macrophages as the factor largely responsible for insulin resistance in obese adipose tissue [65]. Interestingly, obesity causes the loss of microRNAs that characterize fully differentiated and metabolically active adipocytes, including miRs-103 and -143. On the other hand, expression of miRs-221 and -222 decreased during adipogenesis and increased in obese adipocytes. This inverse regulation of microRNAs during adipogenesis vs. obesity is likely mediated by high levels of inflammatory TNFα in obese fat tissue, since miRs-103 and -143 were decreased and miRs-221 and -222 were increased by TNFα [65].

Obese subjects were found to have a unique microRNA expression profile in omental fat as well as blood, when compared with nonobese individuals, with miRs-17-5p and -132 upregulated only in fat and in the circulation of obese subjects [67]. These two microRNAs also correlated significantly with body mass index (BMI) and fasting blood glucose levels. A role for miR-132 in inflammatory processes during obesity was suggested by an *in vitro* study using primary human preadipocytes and *in vitro* differentiated adipocytes, where in response to nutritional availability, induction of miR-132 decreases sirtuin 1-mediated deacetylation of p65 leading to activation of NF-kB and transcription of IL-8 and MCP-1 [68].

Inflammatory microRNAs also have functions in diabetic conditions. MiR-21 and -146 expressions were induced by the proinflammatory cytokines IL-1β and TNFα in pancreatic islets, and inhibition of these microRNAs prevented the reduction in glucose-induced insulin secretion as a result of cytokine exposure [69]. *In vivo*, in diabetic kidney disease, miR-21 levels were increased in renal cortices of type I diabetic mice, leading to decreased levels of phosphatase and tensin homolog (which at normal levels inhibits renal cell hypertrophy and matrix expansion) and increased activity of TOR complex 1, necessary for cellular hypertrophy [70]. MiR-221 was induced by high glucose levels in HUVECs, inhibiting c-kit and consequently impairing EC migration [71]. Finally, exposure of adipocytes to high glucose levels induced miR-222, among others [72].

13.5 CIRCULATING microRNA PROFILES OF CARDIOVASCULAR DISEASES

13.5.1 Circulating microRNA Profiles of Cardiac Disease

Although there is much to discover about the exact function of circulating microRNAs, their presence in the circulation and association with diverse pathologies is now well accepted [73]. MicroRNAs circulate either in cell-shed microparticles—including apoptotic bodies and exosomes—[74], in high-density lipoprotein (HDL) particles [75], or as cell-free miRNAs in serum, recently demonstrated to be carried by Argonaute 2 [76]. The association of circulating microRNAs with cardiovascular disease is reviewed elsewhere in this issue of Cardiovascular Research. Here, we will briefly touch upon the role of circulating microRNAs in cardiovascular diseases with a prominent inflammatory character (Figure 13.1). The most studied in this respect is MI, which seems to induce a very consistent serum presence of the cardiac-specific microRNAs miR-1, -208, and -499 [73,77–79]. Their unique presence following cardiac damage correlates with clinical markers of damage, including troponin T [78] and creatine kinase-MB activity [77], and may even be earlier than troponins [79]. This suggests that serum presence of these microRNAs represents passive leakage from damaged cardiac myocytes.

Another typical form of HF with strong immune involvement is viral myocarditis, and we hypothesized the presence of inflammatory microRNAs in the circulation of these patients to increase [78]. However, serum presence of leucocyte-associated miRs 146a, 146b, 155, and 223 was not changed in patients with viral myocarditis when compared with controls, despite significant leukocytosis in these patients. On the other hand, cardiac miRs 208 and 499 did show a significant increase in acute viral myocarditis, correlating with the degree of myocyte leakage or death indexed by troponin T release.

The cellular origin of miR-423-5p, which was increased in sera of HF patients, is yet unknown [8]. Interestingly, one of the highest detected and upregulated serum microRNAs in this study, besides miR-423-5p, was the inflammation- and EC-associated miR-221 [8].

13.5.2 Circulating microRNA Profiles of Vascular Disease

On the vascular site, CAD and atherosclerosis have a major inflammatory component. An interesting pattern observed in CAD was that highly present circulating microRNAs would decrease in the blood of CAD patients and most of these were EC-derived, including miR-126 and the miR-17-92 cluster (Figure 13.1) [80]. Downregulation of miR-126 in CAD was confirmed by Zampetaki *et al.* [81] Among the downregulated microRNAs in CAD blood were also miRs-21, -146a, -221, and -223 [80]. The overlap with a CAD whole blood profiling study with similar patient numbers [82] is small; of the 46 listed downregulated miRs in Fichtlscherer *et al.* [80], 2 were also found downregulated in whole blood (miRs-20a and -93) and miR-23a even showed an opposite regulation in the two studies. A third study on CAD and microRNAs used PBMCs as source of RNA, and found relatively unknown microRNAs to be involved: miRs-135a and -147 [83]. Note that in the latter study, only 157 microRNAs were studied. Interestingly, miRs-21 and -223 were seen as housekeeping miRs.

A larger study addressed the presence of a selected set of 13 circulating microRNAs in 104 patients with atherosclerosis obliterans when compared with 105 age-matched controls [84]. As opposed to findings in CAD, miR-21 presence was significantly higher in sera of sclerotic patients, while levels of miRs-221 and -222 were unchanged when compared with controls. The presence of circulating microRNAs in atherosclerosis was also studied selectively in HDL particles, whose microRNA content was found to change in subjects with familial hypercholesterolemia, a genetic disorder of the LDL receptor which results in extreme atherosclerosis in homozygous cases [75]. In HDL of healthy subjects, miR-223 was in the top 10 of the highest expressed microRNAs, but its levels increased dramatically in HDL of familial hypercholesterolemia subjects. Also, in atherosclerotic apolipoprotein E and LDL receptor knock-out mice, HDL particles were found to contain increased amounts of miR-223 when compared with controls. No other prominent inflammatory microRNAs, such as miR-21, -221, -155, and -146a/b, stood out, only miR-222 seemed increased in HDL of familial hypercholesterolemia, similar to miR-223 but less dramatic. A preliminary conclusion from these early studies could be that the circulating microRNA repertoire of CAD and atherosclerosis surprisingly diverges.

13.5.3 Circulating microRNA Profiles of Diabetes

Circulating microRNAs have also been proposed as novel biomarkers for diabetes [81,85,86]. In an early study, Chen *et al.* [85] studied not only sera of subjects with DM but also of subjects with nonsmall cell lung carcinoma and compared them to healthy controls. Interestingly, they found that there was considerable overlap in serum microRNA presence for the two diseases and not for controls, and attributed this to the general inflammatory response seen in these diseases.

Although these authors state that the overlapping microRNAs may be related to the body's immune system, unfortunately there is no detailed information on their identity.

Zampetaki *et al.* [81] screened pools of diabetic sera derived from the Bruneck cohort for the expression of 754 microRNAs, detected 130 of them, and found 30 microRNAs to be differentially present between diabetic and control pools. Thirteen topologically unique microRNAs were selected for further validation in a larger group of 80 DM patients vs. 80 matched controls. Here, most miRs decreased significantly in DM, including miR-126, -21, and -223. Interestingly, levels of some miRs including -126 and -223 were already altered before DM manifestation. MiR-126 presence was analyzed in the entire Bruneck cohort of 822 subjects, and emerged as a significant predictor of manifest DM and even correlated negatively with increasing glucose intolerance. These authors propose the use of the five top-ranked microRNAs, miR-15a, -29b, -126, -223, and -150, as highly sensitive biomarkers of DM.

Recently, serum levels of seven microRNAs, selected for their proven involvement in diabetes-related processes, were studied in small groups of patients with established DM type II when compared with patients who either were susceptible (with comparably high BMIs of just above 26) or had prediabetes [86]. Here, all seven studied microRNAs were found increased in diabetic subjects, including miRs -9, -34a, and -146a, which are linked to inflammation and/or aging (Tables 13.1 and 13.2). None of these seven was identified by Zampetaki *et al.* [81] as circulating biomarkers of DM. While Zampetaki *et al.* [81] predominantly found downregulated microRNAs in diabetic sera, with one exception (miR-28-3p), the seven studied microRNAs were all upregulated in sera of DM patients. Finally, Caporali *et al.* [24] found increased levels of the EC miR-503 in sera of diabetic individuals and suggested this microRNA to have a role in angiogenesis.

Recently, hematopoietic cell-derived microRNAs, such as the above-mentioned miRs-146a, -155, and -223, were shown to confound circulating microRNA levels [25]. Also, miRs-21, -221, -222, and -423-5p are among the circulating microRNAs that can be of hematopoietic cell origin. Remarkably, miRs-21, -146a, and -223 are often among the top-detected microRNAs in serum profiling studies [8,75,81]. This warrants the cautious interpretation of circulating inflammation-related microRNA levels. Indeed, as mentioned above, the agreement between different studies on vascular disease is poor, and the field needs larger patient group studies with standardized measurement methods to establish the exact presence, including identity and origin, of microRNAs in the circulation.

13.5.4 Functional Roles for Circulating microRNAs

That circulating microRNAs do more than just being available as disease biomarkers is becoming increasingly clear. Indeed, HDL-bound microRNAs are delivered to recipient cells and were also shown to modulate the target gene expression in these cells [75]. Microvesicles also use microRNAs to mediate intercellular communication, and microRNAs from lung were shown to end up in bone marrow cells in culture and actively repress production of pulmonary epithelial cell mRNAs in these marrow cells [26]. The same principle would hold true for cardiac

TABLE 13.1 MicroRNAs in TLR Signaling: Involvement in HF

MicroRNA	Regulation in HF	Role in HF	Reference	Role In Human Inflammatory Autoimmune Diseases (Dai and Ahmed [3])
let-7	↑	**Inhibits cardiac hypertrophy**	[24]	
miR-9	↓	Targets myocardin to inhibit hypertrophy	[25]	
miR-16	↑/↓	–	[11,12,26,27]	RA
miR-21	↑	**Controls the extent of interstitial fibrosis and cardiac hypertrophy**	[5]	**SLE**
miR-27b	↑	Upregulated early in hypertrophy, role in angiogenesis	[22]	
miR-106	↑	–	[4]	
miR-125b	↑	–	[4]	SLE (125a)
miR-132	↑	**Targets cardiac L-type Ca channel β2 subunit protein**	[9,11–13]	**RA**
miR-145	↑	**Upregulated in human aortic stenosis**	[4]	
miR-146	↑	**Doxorubicin-induced HF**	[15]	**SLE, RA**
miR-155	–	Targets AT1R	[28,29]	RA, MS
miR-199	↑	**Regulates cardiac myocyte hypertrophy**	[16–18]	
miR-221	↑/↓	–	[4]	
miR-223	↑	Diabetic HF	[30]	SLE, RA
miR-4661	↓	–	[4]	

MicroRNAs highlighted in bold are also involved in aging. MiR-105, -348, -579, and -369-3p are also part of the TLR signaling pathway but have no proven regulation in HF. SLE, systemic lupus erythematosus; RA, rheumatoid arthritis; MS, multiple sclerosis.
Based on Table 13.2 in O'Neill et al. [23]

TABLE 13.2 MicroRNAs in Aging Pathways: Involvement in HF

MicroRNA	Regulation in HF	Role in HF	Reference	Role in Human Inflammatory Autoimmune Diseases (Dai and Ahmed [3])
let-7b	↑	–	[4]	
miR-1	↓	Antihypertrophic	[4]	
miR-21	↑	Controls the extent of interstitial fibrosis and cardiac hypertrophy	[5]	**SLE**
miR-24	↑	Suppresses cardiomyocyte and EC apoptosis	[6,7]	
miR34a	↑	Role in endothelial senescence, predisposing to atherosclerosis	[8,9]	MS
miR-100	↑	Involved in the beta-adrenergic receptor-mediated repression of "adult" cardiac genes	[10]	
miR-106	↑	–	[4]	
miR-128	↓	–	[4]	
miR-132	↑	**Targets cardiac L-type Ca channel β2 subunit protein**	[9,11–13]	**RA**
miR-138		Modulates cardiac patterning during embryonic development	[14]	

Continued

TABLE 13.2 MicroRNAs in Aging Pathways: Involvement in HF—Cont'd

MicroRNA	Regulation in HF	Role in HF	Reference	Role in Human Inflammatory Autoimmune Diseases (Dai and Ahmed [3])
miR-140	↑	–	[4]	
miR-145	↑	**Upregulated in human aortic stenosis**	[4]	
miR-146	↑	**Doxorubicin-induced HF**	[15]	**SLE, RA**
miR-199	↑	**Regulates cardiac myocyte hypertrophy**	[16–18]	
miR-206	↑	Associated with protection against cardiac remodeling after MI	[19]	
		Contributes to high glucose-mediated apoptosis in cardiomyocytes	[20]	
miR-217	↑	–	[4]	
miR-302-367	↑	–	[4]	
miR-320	↑	–	[4]	
miR-499	↓	Causes cellular hypertrophy and cardiac dysfunction	[21]	
	(/↑)	Inhibits cardiomyocyte apoptosis	[22]	

MicroRNAs highlighted in bold are also involved in TLR signaling. SLE, systemic lupus erythematosus; RA, rheumatoid arthritis; MS, multiple sclerosis.
Based on Table 13.1 in Chen et al. [2]

tissue-derived microvesicles, and is presumably a mode of communication between systemic organs and the core of the immune system [26]. Vice versa, microparticles secreted from mesenchymal stem cells were readily taken up by H9C2 cardiac myocytes [11].

At a local tissue level, adipocytes were found to communicate with each other using microvesicles, and vesicles were released from large adipocyte-stimulated lipid storage in smaller adipocytes using microRNAs as messengers [27]. In addition, two independent studies showed intercommunication of ECs using miR-126 [12,81]. Endothelial apoptotic bodies were shown to contain microRNAs and high glucose lowered their miR-126 content [81]. MiR-126 was also found as part of microparticles that are produced by apoptotic ECs during atherosclerosis [12]. Importantly, these microparticles conveyed a survival signal to neighboring ECs via miR-126 and its target, regulator of G-protein signaling 16, a G-protein-coupled receptor inhibitor, allowing the production of CXCL12 and thereby antagonizing apoptosis. These studies uniquely show a paracrine signaling function for microRNAs during the process of atherosclerosis.

It is still too early to speculate on whether a distinction can be made between noncell-free microRNAs—that is, microRNAs encapsulated in transport vesicles—and cell-free microRNAs in terms of functionality and possible "second messenger" roles.

13.6 AGING, INFLAMMATION, AND HF: ARE THERE SHARED microRNAs?

It is unclear whether inflammation and cardiovascular morbidity converge into the process of aging, or vice versa, whether aging accelerates inflammation and cardiovascular diseases. Below, we tried to link these together and find common microRNA pathways. Since inflammation is rather a broad concept, we focused on TLR signaling, which is central in HF, the development of the metabolic syndrome, and aging [9,13,87]. A recent review lists microRNAs involved in all levels of this central pathway in inflammation [34], and it is intriguing that most of these microRNAs have proven regulation and/or function in HF (Table 13.1). In addition, most of these microRNAs seem to be upregulated in failing hearts. The same holds true for microRNAs involved in pathways that modulate the aging process, recently reviewed [2], where we see that *all* microRNAs listed are also involved in HF (Table 13.2). In addition, there is some overlap between microRNAs involved in inflammation and aging (almost 50%, as highlighted in both Tables). Therefore, HF-associated microRNAs appear to have central roles in both inflammation and aging.

Our group was the first to study microRNAs in aging-associated HF, and identified miR-18 and -19 to be downregulated in failure-prone aged mice and in cardiac biopsies of HF patients of age [23]. This downregulation was linked to increased expression of the matricellular protein thrombospondin-1 and connective tissue growth factor, and to increased fibrosis in aged failing myocardium. Interestingly, mice on a failure-protected genetic background showed inverse expression patterns of both these microRNAs and matricellular proteins, with miRs-18 and -19 being upregulated with aging, indicating the possibility of their contribution to healthy aging.

MicroRNAs-155, -21, and -146 have established roles in immune and inflammatory pathways, and are central in TLR signaling [16]. MiR-21 has been shown to target the tumor suppressor programmed cell death protein 4 (PDCD4), a proinflammatory protein that promotes activation of the transcription factor NF-kB and suppresses IL-10 [17]. By targeting PDCD4, miR-21 becomes a negative regulator of TLR4 signaling [17]. In addition, miR-21 has been shown to be induced by STAT3 [18] and NF-kB [17], central transcription factors in immune functions but also in cardiac hypertrophy.

EC senescence is also thought to play a role in cardiovascular diseases such as atherosclerosis [1,33]. During endothelial senescence, microRNAs 34a and 217 were found upregulated [6,7]. These endothelial microRNAs triggered endothelial senescence in part through Sirtuin 1, a class III histone deacetylase which is linked to aging and to the regulation of the level of inflammatory responses [88]. MiR-34a was also shown to be upregulated in the heart and spleen of older mice [6]. In addition, miR-20c was induced by oxidative stress in ECs, and its overexpression induced EC growth arrest, apoptosis, and senescence via zinc finger E-box-binding factor 1 [14].

MiR-146a has roles in autoimmunity, and mice lacking this gene show accelerated aging [58]. From 6 to 8 months of age, miR-146a-deficient mice, which are on an aging-susceptible genetic background of 129/Bl6 [23], exhibit signs of immunoproliferative disease, characterized by splenomegaly, lymphadenopathy, and premature death. Liver, kidneys, and lungs had lymphocytic and monocytic infiltrates with some evidence of tissue damage. Recently, miR-146a was found to decrease with aging of ECs, where it targeted NADPH oxidase 4, involved in cell senescence and aging [89]. In agreement with this, miR-146a decreased in models of organismal aging, including foreskin and $CD8^+$ T cells of old vs. young donors [20]. Another microRNA with a role in inflammation, miR-221, increased with aging in ECs [89]. However, this miR decreased in organismal aging [20]. In addition, miR-155 presence in sera of 53 healthy and CAD patients showed a significant negative correlation with age [80].

13.7 CONCLUSIONS AND FUTURE PERSPECTIVES

MicroRNAs have established roles in all aspects of organismal development and disease. In recent years, their roles in inflammation, aging, cancer, and cardiovascular diseases are proven beyond doubt. As reviewed here, microRNAs play major roles in the inflammatory aspects of vascular disease. Given the central role of inflammation in HF, with inflammatory transcription factors including NF-kB and STAT3 and with inflammatory cytokines including TNFα and IL-6 as central mediators of cardiac hypertrophy, a role for inflammatory microRNAs in the development of HF seems ensured but still needs to be proven.

Whether microRNAs will be used some day for the diagnosis of inflammatory involvement in cardiovascular and/or other inflammatory diseases depends on the precise characterization of the role and distribution of microRNAs in the blood. Inflammation-associated microRNAs such as miR-21, -146, and -223 are always detected at high levels in the blood, also in healthy subjects, and to date it is unclear what their function is. Their presence might even be the result of leakage from blood cells, in which these microRNAs have established high levels. Therefore, it is of great importance that standardized techniques are developed for the detection of circulating microRNAs.

Some circulating microRNAs, apparently those contained in some sort of vesicle, appear to function as messengers between cells. This may be a way of tissues, including the heart and vessels, to communicate internally but also with the immune system, possibly changing each other's destiny. In conclusion, the potential of microRNAs for the diagnosis and treatment of human disease seems to be ever growing, and with the discovery of other noncoding RNAs, the field of noncoding RNAs will keep on expanding for the coming years.

Acknowledgments

We thank the colleagues at the Center for Heart Failure Research for fruitful discussions.

Conflict of Interest: none declared.

Funding: B. S. received a Veni grant (016.096.126) from the Netherlands Organization for Scientific Research, a Horizon grant (93519017) from the Netherlands Genomics Initiative, and a research grant from the Netherlands Heart Foundation (NHS 2009B025). S. H. received a Vidi grant from the Netherlands Organization for Scientific Research (91796338) and research grants from the Netherlands Heart Foundation (NHS 2007B036 and 2008B011), Research Foundation—Flanders (FWO 1183211N, 1167610N, G074009N), European Union, FP7-HEALTH-2010, MEDIA, Large scale integrating project, and European Union, FP7-HEALTH-2011, EU-Mascara, Large scale integrating project.

References

[1] Campisi J. Senescent cells, tumor suppression, and organismal aging: good citizens, bad neighbors. Cell 2005;120:513–22.
[2] Chen LH, Chiou GY, Chen YW, Li HY, Chiou SH. microRNA and aging: a novel modulator in regulating the aging network. Ageing Res Rev 2010;9(Suppl 1):S59–66.
[3] Dai R, Ahmed SA. MicroRNA, a new paradigm for understanding immunoregulation, inflammation, and autoimmune diseases. Transl Res 2011;157:163–79.
[4] Schroen B, Heymans S. MicroRNAs and beyond: the heart reveals its treasures. Hypertension 2009;54:1189–94.
[5] Tili E, Croce CM, Michaille JJ. miR-155: on the crosstalk between inflammation and cancer. Int Rev Immunol 2009;28:264–84.
[6] Ito T, Yagi S, Yamakuchi M. MicroRNA-34a regulation of endothelial senescence. Biochem Biophys Res Commun 2010;398:735–40.
[7] Menghini R, Casagrande V, Cardellini M, Martelli E, Terrinoni A, Amati F, et al. MicroRNA 217 modulates endothelial cell senescence via silent information regulator 1. Circulation 2009;120:1524–32.
[8] Tijsen AJ, Creemers EE, Moerland PD, de Windt LJ, van der Wal AC, Kok WE, et al. MiR423–5p as a circulating biomarker for heart failure. Circ Res 2010;106:1035–9.
[9] Konner AC, Bruning JC. Toll-like receptors: linking inflammation to metabolism. Trends Endocrinol Metab 2011;22:16–23.
[10] Sucharov C, Bristow MR, Port JD. miRNA expression in the failing human heart: functional correlates. J Mol Cell Cardiol 2008;45:185–92.
[11] Chen TS, Lai RC, Lee MM, Choo AB, Lee CN, Lim SK. Mesenchymal stem cell secretes microparticles enriched in pre-microRNAs. Nucleic Acids Res 2010;38:215–24.
[12] Zernecke A, Bidzhekov K, Noels H, Shagdarsuren E, Gan L, Denecke B, et al. Delivery of microRNA-126 by apoptotic bodies induces CXCL12-dependent vascular protection. Sci Signal 2009;2:ra81.
[13] Dunston CR, Griffiths HR. The effect of ageing on macrophage toll-like receptor-mediated responses in the fight against pathogens. Clin Exp Immunol 2010;161:407–16.
[14] Magenta A, Cencioni C, Fasanaro P, Zaccagnini G, Greco S, Sarra-Ferraris G, et al. miR-200c is upregulated by oxidative stress and induces endothelial cell apoptosis and senescence via ZEB1 inhibition. Cell Death Differ 2011;18:1628–39.
[15] Zhu W, Yang L, Shan H, Zhang Y, Zhou R, Su Z, et al. MicroRNA expression analysis: clinical advantage of propranolol reveals key microRNAs in myocardial infarction. PLoS One 2011;6:e14736.
[16] Quinn SR, O'Neill LA. A trio of microRNAs that control toll-like receptor signalling. Int Immunol 2011;23:421–5.
[17] Sheedy FJ, Palsson-McDermott E, Hennessy EJ, Martin C, O'Leary JJ, Ruan Q, et al. Negative regulation of TLR4 via targeting of the proinflammatory tumor suppressor PDCD4 by the microRNA miR-21. Nat Immunol 2010;11:141–7.
[18] van der Fits L, van Kester MS, Qin Y, Out-Luiting JJ, Smit F, Zoutman WH, et al. MicroRNA-21 expression in CD4+ T cells is regulated by STAT3 and is pathologically involved in Sezary syndrome. J Invest Dermatol 2011;131:762–8.
[19] Limana F, Esposito G, D'Arcangelo D, Di Carlo A, Romani S, Melillo G, et al. HMGB1 attenuates cardiac remodelling in the failing heart via enhanced cardiac regeneration and miR-206-mediated inhibition of TIMP-3. PLoS One 2011;6:e19845.
[20] Hackl M, Brunner S, Fortschegger K, Schreiner C, Micutkova L, Muck C, et al. miR-17, miR-19b, miR-20a, and miR-106a are down-regulated in human aging. Aging Cell 2010;9:291–6.
[21] Shieh JT, Huang Y, Gilmore J, Srivastava D. Elevated miR-499 levels blunt the cardiac stress response. PLoS One 2011;6:e19481.
[22] Busk PK, Cirera S. MicroRNA profiling in early hypertrophic growth of the left ventricle in rats. Biochem Biophys Res Commun 2010;396:989–93.
[23] van Almen GC, Verhesen W, van Leeuwen RE, van de Vrie M, Eurlings C, Schellings MW, et al. MicroRNA-18 and microRNA-19 regulate CTGF and TSP-1 expression in age-related heart failure. Aging Cell 2011;10:769–79.
[24] Caporali A, Meloni M, Vollenkle C, Bonci D, Sala-Newby GB, Addis R, et al. Deregulation of microRNA-503 contributes to diabetes mellitus-induced impairment of endothelial function and reparative angiogenesis after limb ischemia. Circulation 2011;123:282–91.
[25] Duttagupta R, Jiang R, Gollub J, Getts RC, Jones KW. Impact of cellular miRNAs on circulating miRNA biomarker signatures. PLoS One 2011;6:e20769.
[26] Aliotta JM, Pereira M, Johnson KW, de Paz N, Dooner MS, Puente N, et al. Microvesicle entry into marrow cells mediates tissue-specific changes in mRNA by direct delivery of mRNA and induction of transcription. Exp Hematol 2010;38:233–45.
[27] Muller G, Schneider M, Biemer-Daub G, Wied S. Microvesicles released from rat adipocytes and harboring glycosylphosphatidylinositol-anchored proteins transfer RNA stimulating lipid synthesis. Cell Signal 2011;23:1207–23.
[28] Chen CZ, Li L, Lodish HF, Bartel DP. MicroRNAs modulate hematopoietic lineage differentiation. Science 2004;303:83–6.
[29] Lu H, Buchan RJ, Cook SA. MicroRNA-223 regulates Glut4 expression and cardiomyocyte glucose metabolism. Cardiovasc Res 2010;86:410–20.
[30] Taganov KD, Boldin MP, Chang KJ, Baltimore D. NF-kappaB-dependent induction of microRNA miR-146, an inhibitor targeted to signaling proteins of innate immune responses. Proc Natl Acad Sci U S A 2006;103:12481–6.
[31] Khatami M. Inflammation, aging, and cancer: tumoricidal vs. tumorigenesis of immunity: a common denominator mapping chronic diseases. Cell Biochem Biophys 2009;55:55–79.
[32] Freund A, Orjalo AV, Desprez PY, Campisi J. Inflammatory networks during cellular senescence: causes and consequences. Trends Mol Med 2010;16:238–46.
[33] Wang M, Monticone RE, Lakatta EG. Arterial aging: a journey into subclinical arterial disease. Curr Opin Nephrol Hypertens 2010;19:201–7.
[34] O'Neill LA, Sheedy FJ, McCoy CE. MicroRNAs: the fine-tuners of toll-like receptor signalling. Nat Rev Immunol 2011;11:163–75.
[35] Doench JG, Sharp PA. Specificity of microRNA target selection in translational repression. Genes Dev 2004;18:504–11.
[36] Horie T, Ono K, Nishi H, Nagao K, Kinoshita M, Watanabe S, et al. Acute doxorubicin cardiotoxicity is associated with miR-146a-induced inhibition of the neuregulin-ErbB pathway. Cardiovasc Res 2010;87:656–64.
[37] Lu Y, Zhang Y, Wang N, Pan Z, Gao X, Zhang F, et al. MicroRNA-328 contributes to adverse electrical remodeling in atrial fibrillation. Circulation 2010;122:2378–87.

[38] Ceolotto G, Papparella I, Bortoluzzi A, Strapazzon G, Ragazzo F, Bratti P, et al. Interplay between miR-155, AT1R A1166C polymorphism, and AT1R expression in young untreated hypertensives. Am J Hypertens 2011;24:241–6.
[39] Martin MM, Buckenberger JA, Jiang J, Malana GE, Nuovo GJ, Chotani M, et al. The human angiotensin II type 1 receptor +1166 A/C polymorphism attenuates microRNA-155 binding. J Biol Chem 2007;282:24262–9.
[40] Kumarswamy R, Volkmann I, Thum T. Regulation and function of miRNA-21 in health and disease. RNA Biol 2011;8:706–13.
[41] Thum T, Gross C, Fiedler J, Fischer T, Kissler S, Bussen M, et al. MicroRNA-21 contributes to myocardial disease by stimulating MAP kinase signalling in fibroblasts. Nature 2008;456:980–4.
[42] Dennert R, Crijns HJ, Heymans S. Acute viral myocarditis. Eur Heart J 2008;29:2073–82.
[43] Mestas J, Ley K. Monocyte-endothelial cell interactions in the development of atherosclerosis. Trends Cardiovasc Med 2008;18:228–32.
[44] Kuehbacher A, Urbich C, Zeiher AM, Dimmeler S. Role of Dicer and Drosha for endothelial microRNA expression and angiogenesis. Circ Res 2007;101:59–68.
[45] Suarez Y, Fernandez-Hernando C, Pober JS, Sessa WC. Dicer dependent microRNAs regulate gene expression and functions in human endothelial cells. Circ Res 2007;100:1164–73.
[46] Harris TA, Yamakuchi M, Ferlito M, Mendell JT, Lowenstein CJ. MicroRNA-126 regulates endothelial expression of vascular cell adhesion molecule 1. Proc Natl Acad Sci U S A 2008;105:1516–21.
[47] Zhou J, Wang KC, Wu W, Subramaniam S, Shyy JY, Chiu JJ, et al. MicroRNA-21 targets peroxisome proliferators-activated receptor-{alpha} in an autoregulatory loop to modulate flow-induced endothelial inflammation. Proc Natl Acad Sci U S A 2011;108:10355–60.
[48] Fang Y, Shi C, Manduchi E, Civelek M, Davies PF. MicroRNA-10a regulation of proinflammatory phenotype in athero-susceptible endothelium *in vivo* and *in vitro*. Proc Natl Acad Sci U S A 2010;107:13450–5.
[49] Ni CW, Qiu H, Jo H. MicroRNA-663 upregulated by oscillatory shear stress plays a role in inflammatory response of endothelial cells. Am J Physiol Heart Circ Physiol 2011;300:H1762–9.
[50] Zhu N, Zhang D, Chen S, Liu X, Lin L, Huang X, et al. Endothelial enriched microRNAs regulate angiotensin II-induced endothelial inflammation and migration. Atherosclerosis 2011;215:286–93.
[51] Minami Y, Satoh M, Maesawa C, Takahashi Y, Tabuchi T, Itoh T, et al. Effect of atorvastatin on microRNA 221/222 expression in endothelial progenitor cells obtained from patients with coronary artery disease. Eur J Clin Invest 2009;39:359–67.
[52] Dentelli P, Rosso A, Orso F, Olgasi C, Taverna D, Brizzi MF. microRNA-222 controls neovascularization by regulating signal transducer and activator of transcription 5A expression. Arterioscler Thromb Vasc Biol 2010;30:1562–8.
[53] Chen T, Huang Z, Wang L, Wang Y, Wu F, Meng S, et al. MicroRNA-125a-5p partly regulates the inflammatory response, lipid uptake, and ORP9 expression in oxLDL-stimulated monocyte/macrophages. Cardiovasc Res 2009;83:131–9.
[54] Yang K, He YS, Wang XQ, Lu L, Chen QJ, Liu J, et al. MiR-146a inhibits oxidized low-density lipoprotein-induced lipid accumulation and inflammatory response via targeting toll-like receptor 4. FEBS Lett 2011;585:854–60.
[55] Huang RS, Hu GQ, Lin B, Lin ZY, Sun CC. MicroRNA-155 silencing enhances inflammatory response and lipid uptake in oxidized low-density lipoprotein-stimulated human THP-1 macrophages. J Investig Med 2010;58:961–7.
[56] Guo M, Mao X, Ji Q, Lang M, Li S, Peng Y, et al. miR-146a in PBMCs modulates Th1 function in patients with acute coronary syndrome. Immunol Cell Biol 2010;88:555–64.
[57] Takahashi Y, Satoh M, Minami Y, Tabuchi T, Itoh T, Nakamura M. Expression of miR-146a/b is associated with the toll-like receptor 4 signal in coronary artery disease: effect of renin-angiotensin system blockade and statins on miRNA-146a/b and Toll-like receptor 4 levels. Clin Sci (Lond) 2010;119:395–405.
[58] Boldin MP, Taganov KD, Rao DS, Yang L, Zhao JL, Kalwani M, et al. miR-146a is a significant brake on autoimmunity, myeloproliferation, and cancer in mice. J Exp Med 2011;208:1189–201.
[59] Pauley KM, Stewart CM, Gauna AE, Dupre LC, Kuklani R, Chan AL, et al. Altered miR-146a expression in Sjogren's syndrome and its functional role in innate immunity. Eur J Immunol 2011;41:2029–39.
[60] Nahid MA, Satoh M, Chan EK. Mechanistic role of microRNA-146a in endotoxin-induced differential cross-regulation of TLR signaling. J Immunol 2011;186:1723–34.
[61] Chassin C, Kocur M, Pott J, Duerr CU, Gutle D, Lotz M, et al. miR-146a mediates protective innate immune tolerance in the neonate intestine. Cell Host Microbe 2010;8:358–68.
[62] Godshalk SE, Bhaduri-McIntosh S, Slack FJ. Epstein-Barr virus-mediated dysregulation of human microRNA expression. Cell Cycle 2008;7:3595–600.
[63] Dai R, Phillips RA, Zhang Y, Khan D, Crasta O, Ahmed SA. Suppression of LPS-induced Interferon-gamma and nitric oxide in splenic lymphocytes by select estrogen-regulated microRNAs: a novel mechanism of immune modulation. Blood 2008;112:4591–7.
[64] Sheedy FJ, O'Neill LA. Adding fuel to fire: microRNAs as a new class of mediators of inflammation. Ann Rheum Dis 2008;67(Suppl 3):iii50–5.
[65] Xie H, Lim B, Lodish HF. MicroRNAs induced during adipogenesis that accelerate fat cell development are downregulated in obesity. Diabetes 2009;58:1050–7.
[66] Heneghan HM, Miller N, Kerin MJ. Role of microRNAs in obesity and the metabolic syndrome. Obes Rev 2010;11:354–61.
[67] Heneghan HM, Miller N, McAnena OJ, O'Brien T, Kerin MJ. Differential miRNA expression in omental adipose tissue and in the circulation of obese patients identifies novel metabolic biomarkers. J Clin Endocrinol Metab 2011;96:E846–50.
[68] Strum JC, Johnson JH, Ward J, Xie H, Feild J, Hester A, et al. MicroRNA 132 regulates nutritional stress-induced chemokine production through repression of SirT1. Mol Endocrinol 2009;23:1876–84.
[69] Roggli E, Britan A, Gattesco S, Lin-Marq N, Abderrahmani A, Meda P, et al. Involvement of microRNAs in the cytotoxic effects exerted by proinflammatory cytokines on pancreatic beta-cells. Diabetes 2010;59:978–86.
[70] Dey N, Das F, Mariappan MM, Mandal CC, Ghosh-Choudhury N, Kasinath BS, et al. MicroRNA-21 orchestrates high glucose-induced signals to TOR complex 1, resulting in renal cell pathology in diabetes. J Biol Chem 2011;286:25586–603.
[71] Li Y, Song YH, Li F, Yang T, Lu YW, Geng YJ. MicroRNA-221 regulates high glucose-induced endothelial dysfunction. Biochem Biophys Res Commun 2009;381:81–3.
[72] Herrera BM, Lockstone HE, Taylor JM, Ria M, Barrett A, Collins S, et al. Global microRNA expression profiles in insulin target tissues in a spontaneous rat model of type 2 diabetes. Diabetologia 2010;53:1099–109.

[73] Reid G, Kirschner MB, van Zandwijk N. Circulating microRNAs: association with disease and potential use as biomarkers. Crit Rev Oncol Hematol 2011;80:193–208.
[74] Mause SF, Weber C. Microparticles: protagonists of a novel communication network for intercellular information exchange. Circ Res 2010;107:1047–57.
[75] Vickers KC, Palmisano BT, Shoucri BM, Shamburek RD, Remaley AT. MicroRNAs are transported in plasma and delivered to recipient cells by high-density lipoproteins. Nat Cell Biol 2011;13:423–33.
[76] Arroyo JD, Chevillet JR, Kroh EM, Ruf IK, Pritchard CC, Gibson DF, et al. Argonaute 2 complexes carry a population of circulating microRNAs independent of vesicles in human plasma. Proc Natl Acad Sci U S A 2011;108:5003–8.
[77] Adachi T, Nakanishi M, Otsuka Y, Nishimura K, Hirokawa G, Goto Y, et al. Plasma microRNA 499 as a biomarker of acute myocardial infarction. Clin Chem 2010;56:1183–5.
[78] Corsten MF, Dennert R, Jochems S, Kuznetsova T, Devaux Y, Hofstra L, et al. Circulating MicroRNA-208b and MicroRNA-499 reflect myocardial damage in cardiovascular disease. Circ Cardiovasc Genet 2010;3:499–506.
[79] Wang GK, Zhu JQ, Zhang JT, Li Q, Li Y, He J, et al. Circulating microRNA: a novel potential biomarker for early diagnosis of acute myocardial infarction in humans. Eur Heart J 2010;31:659–66.
[80] Fichtlscherer S, De Rosa S, Fox H, Schwietz T, Fischer A, Liebetrau C, et al. Circulating microRNAs in patients with coronary artery disease. Circ Res 2010;107:677–84.
[81] Zampetaki A, Kiechl S, Drozdov I, Willeit P, Mayr U, Prokopi M, et al. Plasma microRNA profiling reveals loss of endothelial miR-126 and other microRNAs in type 2 diabetes. Circ Res 2010;107:810–7.
[82] Taurino C, Miller WH, McBride MW, McClure JD, Khanin R, Moreno MU, et al. Gene expression profiling in whole blood of patients with coronary artery disease. Clin Sci (Lond) 2010;119:335–43.
[83] Hoekstra M, van der Lans CA, Halvorsen B, Gullestad L, Kuiper J, Aukrust P, et al. The peripheral blood mononuclear cell microRNA signature of coronary artery disease. Biochem Biophys Res Commun 2010;394:792–7.
[84] Li T, Cao H, Zhuang J, Wan J, Guan M, Yu B, et al. Identification of miR-130a, miR-27b and miR-210 as serum biomarkers for atherosclerosis obliterans. Clin Chim Acta 2011;412:66–70.
[85] Chen X, Ba Y, Ma L, Cai X, Yin Y, Wang K, et al. Characterization of microRNAs in serum: a novel class of biomarkers for diagnosis of cancer and other diseases. Cell Res 2008;18:997–1006.
[86] Kong L, Zhu J, Han W, Jiang X, Xu M, Zhao Y, et al. Significance of serum microRNAs in pre-diabetes and newly diagnosed type 2 diabetes: a clinical study. Acta Diabetol 2011;48:61–9.
[87] Frantz S, Ertl G, Bauersachs J. Mechanisms of disease: toll-like receptors in cardiovascular disease. Nat Clin Pract Cardiovasc Med 2007;4:444–54.
[88] Salminen A, Huuskonen J, Ojala J, Kauppinen A, Kaarniranta K, Suuronen T. Activation of innate immunity system during aging: NF-kB signaling is the molecular culprit of inflamm-aging. Ageing Res Rev 2008;7:83–105.
[89] Vasa-Nicotera M, Chen H, Tucci P, Yang AL, Saintigny G, Menghini R, et al. miR-146a is modulated in human endothelial cell with aging. Atherosclerosis 2011;217:326–30.

CHAPTER

14

The Role of Cytokines in Clinical Heart Failure

Douglas L. Mann

Cardiovascular Division, Department of Medicine, Center for Cardiovascular Research, Washington University School of Medicine, St. Louis, MO, USA

The first study that linked clinical heart failure (HF) with the expression of inflammatory cytokines was by Levine and colleagues [1] who reported elevated circulating levels of tumor necrosis factor (TNF) in HF patients with a reduced ejection fraction. Since this original report, there has been an exponential rise in the number of cytokines and chemokines that have been identified in the setting of HF with a reduced ejection fraction. Moreover, elevated levels of inflammatory mediators have also been identified in acute decompensated HF as well as in HF patients with a preserved ejection. Thus, there is evidence of an ongoing inflammatory response in all manifestations of clinical HF.

The early clinical observations with respect to TNF prompted a series of experimental studies, which demonstrated that the sustained expression of TNF at levels that were observed in HF patients was sufficient to provoke left ventricular (LV) dysfunction and LV remodeling [2]. These and other preclinical studies formed the basis for several multicenter clinical trials that utilized "targeted" approaches to neutralize TNF in patients with moderate to advanced HF. Unfortunately, these targeted anti-TNF clinical trials were negative with respect to the primary end points of the trial and in some patients resulted in worsening HF and/or death [3,4]. The negative outcomes of these clinical trials had a chilling effect on further clinical efforts to develop therapeutics in this area. Fortunately, progress in the field over the past decade has led to a clearer appreciation of the role of innate immunity and adaptive immunity in the heart and a renewed interest in the clinical role of inflammation in HF. In this chapter, we will discuss the role of cytokines in clinical HF, as well re-evaluate the clinical trials that have been conducted in this area, in light of the new information that has been obtained.

14.1 ROLE OF INFLAMMATION IN THE PATHOGENESIS OF HF

The interest in understanding the role of inflammation in HF was sparked by observations from multiple laboratories that many aspects of the clinical syndrome of HF can be explained by the known biological effects of proinflammatory cytokines (Table 14.1). The "cytokine hypothesis" [5] for HF postulates that HF progresses, at least in part, as a result of the deleterious effects exerted by endogenous cytokine cascades on the heart and the peripheral circulation. It bears emphasis that the cytokine hypothesis does not imply that cytokines cause "heart failure" *per se*, but rather that the overexpression of cytokine cascades contributes to disease progression of HF. Much like the elaboration of neurohormones, the elaboration of cytokines may represent a biological mechanism that is responsible for worsening HF.

14.1.1 Effects of Cytokines on LV Function

Proinflammatory cytokines were first shown to cause LV dysfunction in the setting of the systemic inflammatory response during sepsis. Direct injections of TNF were shown to produce hypotension, metabolic acidosis, and rapid death within minutes, whereas injections of anti-TNF antibodies were sufficient to attenuate the hemodynamic

Inflammation in Heart Failure
http://dx.doi.org/10.1016/B978-0-12-800039-7.00014-1

© 2015 Elsevier Inc. All rights reserved.

TABLE 14.1 Deleterious Effects of Inflammatory Mediators in Heart Failure

Left ventricular dysfunction
Pulmonary edema in humans
Cardiomyopathy in humans
Reduced skeletal muscle blood flow
Endothelial dysfunction
Anorexia and cachexia
Receptor uncoupling from adenylate cyclase experimentally
Activation of the fetal gene program experimentally
Cardiac myocyte apoptosis experimentally

Reproduced with permission Mann DL. Activation of inflammatory mediators in heart failure. In: Mann DL, editor. Heart failure: A companion to Braunwald's heart disease. 2nd ed. Philadelphia: Elsevier/Saunders; 2011. p. 163-84.

collapse observed during endotoxin shock. Subsequent studies in dogs and rats showed that circulating levels of TNF produced negative inotropic effects *in vivo* and *in vitro* (reviewed in Ref. [6]). More recent studies in transgenic mice with cardiac restricted overexpression of TNF showed that forced overexpression of TNF resulted in depressed LV ejection performance that was dependent on TNF "gene dosage" [7].

With respect to the potential mechanisms for the deleterious effects of TNF on LV function, the literature suggests that TNF modulates myocardial function through an immediate pathway that manifests within minutes and is mediated by activation of the neutral sphingomyelinase pathway, by a delayed pathway that requires hours to days to develop, and by nitric oxide mediated blunting of β-adrenergic signaling (reviewed in Ref. [6]). Whereas the negative inotropic effects of IL-1 appear to be mediated, at least in part, through the production of nitric oxide (i.e., the delayed pathway), the negative inotropic effects of IL-6 are less well understood. Recent studies suggested that TNF and IL-1 may produce negative inotropic effects *indirectly* through activation and/or release of IL-18. Remarkably, blockade of IL-18 using neutralizing IL-18 binding protein leads to an improvement in myocardial contractility in atrial tissue following ischemia reperfusion injury [8]. Although the signaling pathways that are responsible for the IL-18 induced negative inotropic effects that have not been delineated thus far, it is likely that they will overlap those for IL-1, given that the IL-18 receptor complex utilizes components of the IL-1 signaling chain.

14.1.2 Effects of Proinflammatory Cytokines on LV Remodeling

LV ventricular remodeling refers to the multitude of changes that occur in cardiac shape, size, and composition in response to myocardial injury. Inflammatory mediators have a number of important biological effects that may play an important role in the process of LV remodeling, including cardiac myocyte hypertrophy, alterations in fetal gene expression, activation of collagenolytic matrix metalloproteinase (MMP), myocardial fibrosis, as well as progressive myocyte loss through apoptosis [6]. Antagonism of innate immune receptors (TLR2, TLR4), innate immune signaling pathways (MyD88, IRAK-4, IRAK-M, and NLRP3), and the proinflammatory cytokines downstream from these pathways (TNF, IL-1β, IL-18) has been shown to attenuate adverse LV remodeling following acute myocardial infarction (reviewed in Ref. [9]). Studies in chimeric mice, wherein it has been possible to separate the role of innate immune signaling in cells derived from the bone marrow from the effects in the myocardium, have demonstrated that activation of innate immune signaling pathways in bone marrow derived neutrophils and monocytes contributes to tissue damage, progressive fibrosis, and adverse cardiac remodeling. Activation of the same pathways in cardiac myocytes is beneficial in the short-term through mitochondrial stabilization, enhanced sarcolemma membrane integrity, as well as through conservation of energy secondary to the development of reversible LV dysfunction (reviewed in Ref. [9]). Studies in experimental models wherein the inflammatory signaling is sustained have also provided important insights in the mechanisms for inflammation-induced adverse LV remodeling. For example, a study in rats showed that infusion of concentrations of TNF that overlap those observed in HF patients led to a time-dependent change in LV dimension that was associated with progressive degradation of the extracellular matrix [2]. Studies in transgenic mice with targeted overexpression of TNF have shown that these mice develop progressive LV dilation, and that TNF-induced activation of MMPs is responsible for collagen degradation and progressive LV dilation [10]. These studies demonstrated that sustained myocardial inflammation leads to temporal changes in the balance between MMP activity and tissue inhibitor of matrix metalloproteinases (TIMPs) and mast cell-mediated TGF-β signaling [11]. These time-dependent changes favor degradation of the extracellular matrix during the onset of inflammation and progressive myocardial fibrosis following sustained inflammation. Thus, the sustained activation of inflammatory signaling contributes to LV remodeling through a variety of different mechanisms that involve both the myocyte and nonmyocyte components of the myocardium.

14.2 INFLAMMATION AS A THERAPEUTIC TARGET IN HF

Given that elevated levels of proinflammatory cytokines mimic many aspects of the HF phenotype and that the deleterious effects of inflammatory mediators are potentially reversible once inflammation subsides, investigators have used a variety of different approaches to antagonize inflammatory mediators in HF (Table 14.2). These fall into one of three broad categories: anti-inflammatory therapies, immunomodulatory therapies, and autoimmune strategies. Because the topic of autoimmunity is not the focus of this chapter, it will not be discussed further herein.

14.2.1 Anti-inflammatory Therapies

The biological effects of proinflammatory mediators can be antagonized through transcriptional or translational approaches, or by so-called biological response modifiers that bind and/or neutralize soluble mediators (e.g., TNF or IL-1β). Many of these strategies have been explored in phase II-III clinical trials, as described below.

14.2.1.1 Transcriptional Suppression of Proinflammatory Cytokines

Pentoxyfilline is a xanthine-derived agent that is known to inhibit TNF transcription and translation. Pentoxyfilline has been studied in a number of small randomized trials in patients with ischemic and dilated cardiomyopathy (Table 14.2) [17,18,20,26]. The use of pentoxifylline was associated with significant improvement in NYHA functional class and/or LV ejection fraction in each of these studies. Importantly, the beneficial effects were seen in all NYHA classes of HF, in patients with ischemic and nonischemic cardiomyopathy, as well as in patients treated with ACE inhibitors and beta-blockers. Apposite to the present discussion, the beneficial effects on cardiac function in some of the studies were accompanied by decreased circulating plasma levels of TNF [27]. Given that pentoxifylline is a nonspecific phosphodiesterase inhibitor, it is possible that the salutary effects of this agent might be unrelated to its anti-inflammatory properties.

Thalidomide (α-*N*-phthalimidoglutarimide) has also been used to suppress TNF production. The mechanism of action of thalidomide with respect to reducing TNF levels appears to be through enhancing mRNA degradation; [28] however, the precise mechanism of action of thalidomide is unclear, and contradictory results have been reported regarding its effects on cytokine levels *in vivo*. Thalidomide was safe and potentially effective in a small open-label dose escalation study in patients with HF. There was a significant increase in the 6-min walk distance and a trend ($p=0.16$) toward improvement in LV ejection fraction and quality of life (QoL) after 12 weeks of maintenance therapy with thalidomide [29]. However, dose-limiting toxicity was observed in two patients (50 and 200 mg/day). In a larger placebo-controlled study of 56 patients with NYHA class II-III HF secondary to an ischemic and nonischemic cardiomyopathy and an LV ejection fraction ≤40%, treatment with up to 200 mg/day of thalidomide for 12 weeks resulted in an increase in LV ejection fraction and a decrease in LV end-diastolic volume [14]. These salutary changes were accompanied by a decrease in circulating levels of MMP 2, but an increase in circulating levels of TNF. The effect of thalidomide on LV ejection fraction was observed to a greater degree in patients with dilated cardiomyopathy, who were able to tolerate higher doses of thalidomide [14].

14.2.1.2 Translational Suppression of Proinflammatory Cytokines

Dexamethasone, thought to suppress TNF biosynthesis at the translational level, may also block TNF biosynthesis at the transcriptional level. In an early study, Parrillo and colleagues [16] randomized 102 patients with dilated cardiomyopathy to treatment with prednisone (60 mg/day) or placebo. After 3 months of therapy, these investigators observed a >5% increase in EF in ~50% of the prednisone-treated patients, whereas ~25% of the controls had a significant improvement in LV EF ($p=0.005$). However, the mean increase in LV ejection fraction was not significantly ($p=0.054$) different between the prednisone-treated group (4.3 ± 1.5%) when compared to controls (2.1 ± 0.8%). When patients were divided into a "reactive" group (defined and prespecified as a fibroblastic/lymphocytic infiltration or immunoglobulin deposition on endomyocardial biopsy, a positive gallium scan, or an elevated erythrocyte and nonreactive) and a "nonreactive" group, the authors noted that ~65% of "reactive" patients had an improved LV ejection fraction at 3 months, whereas ~25% of the "reactive" control patients had an improved LV ejection fraction ($p=0.004$). The prednisone-treated nonreactive patients did not have significantly improved LV function ($p=0.51$). This is the first study that demonstrated that patients with dilated cardiomyopathy benefited from an anti-inflammatory therapy.

TABLE 14.2 Randomized Clinical Trials that have Targeted Inflammation in Heart Failure

Study	Number Patients	NYHA Class	Agent	Category	Follow-up (Months)	Mean Age	Mean LVEF	% ACE-ARB/BB	Primary End Point	Outcome
Randomized anti-cytokine clinical trials in heart failure										
ATTACH [4]	150	II-IV III, IV	Infliximab	DCM, IHD	7	61	24	100/73	Clinical composite score	High dose had adverse effect on clinical outcomes
CORONA [12]	5011	II-IV	Rosuvastatin	IHD	32.8	73	ns	91/75	CV death, nonfatal MI and stroke	No effect on CV death, nonfatal MI and stroke, decreased HF hospitalizations
GISSI-HF [13]	4574	II-IV	Rosuvastatin	DCM, IHD	46.8	68	33*	95/62	Death, Death and CV hospitalization	No effect on Death, Death and CV hospitalization
Gullestad *et al.* [14]	56	II, III	Thalidomide	DCM, IHD	3	66	25	100/91	LVEF, LV volumes, symptoms	Improved LVEF and LV remodeling
Hare *et al.* [15]	405	III, IV	Oxypurinol	DCM, IHD		65	26	95/01	Composite of HF mortality + morbidity + QoL	No overall effect; effect in those with elevated uric acid
Parillo *et al.* [16]	102/RCT	Ns	Prednisone	DCM	3	43	17	Na/na	LVEF	Improved LVEF
RECOVER/ RENAISSANCE/ RECOVER [3]	1500	II-IV	Etanercept	DCM, IHD	5.7/12.9	63	23	98/62	Clinical composite score/ Death, or heart failure hospitalization	No effect on clinical status, death, or heart failure hospitalization
Skudicky *et al.* [17]	39	II, III	Pentoxifylline	DCM	6	49	24	100/100	NYHA class, exercise tolerance, and LVEF	Improved symptoms and LVEF
Sliwa *et al.* [18]	28	II, III	Pentoxifylline	DCM	6	53	24	100/na	NYHA class and LVEF	Improved symptoms and LVEF
Sliwa *et al.* [19]	18	IV	Pentoxifylline	DCM	1	46	15	100/na	Symptoms, cytokines, and LVEF	Improved symptoms and LVEF
Sliwa *et al.* [20]	38	I-IV	Pentoxifylline	IHD	6	55	25	100/100	NYHA class and LVEF	Improved symptoms and LVEF
UNIVERSE [21]	87	II-IV	Rosuvastatin	DCM, IHD	6.5	62	29	98/85	LVEF	No effect on LVEF

Randomized immunomodulation clinical trials in heart failure										
ACCLAIM [22]	2426	II-IV	Celacade	DCM, IHD	10.2	64	23	94/87	Death, or CV hospitalization	No effect on death, or CV hospitalization
Gullestad *et al.* [23]	40	II-IV	IVIG	DCM, IHD	6	61	27	100/75	NYHA class, and LVEF	Improved clinical status and LVEF
IMAC [24]	62/RCT	I-IV	IVIG	DCM	12	43	25	90/18	LVEF and symptoms	No effect
METIS [25]	50/RCT	II-IV	Methotrexate	IHD	3	59	35	85/84	6MWT	No effect

ACE, angiotensin converting enzyme; ARB, angiotensin II receptor blocker; BB, beta-adrenergic receptor blocker; IVIG, intravenous immunoglobulin; LVEF, left ventricular ejection fraction; Mo, months; Na, not available; Ns, not specified; NYHA, New York Heart Association; QoL, quality of life.

** 10% of the patients in GISSI-HF has an EF >40%.*

Reproduced with permission Mann DL. Activation of inflammatory mediators in heart failure. In: Mann DL, editor. Heart failure: a companion to Braunwald's heart disease. 2nd ed. Philadelphia: Elsevier/Saunders; 2011. p. 163-84. Modified from Aukrust Gullestad L, Ueland T, Vinge LE, Finsen A, Yndestad A, Aukrust P. Inflammatory cytokines in heart failure: mediators and markers. Cardiology 2012;122:23-35.

14.2.1.3 Targeted Anti-cytokine Approaches Using Biological Response Modifiers

Two different targeted approaches have been taken to selectively antagonize proinflammatory cytokines in the setting of HF (Table 14.2). The first approach employed a genetically engineered TNF receptor (etanercept) that acts as a "decoy" to prevent TNF from binding to its TNF receptors on target cells, whereas the second approach employed a chimeric monoclonal antibody that neutralizes circulating TNF.

14.2.1.3.1 SOLUBLE TNF RECEPTORS

Etanercept (Enbrel™) is genetically engineered humanized protein consisting of two human TNF p75 receptors coupled to a human IgG_1:Fc fragment. Two small short-term studies in patients with stable HF showed that treatment with 25 mg biw etanercept resulted in improved QoL, increased 6-min walk distance, and improved LV ejection performance after 3 months of treatment [6]. These early trials formed the basis for two moderate-size multicenter trials that employed parallel study designs but differed in the dose of etanercept that was used. The *R*andomized *E*tanercept *N*orth *A*mer*I*can *S*trategy to *S*tudy *A*ntago*N*ism of *C*ytokin*E*s (RENAISSANCE; $n=900$) trial was conducted in North America, whereas the *R*esearch into *E*tanercept *C*yt*o*kine Antagonism in *Ve*ntricula*r* Dysfunction (RECOVER; $n=900$) trial was held in Europe and Australia. The primary end point for both trials was a clinical composite score wherein patients were assessed as improved, unchanged, or worsened at 24 weeks. In RENAISSANCE, patients were treated with placebo or subcutaneous etanercept 25 mg biw or 25 mg tiw, whereas the RECOVER trial employed doses of 25 mg qw or 25 mg biw of subcutaneous etanercept. A third prespecified trial, the *R*andomized *E*ta*ne*rce*p*t *W*orldwide Ev*al*uation (RENEWAL; $n=1500$) trial, utilized the pooled data from the RENAISSANCE (biw and tiw dosing) and RECOVER (biw dosing only). The primary end point for RENEWAL was all-cause mortality and hospitalization for HF. On the basis of prespecified stopping rules, both trials were terminated prematurely because of the lack of benefit of etanercept on the clinical composite in RENAISSANCE ($p=0.17$) and RECOVER ($p=0.34$) (Figure 14.1a). The prespecified analysis of RENEWAL showed that there was no effect of etanercept on the primary end point (Figure 14.1b) of death or chronic HF hospitalization (hazard ratio = 1.1, 95% CI 0.91-1.33, $P=0.33$) [3]. However, in a post hoc analysis of the RENAISSANCE trial, patients receiving biw and tiw etanercept experienced, respectively, an increased 1.21 ($p=0.17$) and 1.23 ($p=0.13$) risk of death/HF hospitalization when compared with the placebo group. Further analysis of the components of the clinical composite score in the RENAISSANCE

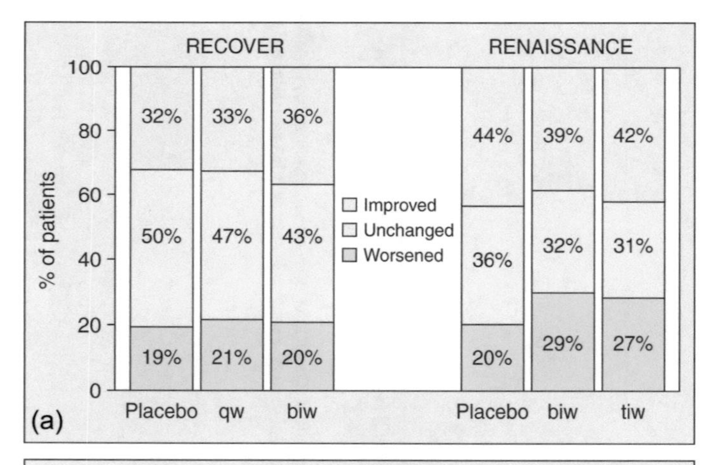

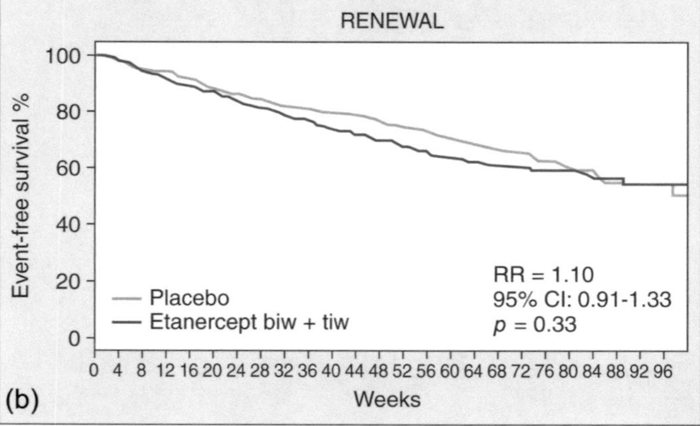

FIGURE 14.1 Results of the RENAISSANCE, RECOVER, AND RENEWAL trials. (a) analysis of the "clinical status" composite score for the RECOVER and RENAISSANCE trials in the placebo and etanercept groups. (b) Kaplan-Meier analysis of the time to death or heart failure hospitalizations in the placebo and etanercept group (biw and tiw) in the RENEWAL analysis. Reproduced with permission Mann DL. Activation of inflammatory mediators in heart failure. In: Mann DL, editor. Heart failure: a companion to Braunwald's heart disease. 2nd ed. Philadelphia: Elsevier/Saunders; 2011. p. 163-84.

trial revealed that there was a significantly greater proportion of etanercept-treated patients (29%, $p<0.04$) in the worsened category at 24 weeks when compared to placebo-treated patients (20%). Increases in the risk of death/HF hospitalization and a worsening clinical composite were not observed in RECOVER, wherein the dose and duration of etanercept dosing was less. Patients in RECOVER received etanercept for a median time of 5.7 months, whereas patients in RENAISSANCE received etanercept for 12.7 months. Had these trials not been stopped prematurely for futility, the hazard ratios for increased death/HF hospitalization may have also been worse in the RECOVER trial. On the basis of these findings, the prescribing information for etanercept has been updated and now suggests that physicians exercise caution in the use of etanercept in patients with HF.

Although the precise explanation for the worsening HF in the RENAISSANCE trial is not known, it bears emphasis that TNF receptor antagonists have intrinsic biological activity and, in certain settings, can act as agonists (referred to as a stimulating antagonist [30]). We and others have reported that in some settings etanercept can stabilize TNF and increase its bioactivity (see Ref. [6] for further discussion). Although the stabilizing effects of etanercept might not be problematic in rheumatoid arthritis, wherein TNF is encapsulated within a joint space and peripheral circulating TNF levels are relatively low (compared to HF) or are nonexistent, it is possible that an increase in the circulating levels of biologically active TNF in a HF patients might contribute to worsening HF.

14.2.1.3.2 MONOCLONAL ANTIBODIES

Infliximab (Remicade™) is a chimeric monoclonal antibody consisting of a genetically engineered anti-TNF murine Fab fragment fused to a human FC portion of human IgG_1. Although infliximab had been shown to be effective in effective in Crohn's disease and rheumatoid arthritis, infliximab had never been tested in preclinical nor early phase I clinical studies in HF patients. The *Anti-TNFα Therapy Against CHF* (ATTACH) trial was a phase II study in 150 patients with moderate to advanced HF (NYHA class III, IV). The primary end point of the ATTACH trial was the clinical composite score that was also employed in RENAISSANCE and RECOVER [3]. Patients were randomized to receive three separate intravenous infusions of infliximab (5 or 10 mg/kg) at baseline and at 2 and 4 weeks, followed by an assessment of the clinical composite score at 14 and 28 weeks. Because ATTACH was a pilot phase II study, there was no requirement for a formal Data Safety Monitoring Board to monitor ongoing clinical outcomes during the trial. Analysis of the completed data set revealed that there were increased rates of mortality and HF hospitalization, particularly in the group that was receiving the highest dose of infliximab (Figure 14.2). On the basis of these

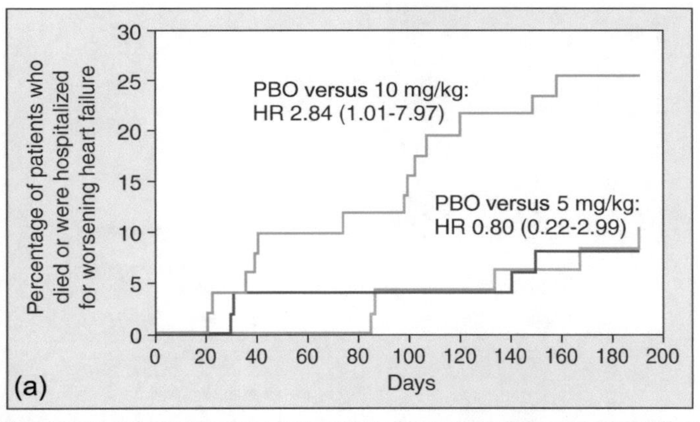

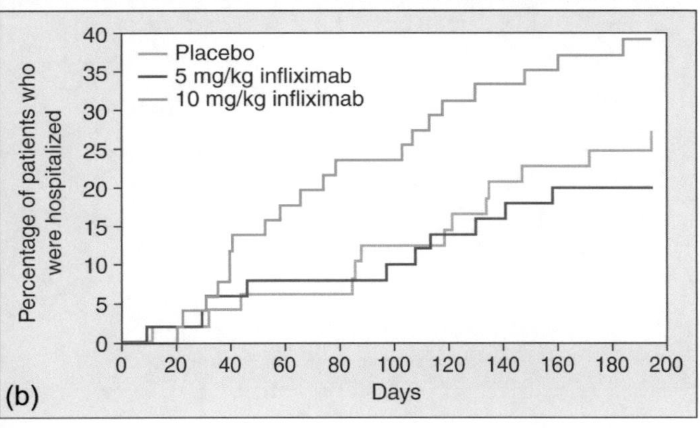

FIGURE 14.2 Results of the ATTACH trial. (a) Kaplan-Meier rates of death and hospitalization for heart failure. (b) Kaplan-Meier rates of hospitalization for any reason. (Key: PBO, placebo; HR, hazard ratio). Reproduced with permission Mann DL. Activation of inflammatory mediators in heart failure. In: Mann DL, editor. Heart failure: a companion to Braunwald's heart disease. 2nd ed. Philadelphia: Elsevier/Saunders; 2011. p. 163-84.

findings, the prescribing information for infliximab has been changed, and it is now recommended that treatment with infliximab be discontinued in patients with worsening HF and that treatment with infliximab should not be initiated in patients with HF.

Analogous to the discussion above for the RENAISSANCE trial, it is not possible to precisely identify the mechanism for the untoward outcomes in ATTACH. However, the publication of the full trial results from ATTACH has allowed for some potential mechanistic insights that were not previously available [6]. As shown in Figure 14.3a, one of the mechanisms of action of infliximab is to bind to cells expressing TNF on their membrane, and to lyse these cells through complement fixation. Although this type of biological activity is beneficial in eliminating clones of activated T-cells in Crohn's disease, it is predictable that infliximab might be deleterious in HF if infliximab bound to TNF that was expressed on the sarcolemma of failing cardiac myocytes (TNF is not expressed in the nonfailing heart), which would lead to complement fixation, lysis of cardiac myocyte cell membranes, and cell death [31]. Analysis of the ATTACH trial indirectly supports this point of view. As shown in Figure 14.3b, plasma levels of immunoreactive TNF increased at the time of treatment with infliximab at 2 and 6 weeks, as well as after the last dose in infliximab at 6 weeks. Although the increase in TNF levels was attributed to TNF that was bound to infliximab (and hence presumably neutralized), this explanation does not explain the striking 25-fold increase in TNF levels at 10-28 weeks, when the infliximab levels were declining below detectable levels (see Figure 14.3b). Moreover, there was a progressive and paradoxical rise in CRP and IL-6 levels over the course of the ATTACH trial, consistent with ongoing tissue injury. Accordingly, one biologically plausible explanation for the increase in patient morbidity and mortality in the ATTACH trial is that infliximab was overtly toxic through complement fixation in the heart.

14.2.1.3.3 IL-1 RECEPTOR ANTAGONIST

Anakinra (Kineret™) is an interleukin-1 (IL-1) receptor antagonist that blocks the biologic activity of naturally IL-1 by competitively inhibiting the binding of IL-1 to the Interleukin-1 type receptor. Anakinra has been shown to prevent adverse cardiac remodeling following LAD ligation in mice [32] but did not have a significant effect on LV remodeling in small randomized study in patients with acute myocardial infarction [33]. Although the experience

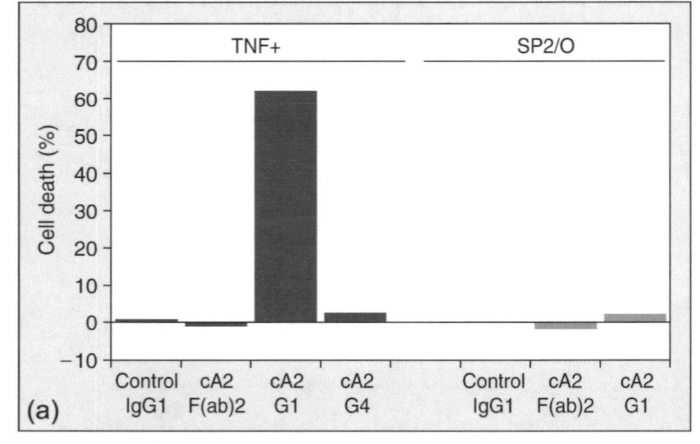

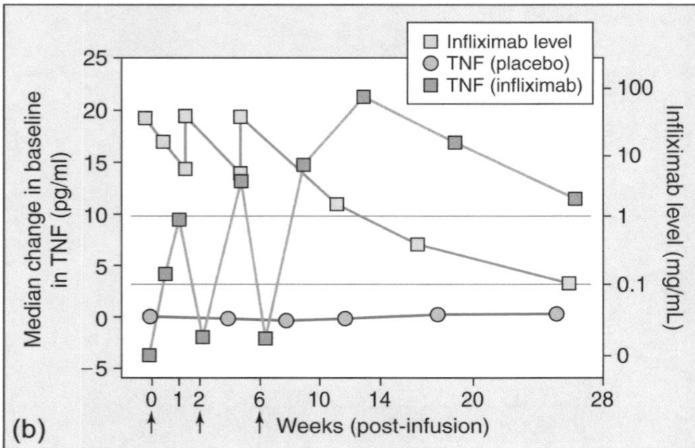

FIGURE 14.3 Biological properties of infliximab. (a) Infliximab (cA2 G1) is cytotoxic for cells that express TNF on their cell membranes (TNF+), whereas it is not cytotoxic for cells that do not express TNF on their membranes (SP2/O). The mechanism for the cytotoxic effects of infliximab was demonstrated using F(ab)2 fragments of infliximab, which lack the Fc domain and therefore cannot fix complement. As shown, the F(ab)2 fragment of infliximab was not cytotoxic for TNF+ cells. (b) Levels of immunoreactive TNF in patients who received placebo and infliximab (10 mg/kg) are displayed in relation to the circulating levels of infliximab (data are redrawn from Figures 2 and 4 in reference [4]). The dotted horizontal lines depict the upper and lower limits of the therapeutic window for infliximab. Reproduced with permission Mann DL. Activation of inflammatory mediators in heart failure. In: Mann DL, editor. Heart failure: a companion to Braunwald's heart disease. 2nd ed. Philadelphia: Elsevier/Saunders; 2011. p. 163-84.

with anakinra in HF has been limited, two small studies have shown significant improvements in exercise performance in patients with HF with a depressed ejection fraction ($n=7$) and a preserved ejection fraction ($n=12$) [34,35].

14.2.1.4 Other Anti-inflammatory Agents

The biological effects of proinflammatory mediators can also be antagonized using pleiotropic drugs that have anti-inflammatory properties. Three of these therapeutics, statins, N-3 polyunsaturated fatty acids (PUFAs), and oxypurinol, have been tested in phase III clinical trials.

14.2.1.4.1 STATINS

Stains have a variety of pleiotropic effects, including inhibition of inflammatory responses, increased nitric oxide bioavailability, improved endothelial function, as well as antioxidant properties (reviewed in Ref. [36]). Based upon several retrospective analyses, several large clinical trials were performed with rosuvastatin in patients with HF, as discussed in Chapter 12.

14.2.1.4.2 N-3 POLYUNSATURATED FATTY ACIDS

There is a large body of experimental evidence suggesting that N-3 PUFA have favorable effects on inflammation, including a reduction of endothelial activation and production of inflammatory cytokines, platelet aggregation, autonomic tone, blood pressure, heart rate, and LV function. In a parallel arm of the GISSI-HF study, patients with NYHA class II-IV HF were randomized to receive PUFAs or placebo. The GISSI-HF trial showed that long-term administration of 1g/day of omega N-3 PUFA resulted in a significant reduction in both all-cause mortality (adjusted HR 0.91 [95.5% CI 0.83-0.99], $p=0.041$) and all-cause mortality and cardiovascular admissions (adjusted HR 0.92 [99% CI 0.849-0.999], $p=0.009$), in all the predefined subgroups, including HF patients in nonischemic cardiomyopathy group [37]. Although N-3 PUFA are not endorsed by current practice guidelines, the use of N-3 PUFA may be considered in patients who remain symptomatic despite optimal medical therapy.

14.2.1.4.3 OXYPURINOL

Elevated levels of uric acid (UA) are known to predict mortality and the need for heart transplantation in patients with HF [38]. UA is a byproduct of the purine metabolism via the xanthine oxidase (XO) pathway. Serum uric acid (SUA) levels may increase in the HF because of increased generation, decreased excretion, or a combination of both factors. Recent studies have shown that UA can trigger interleukin-1β-mediated inflammation via activation of the NOD-like receptor protein (NLRP)3 inflammasome, which is a large multimolecular complex that plays a critical role in the processing of immature interleukin-1β to mature "secretable" form of interleukin-1β. Monosodium urate crystals have can also activate the innate immune system through engagement of TLR2 and TLR4 [39].

The OPTIME-HF trial was a prospective randomized clinical trial that evaluated the effects of the XO inhibitor oxypurinol in patients with New York Heart Association functional class III to IV HF with a LV ejection fraction <40% [15]. The end point of the trial was a clinical composite comprising morbidity, mortality, and QoL evaluated at 24 weeks. The percentage of patients characterized as improved, unchanged, or worsened did not differ between those receiving oxypurinol or placebo. In a subgroup analysis, patients with elevated SUA level of >9.5mg/dl responded favorably to oxypurinol, whereas oxypurinol patients with SUA <9.5mg/dl exhibited a trend toward worsening. The ongoing NIH-sponsored EXACT (Using Allopurinol to Relieve Symptoms in Patients with Heart Failure and High Uric Acid Levelstrial; NCT00987415) tested the hypothesis that treatment with allopurinol will lead to improvements in a composite clinical in HF patients with a reduced ejection fraction and a SUA of >9.5mg. The EXACT trial did not show a benefit for allopurinol on clinical outcomes in patients with HF.

14.2.2 Immunomodulation

An alternative approach to targeting specific components of the inflammatory cascade is to employ strategies that dampen the various components systemic inflammatory response. Given the increasing recognition that cross talk between innate and adaptive immune systems leads to progressive LV remodeling following acute myocardial infarction, and that adverse LV remodeling is driven by activation of macrophages, dendritic cells, and $CD4^+$ T-cells [40–42], there has been interest in developing broad-based immunomodulatory strategies for patients with HF. Thus far, three different approaches have been employed in HF studies: intravenous immunoglobulin (IVIG), methotrexate, and immune modulation therapy (IMT).

14.2.2.1 Intravenous Immuoglobulin

Therapy with IVIG has been tried in a wide range of immune-mediated disorders, such as Kawasaki's syndrome, dermatomyositis, multiple sclerosis, and, most recently, dilated cardiomyopathy, wherein the initial results have been encouraging. In a double-blind, placebo-controlled study of 20 ischemic and nonischemic NYHA class II-IV HF patients with an LV ejection fraction <40%, monthly IVIG treatment for 6 months resulted in a significant increase in LV ejection fraction from 26% to 31%, independent of HF etiology [23]. These improvements in functional class and LV function were accompanied by an increase in the anti-inflammatory mediators IL-10, IL-1 receptor antagonist (IL-1Ra), and soluble TNF receptors, as well as a slight decrease in plasma TNF suggesting that IVIG evoked a net anti-inflammatory effect. In contrast to these encouraging results, induction therapy with IVIG in the IMAC (Intravenous in Myocarditis and Acute Cardiomyopathy) trial in patients with recent-onset cardiomyopathy (<6 months) and an LV ejection fraction <40% demonstrated no significant effect on LV ejection fraction when compared to placebo [24]. However, it bears emphasis that there was also an increase in LV ejection fraction from 23% to 42% in the placebo arm, which would have made it difficult to show a statistically significant increase in LV ejection fraction in the treatment arm. Moreover, there were important differences in the IVIG dosing strategies in IMAC and the study by Gullestad and colleagues. That is, while both studies used induction therapy (a total of 2 g/kg IVIG), in the study by Gullestad *et al.* maintenance therapy (monthly infusions [0.4 g/kg] for a total of 5 months) was also given. Thus, one possible reason for the different outcomes in these two studies is that IVIG maintenance therapy is required for an extended period of time, as has been observed in other chronic inflammatory disorders.

14.2.2.2 Methotrexate

Epidemiological studies have shown that patients with rheumatoid arthritis have an increased incidence of HF [43], and that the HF that develops in elderly rheumatoid arthritis patients cannot be explained entirely by traditional cardiovascular risk factors nor by the presence of ischemic heart disease [44]. Notably, the HF that develops in these patients is associated with concomitant increase in circulating levels of TNF [43]. Methotrexate, which was originally developed as a folate antagonist for the treatment of cancer, has become a mainstay of therapy in rheumatoid arthritis. Several mechanisms have been proposed including inhibition of T cell proliferation via its effects on purine and pyrimidine metabolism, inhibition of transmethylation reactions required for the prevention of T cell cytotoxicity, interference with glutathione metabolism leading to alterations in recruitment of monocytes and other cells to the inflamed joint, and promotion of the release of the endogenous anti-inflammatory mediator adenosine [45]. Of note, the use of methotrexate in rheumatoid arthritis has also been associated with reduced cardiovascular events, including HF hospitalization, especially in patients 65-years old or older [46]. Methotrexate was evaluated in a small ($n = 71$) prospective randomized clinical trial of HF patients treated with 7.5 mg qw for 12 weeks [47]. Compared to patients on optimal medical therapy, addition of low-dose methotrexate resulted in a significant reduction in the circulating levels of proinflammatory cytokines (TNF, IL-6, and MCP-1) and upregulation of the anti-inflammatory cytokines (IL-10 and soluble IL-1 receptor antagonist). There were also improvements in NYHA classification, 6-min walk test distance, and QoL when compared with baseline values. However, methotrexate had no effect of LV remodeling nor LV ejection fraction after 12 weeks of therapy. The main adverse effects reported for low-dose methotrexate were related to gastrointestinal symptoms; there were no severe drug toxicities such as bone marrow suppression or alopecia recorded. The METIS (Methotrexate Therapy Effects in the Physical Capacity of Patients with Ischemic Heart Failure) trial evaluated low-dose methotrexate in 50 patients with chronic ischemic heart disease. Patients were given methotrexate (7.5 mg) or placebo, plus folic acid (5 mg), for 12 weeks. The primary end point was the difference in 6-min walk test (6MWT) distance before and after treatment. There was no significant difference between groups in distance covered in the 6-min walk test, nor NYHA classification [25]. The effects of methotrexate on the rate of HF hospitalization (secondary outcome measure) are being evaluated in the ongoing CIRT (Cardiovascular Inflammation Reduction Trial [NCT 1594333]) trial, which examines whether low-dose methotrexate reduces heart attacks, strokes, or death in people with type 2 diabetes or metabolic syndrome that have had a heart attack or known coronary artery disease.

14.2.2.3 Immune Modulation Therapy

IMT (Celacade™; Vasogen, Inc) utilized a medical device that exposes a sample of blood to a combination of physiochemical stressors *ex vivo*. The treated blood sample is administered intramuscularly along with local anesthetic into the same patient from whom the sample is obtained. The physiochemical stresses to which the autologous blood sample is subjected are known to initiate or facilitate apoptotic cell death. The uptake of apoptotic cells by macrophages results in a downregulation of proinflammatory cytokines, including TNF, IL-1β, and IL-8, and an increase in production of the anti-inflammatory cytokines, including TGF-β and IL-10 [48]. Given the imbalance between pro- and anti-inflammatory cytokines in patients with HF [49], it was hypothesized that IMT would restore this balance toward normal. In a pilot study employing Celacade™ in 73 patients with moderate HF, the investigators noted that the group receiving Celacade™ experienced significantly fewer hospitalizations or deaths when compared to the placebo group. The decrease in event rate in the treatment arm was accompanied by improvements in QoL and NYHA clinical classification [50]. Based on the encouraging results of the early studies the ACCLAIM (Advance Chronic Heart Failure Clinical Assessment of Immune Modulation) pivotal study was conducted in 2426 patients with NYHA class II-IV HF patients with ischemic and nonischemic dilated cardiomyopathy. Patients were randomly assigned to receive Celacade ($n = 1213$) or placebo ($n = 1213$) by intragluteal injection on days 1, 2, 14, and every 28 days thereafter [22].The primary end point was an event-driven composite of time to death from any cause or first hospitalization for cardiovascular reasons. There was no significant difference between the Celacade™ and placebo-treated patients with respect to the primary end point of the trial, which was death from any cause or cardiovascular hospitalization (HR 0.92; 95% CI 0.80-1.05; $p = 0.22$). However, in a prespecified subgroup analysis of patients with NYHA II HF and patients without a history of previous myocardial infarction, it was noted that treatment with Celacade™ was associated with a 39% (0.61; 95% CI 0.46-0.80; $p = 0.0003$) and 26% (0.74; 0.57-0.95; $p = 0.02$) reduction in the risk death from any cause or first hospitalization for cardiovascular reasons, respectively, suggesting that IMT may have benefited patients with nonischemic cardiomyopathy and/or patients with milder HF (NYHA class II).

14.3 SUMMARY AND FUTURE DIRECTIONS

As summarized in the foregoing review, the experimental evidence linking activation of the innate immune system to the pathogenesis of HF has grown exponentially since the original description in 1990. Unfortunately, the ability to translate this information to HF patients has not met with success in phase III clinical trials, and in some cases has led to worsening HF. Although the reasons for the inability to identify a safe and effective anti-inflammatory in HF are not known, the information gleaned over the past two decades may prove useful in identifying future therapeutic inflammatory targets. For example, the intrinsic complexity of innate immune biological signaling pathways was simply not known during the planning and implantation of the two targeted anti-TNF trials. As noted by Bruce Beutler there is a "price to be paid" for paralyzing innate immunity in an attempt to modulate the inflammation that arises from infectious and noninfectious etiologies [51]. It is also important to recognize that ACE inhibitors, β-blockers, and aldosterone antagonists, which are the mainstays of treatment in HF, are also "anti-inflammatory" (see also Chapter 10), insofar as the downstream signal transduction pathways from these classical neurohormonal pathways converge on NF-κB signaling, and are thus inherently proinflammatory. Moreover, demonstrating the benefit of additional add-on therapies on top of conventional triple therapy (ACE, β-blocker, aldosterone antagonist) is difficult because the annual mortality for patients with moderate HF is now ~5-7%. Accordingly, in future studies it may be necessary to use biomarkers to select HF patients who have ongoing inflammation despite optimal medical therapy. Indeed, a recent consensus statement from the Translation Research Committee of the Heart Failure Association of the European Society of Cardiology suggested that there may not be a common inflammatory pathway that characterizes all of the different forms of HF, and that going forward it would be important to design specific anti-inflammatory approaches for different types and stages of HF, as well as to determine the specific inflammatory pathways that are activated in different forms of HF [52]. *Given, the inherent difficulties in developing new heart failure therapies in general, as well as the specific difficulties in targeting innate immunity mentioned above, is there a foreseeable future for developing anti-inflammatory strategies in heart failure?* Despite the inauspicious beginning with targeted anti-inflammatory approaches, the expanding body of knowledge in the field of innate immunity and the development of new therapeutic targets in this area, coupled with the ability to utilize inflammatory biomarkers to identify subsets of HF patients who have ongoing inflammation despite optimal medical and device therapy, raises the exciting possibility that we ultimately will be able to identify subsets of HF patients who will benefit from anti-inflammatory strategies in HF.

Acknowledgments

The author would like to apologize in advance to colleagues whose work was not directly cited in this review because of the imposed space limitations.

Funding Sources: This research was supported by research funds from the N.I.H. (RO1 HL89543, RO1 111094).

Reference

[1] Levine B, Kalman J, Mayer L, et al. Elevated circulating levels of tumor necrosis factor in severe chronic heart failure. N Engl J Med 1990;223:236–41.
[2] Bozkurt B, Kribbs S, Clubb Jr FJ, et al. Pathophysiologically relevant concentrations of tumor necrosis factor-a promote progressive left ventricular dysfunction and remodeling in rats. Circulation 1998;97:1382–91.
[3] Mann DL, McMurray JJV, Packer M, et al. Targeted anti-cytokine therapy in patients with chronic heart failure: results of the Randomized EtaNcercept Worldwide evALuation (RENEWAL). Circulation 2004;109:1594–602.
[4] Chung ES, Packer M, Lo KH, et al. Randomized, double-blind, placebo-controlled, pilot trial of infliximab, a chimeric monoclonal antibody to tumor necrosis factor-{alpha}, in patients with moderate-to-severe heart failure: results of the anti-TNF Therapy Against Congestive Heart failure (ATTACH) trial. Circulation 2003;107:3133–40.
[5] Seta Y, Shan K, Bozkurt B, et al. Basic mechanisms in heart failure: the cytokine hypothesis. J Card Fail 1996;2:243–9.
[6] Mann DL. Inflammatory mediators and the failing heart: past, present, and the foreseeable future. Circ Res 2002;91:988–98.
[7] Franco F, Thomas GD, Giroir BP, et al. Magnetic resonance imaging and invasive evaluation of development of heart failure in transgenic mice with myocardial expression of tumor necrosis factor-alpha. Circulation 1999;99:448–54.
[8] Pomerantz BJ, Reznikov LL, Harken AH, et al. Inhibition of caspase 1 reduces human myocardial ischemic dysfunction via inhibition of IL-18 and IL-1beta. Proc Natl Acad Sci U S A 2001;98:2871–6.
[9] Mann DL. The emerging role of innate immunity in the heart and vascular system: for whom the cell tolls. Circ Res 2011;108:1133–45.
[10] Sivasubramanian N, Coker ML, Kurrelmeyer K, et al. Left ventricular remodeling in transgenic mice with cardiac restricted overexpression of tumor necrosis factor. Circulation 2001;104:826–31.
[11] Zhang W, Chancey AL, Tzeng HP, et al. The development of myocardial fibrosis in transgenic mice with targeted overexpression of tumor necrosis factor requires mast cell-fibroblast interactions. Circulation 2011;124:2116.
[12] Kjekshus J, Apetrei E, Barrios V, et al. Rosuvastatin in older patients with systolic heart failure. N Engl J Med 2007;357:2248–61.
[13] Gissi-HF Investigators. Effect of rosuvastatin in patients with chronic heart failure (the GISSI-HF trial): a randomised, double-blind, placebo-controlled trial. Lancet 2008;372:1231–9.
[14] Gullestad L, Ueland T, Fjeld JG, et al. Effect of thalidomide on cardiac remodeling in chronic heart failure: results of a double-blind, placebo-controlled study. Circulation 2005;112:3408–14.
[15] Hare JM, Mangal B, Brown J, et al. Impact of oxypurinol in patients with symptomatic heart failure. Results of the OPT-CHF study. J Am Coll Cardiol 2008;51:2301–9.
[16] Parrillo JE, Cunnion RE, Epstein SE, et al. A prospective randomized controlled trial of prednisone for dilated cardiomyopathy. N Engl J Med 1989;321:1061–8.
[17] Skudicky D, Bergemann A, Sliwa K, et al. Beneficial effects of pentoxifylline in patients with idiopathic dilated cardiomyopathy treated with angiotensin-converting enzyme inhibitors and carvedilol: results of a randomized study. Circulation 2001;103:1083–8.
[18] Sliwa K, Skudicky D, Candy G, et al. Randomized investigation of effects of pentoxifylline on left ventricular performance in idiopathic dilated cardiomyopathy. Lancet 1998;351:1091–3.
[19] Gordon S. Pattern recognition receptors doubling up for the innate immune response. Cell 2002;111:927–30.
[20] Sliwa K, Woodiwiss A, Kone VN, et al. Therapy of ischemic cardiomyopathy with the immunomodulating agent pentoxifylline: results of a randomized study. Circulation 2004;109:750–5.
[21] Krum H, Ashton E, Reid C, et al. Double-blind, randomized, placebo-controlled study of high-dose HMG CoA reductase inhibitor therapy on ventricular remodeling, pro-inflammatory cytokines and neurohormonal parameters in patients with chronic systolic heart failure. J Card Fail 2007;13:1–7.
[22] Torre-Amione G, Anker SD, Bourge RC, et al. Results of a non-specific immunomodulation therapy in chronic heart failure (ACCLAIM trial): a placebo-controlled randomised trial. Lancet 2008;371:228–36.
[23] Gullestad L, Aass H, Fjeld JG, et al. Immunomodulating therapy with intravenous immunoglobulin in patients with chronic heart failure. Circulation 2001;103:220–5.
[24] McNamara DM, Holubkov R, Starling RC, et al. Controlled trial of intravenous immune globulin in recent-onset dilated cardiomyopathy. Circulation 2001;103:2254–9.
[25] Moreira DM, Vieira JL, Gottschall CA. The effects of METhotrexate therapy on the physical capacity of patients with ISchemic heart failure: a randomized double-blind, placebo-controlled trial (METIS trial). J Card Fail 2009;15:828–34.
[26] Sliwa K, Woodiwiss A, Candy G, et al. Effects of pentoxifylline on cytokine profiles and left ventricular performance in patients with decompensated congestive heart failure secondary to idiopathic dilated cardiomyopathy. Am J Cardiol 2002;90:1118–22.
[27] Skudicky D, Sliwa K, Bergemann A, et al. Reduction in Fas/APO-1 plasma concentrations correlates with improvement in left ventricular function in patients with idiopathic dilated cardiomyopathy treated with pentoxifylline. Heart 2000;84:438–9.
[28] Moreira AL, Sampaio EP, Zmuidzinas A, et al. Thalidomide exerts its inhibitory action on tumor necrosis factor-alpha by enhancing messenger RNA degradation. J Exp Med 1993;177:1675–80.
[29] Agoston I, Dibbs ZI, Wang F, et al. Preclinical and clinical assessment of the safety and potential efficacy of thalidomide in heart failure. J Card Fail 2002;8:306–14.
[30] Klein B, Brailly H. Cytokine-binding proteins: stimulating antagonists. Immunol Today 1995;16:216–20.
[31] Homeister JW, Lucchesi BR. Complement activation and inhibition in myocardial ischemia and reperfusion injury. Annu Rev Pharmacol Toxicol 1994;34:17–40.
[32] Abbate A, Salloum FN, Van Tassell BW, et al. Alterations in the interleukin-1/interleukin-1 receptor antagonist balance modulate cardiac remodeling following myocardial infarction in the mouse. PLoS One 2011;6:e27923.
[33] Abbate A, Van Tassell BW, Biondi-Zoccai G, et al. Effects of interleukin-1 blockade with anakinra on adverse cardiac remodeling and heart failure after acute myocardial infarction [from the Virginia Commonwealth University-Anakinra Remodeling Trial (2) (VCU-ART2) pilot study]. Am J Cardiol 2013;111:1394–400.

[34] Van Tassell BW, Arena R, Biondi-Zoccai G, et al. Effects of interleukin-1 blockade with anakinra on aerobic exercise capacity in patients with heart failure and preserved ejection fraction (from the D-HART pilot study). Am J Cardiol 2014;113:321–7.
[35] Van Tassell BW, Arena RA, Toldo S, et al. Enhanced interleukin-1 activity contributes to exercise intolerance in patients with systolic heart failure. PLoS One 2012;7:e33438.
[36] Ramasubbu K, Estep J, White DL, et al. Experimental and clinical basis for the use of statins in patients with ischemic and nonischemic cardiomyopathy. J Am Coll Cardiol 2008;51:415–26.
[37] Yancy CW, Jessup M, Bozkurt B, et al. 2013 ACCF/AHA guideline for the management of heart failure: a report of the American College of Cardiology Foundation/American Heart Association Task Force on Practice Guidelines. Circulation 2013;128:e240–e327.
[38] Anker SD, Doehner W, Rauchhaus M, et al. Uric acid and survival in chronic heart failure: validation and application in metabolic, functional, and hemodynamic staging. Circulation 2003;107:1991–7.
[39] Ghaemi-Oskouie F, Shi Y. The role of uric acid as an endogenous danger signal in immunity and inflammation. Curr Rheumatol Rep 2011;13:160–6.
[40] Ismahil MA, Hamid T, Bansal SS, et al. Remodeling of the mononuclear phagocyte network underlies chronic inflammation and disease progression in heart failure: critical importance of the cardiosplenic axis. Circ Res 2014;114:266–82.
[41] Hofmann U, Beyersdorf N, Weirather J, et al. Activation of CD4+ T lymphocytes improves wound healing and survival after experimental myocardial infarction in mice. Circulation 2012;125:1652–63.
[42] Epelman S, Lavine KJ, Beaudin AE, et al. Embryonic and adult-derived resident cardiac macrophages are maintained through distinct mechanisms at steady state and during inflammation. Immunity 2014;40:91–104.
[43] Nicola PJ, Maradit-Kremers H, Roger VL, et al. The risk of congestive heart failure in rheumatoid arthritis: a population-based study over 46 years. Arthritis Rheum 2005;52:412–20.
[44] Crowson CS, Nicola PJ, Kremers HM, et al. How much of the increased incidence of heart failure in rheumatoid arthritis is attributable to traditional cardiovascular risk factors and ischemic heart disease? Arthritis Rheum 2005;52:3039–44.
[45] Nicola PJ, Crowson CS, Maradit-Kremers H, et al. Contribution of congestive heart failure and ischemic heart disease to excess mortality in rheumatoid arthritis. Arthritis Rheum 2006;54:60–7.
[46] Setoguchi S, Schneeweiss S, Avorn J, et al. Tumor necrosis factor-alpha antagonist use and heart failure in elderly patients with rheumatoid arthritis. Am Heart J 2008;156:336–41.
[47] Gong K, Zhang Z, Sun X, et al. The nonspecific anti-inflammatory therapy with methotrexate for patients with chronic heart failure. Am Heart J 2006;151:62–8.
[48] Fadok VA, Bratton DL, Konowal A, et al. Macrophages that have ingested apoptotic cells in vitro inhibit proinflammatory cytokine production through autocrine/paracrine mechanisms involving TGF-beta, PGE2, and PAF. J Clin Invest 1998;101:890–8.
[49] Aukrust P, Ueland T, Lien E, et al. Cytokine network in congestive heart failure secondary to ischemic or idiopathic dilated cardiomyopathy. Am J Cardiol 1999;83:376–82.
[50] Torre-Amione G, Sestier F, Radovancevic B, et al. Effects of a novel immune modulation therapy in patients with advanced chronic heart failure: results of a randomized, controlled, phase II trial. J Am Coll Cardiol 2004;44:1181–6.
[51] Beutler B. Inferences, questions and possibilities in toll-like receptor signalling. Nature 2004;430:257–63.
[52] Heymans S, Hirsch E, Anker SD, et al. Inflammation as a therapeutic target in heart failure? A scientific statement from the Translational Research Committee of the Heart Failure Association of the European Society of Cardiology. Eur J Heart Fail 2009;11:119–29.

Index

Note: Page numbers followed by *f* indicate figures and *t* indicate tables.

Printed in the United States
By Bookmasters